T0329541

ENGINEERING THE CMOS LIBRARY

ENGINEERING THE CMOS LIBRARY
Enhancing Digital Design Kits for Competitive Silicon

DAVID DOMAN

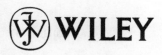

A JOHN WILEY & SONS, INC., PUBLICATION

For general information on our other products and services or for technical support, please contact our Customer Care Department within the United States at (800) 762-2974, outside the United States at (317) 572-3993 or fax (317) 572-4002.

Wiley also publishes its books in a variety of electronic formats. Some content that appears in print may not be available in electronic formats. For more information about Wiley products, visit our web site at www.wiley.com.

Library of Congress Cataloging-in-Publication Data:

Doman, David.
    Engineering the CMOS library : enhancing digital design kits for competitive silicon / David Doman.
        p.   cm.
    ISBN 978-1-118-24304-6
        1. Digital integrated circuits—Design and construction.   2. Metal oxide semiconductors, Complementary.   3. Industrial efficiency.   I. Title.
    TK7874.65.D66   2012
    621.3815—dc23
                                                                                                2011043303

10 9 8 7 6 5 4 3 2 1

CONTENTS

PREFACE

In many ways, it is correct to view engineering as an inexact science. Yes, as with mathematics and the hard sciences, there is often a need for proofs and exacting arguments. HOWEVER, there is often, also, the need to produce, in a timely manner, just enough. That is to say, there are circumstances when a "sufficient" solution today is better than a "perfect" solution tomorrow. This mind-set is often the case for integrated circuit design, in general, and for stdcell IP design, in particular. As just one example, a classical way of implementing a flip-flop that has the added functionality of both an "asynchronous set" and an "asynchronous reset" is to add just enough circuitry to the flip-flop for each of those two additional functions to operate independently. Doing so assures a minimal sized stdcell in both net-list size and in layout area. However, it means that the flip-flop operates un-tractably when the controls for those two asynchronous functions are operated together. This especially causes issues if the control signals are de-asserted simultaneously. Dependent on the race between the two control signals, operating in a real world parasitic environment, such a flip-flop might settle either into a "set" configuration, or into a "reset" configuration, or into a nonsensical combination of those two states, or even settle into some meta-stable oscillation mode. The "perfect" solution would be to define a flip-flop circuit that assures some a priori tractable resolution in such an event. That means added transistors to the flip-flop design, further meaning that the flip-flop is larger, probably slower, and definitely more power hungry. In addition, while the addition of such circuitry can certainly reduce the chances of the cell going into some improper operation upon the simultaneous removal of both asynchronous signals, it cannot totally eliminate the chance of this occurring. Therefore, since the "perfect design" produces a larger, slower, more power hungry device and one that is doomed to failure anyway, the "sufficient design" is just to use the minimal circuit additions and to note, in a release document, that the flip-flop operation is not guaranteed under such conditions. This assures a smaller, faster design and allows the user of the flip-flop to develop proper control over the signals that operate the two asynchronous functions (or to ignore the warning at the user's own risk). In other words, the stdcell library designer should trust that the integrated-circuit designer is competent to do his/her job.

While it might be the concerted wish of the integrated-circuit design community that the typical stdcell library (or other IP) offerings that they might be required to use in their design work is of the "perfect" mode, it is far more likely, and even desirable, for those offerings to be of the "sufficient" mode. The knowledge of the existence of the inexactitudes in those "sufficient" mode libraries, the ability to spot where in those libraries such inexactitudes might be, and the ability to take advantage of those inexactitudes in order to produce smaller, faster, cheaper designs, is the underlying cause for the writing of this book. This book is based on personal experience gained while producing stdcell libraries, for multiple companies, both on the integrated-circuit design side and on the IP development side, for technology nodes ranging from 1.25um in the early 1980s to 14nm today. Along with this being a textbook, I believe that it holds grains of wisdom for the integrated-circuit design engineer, the IP or stdcell-library supply engineer, the IP maintenance engineer, the EDA engineer, and the first line managers of all of those realms of engineering. Enjoy.

ACKNOWLEDGMENTS

I gratefully acknowledge the assistance of the many people who have had a hand at allowing me the chance to write this book.

First, I have had the pleasure of knowing Bernard Goodwin, a technical editor, for more than 10 years. I met him as he was finishing the publication of a book written by a co-worker of mine at the time, and I mentioned that I might have an idea for a book. Bernard took the opportunity to challenge me to flesh out the idea. At the time, I little realized that it would take me all of those 10 years to accomplish that fleshing out. Each year, at the Design Automation Conference (DAC), I would make the trip over to the Prentice-Hall booth and tell Bernard that I was still thinking. He never gave up on me. Finally, in November 2009, I decided that I was going to have to start the writing of the book soon or I would never finish and would always look backward wondering why. I contacted Bernard, told him that I was going to send him the first couple of chapters by the end of the year. Based on this, Bernard rounded up a few reviewers and the effort got underway.

After I delivered these first two chapters, Bernard took them to the reviewers: Thomas Dillinger, Faizul Alum, Don Bouldin, and especially Charles Dancak. Their reviews, although sometimes painful to read, were of incredible value to me. I have taken their advice and made many changes as a result. In my opinion, this book is the better because of their advice. Hence, second to Bernard, I would like to thank these four gentlemen.

Third, and most important, I should like to thank my wife, Susan, for putting up with me during my engineering career. She always listened to me explain, repeatedly, how the experiences that I was having at the various companies I worked at would be good fodder for an engineering book. Further, she listened to me as I lamented at not working on that very same book for the past 10 years. Finally, she stood by me as I decided to push through the effort. Susan, I could not do this without you.

Finally, I thank the senior-level management at my last company for giving me the time to write this book and the incentive to get off the starting line and actually do it.

Bernard, Thomas, Faizul, Don, Charles, Susan, and everybody else whom I have forgotten, thank you!

CHAPTER 1

INTRODUCTION

In a positive sense, this book is about gaining a competitive edge in the integrated circuit marketplace. What it will suggest to the reader is that there is unrecognized value hidden in the safety-net margins that exist in the various descriptive views of any piece of intellectual property. This hidden value, which might be useful on any given integrated circuit design, can normally be left "on the table." However, this hidden value can and should be used by the aggressive design engineer (or manager) to beat market competitors. The pages ahead will reveal how the typical design house can enhance the performance, reduce the power, and improve the density of standard-cell logic. It will show how to add value to the generic, foundry-provided standard-cell library that many companies use "out of the box." It will identify low-risk opportunities where aggressive designers and managers can shave off margin from overdesigned standard cells.

However, the other side of the preceding is also true. That is to say, this is a dangerous book. The reason it is dangerous is that no engineer or manager has ever been fired because of not following the herd and either accomplished or failed to accomplish exactly what any other engineer or manager accomplished or failed to accomplish. The not-so-hidden message of this book is that by breaking from the herd and attempting to use that safety net margin, wherever it is found, the results, *when successful*, will reap market benefit. The dangerous downside of this activity, however, is the risk of taking a little too much of the perceived margin and actually forcing a failure on silicon. The cost of such aggressiveness could easily be career limitations. Now, before you put this book back down, let me point out that taking advantage of such margin is not a new idea. In fact, it is rather old and it used to be rather common for the design engineer to take advantage of

Engineering the CMOS Library: Enhancing Digital Design Kits for Competitive Silicon, First Edition. David Doman.
© 2012 John Wiley & Sons, Inc. Published 2012 by John Wiley & Sons, Inc.

it. It used to be called "adding value" by *not* doing it the same way as everybody else. However, over time, as the cost of development of integrated circuits and the market cost of failure of those integrated circuits have increased, this concept has been replaced by "how not to mess up" by doing it the same way as everybody else. The result is that most integrated circuits are now designed in an extremely conservative manner.

About twenty years ago, as of the writing of this book, I was working for a microprocessor design company in the data communications integrated circuit design center. At that time, the design center was working on some Fiber Distributed Data Interface (FDDI) devices. During the development of some of those integrated circuits, I was asked if it was possible to decrease the cost by reducing the number of layers that the router that was being used on that particular design could use. Specifically, the design center was using a near-first-generation sea-of-gates router, and I was asked if it were possible to route the design as a channeled route using the sea-of-gates router that was available. I told the design manager that I could do channeled routing with the sea-of-gates router but that the sea-of-gates router was not a channeled router. My answer perplexed the design manager until I told him that you can use a wrench to hammer a nail into a wall, but that does not make a wrench a hammer.

The preceding illustration is what this book is basically about. Too often in the current world of integrated circuit design, we look for the perfect hammer in order to drive the perfect nail into the perfect wall exactly as the hammer and nail and wall manufacturer has instructed us. This is the previously mentioned safe "I won't mess up" follow-the-herd method of integrated circuit design. However, doing that will mean your results will be exactly like the results of the competing design center that set up shop down the block. This book is about looking for the ways of using the tools that we have in creative ways. It is about using the wrench because the hammer is too expensive. It is also about using the hammer—but in ways that the design center down the block will not.

What this book is not about is device physics. You will not find any physics-based or electronics-based equation anywhere in it. There is no consideration herein about ion currents or oxide thickness. There already are more than enough books available on all of those subjects (and others as well). Rather, this book explores the everyday consider-ations of the library design and support engineer. That library design and support engineer may be in a specific library function in a separate organization or be a design engineer who is integrated within the design center that is using the library that is under consideration and is doing exploration of the library as an additional function to his or her regular design activities. This book is meant to be used as an instigator of thought. As mentioned, there are no physics or electronics equations in this book (although there are a few Boolean logic equations). Indeed, I have taken the tack of explaining in either words or illustrations as opposed to hard physics. An outcome of this book should be that design centers might push technologies and libraries in ways that the fabrication houses might be nervous about but that would (or should or could) be covered sufficiently through the validation and verification efforts of the design centers to be rendered viable.

A digital design kit, sometimes referred to as a DDK, is a collection of small-scale functions together with the various views that describe these functions for various pro-cess, design, and test (and other) tools. They are used by design-center engineers to design modern application-specific integrated circuits (ASICs) that can be processed success-fully in the external fabrication house that supplied the original digital design kit.

In a sense, a DDK can be thought of as a collection of various bricks, each brick different, together with an assortment of descriptions of each type of brick that allows them to be placed together in such a manner as to further allow somebody to build a structure. Another way of looking at a digital design kit is as an alphabet, together with the rules of usage of this alphabet for word and sentence construction that allow somebody to write a book. In either of these two cases, the bricks or the alphabets, if you can keep your supply separate from those available to anybody else, then you can build either a unique structure or better book. Otherwise, if you can somehow adjust your bricks or alphabet in a manner not available to others, you can still build the better structure or write the better book.

For some design centers—those that belong to companies that have their own fabrication facilities—these DDKs can be viewed as private in-house tools that are not available to design centers for competing companies. This is good because it allows the company to adjust the fabrication and hence the views of the digital design kit in ways that generate a competitive advantage for their designs over those of competing companies. For many or even most companies around today, however, this is not the case. These "fabrication-less" companies subcontract the processing of their designs to external fabrication houses and have no real control of the processing at the external fabrication houses and must accept the process "as is"—unless they are willing to pay significant pricing and significant time and engineering resources to convince the external fabrication house that it is worth their time and effort and cost to allow special handling of the design centers designs. As a result, few such design centers are willing to expend the energy to do so. Hence, the DDK they use to do their designs is the same DDK that the next design center down the street is using. These design centers end up using the digital design kit "straight out of the cellophane" with no adjustments.

There are advantages for having common views and tools, such as the ability to hire workers already familiar with such tools and views. However, the real result of this is that if two or more such companies decide to do the same sort of design for the same market while using the same digital design kit, they tend to produce nearly the same design. This is further forced because most design centers use the same Engineering Design Automation (EDA) tools or at least very similar ones. Yes, there may be Input-Output Structure (IO) order differences, perhaps scan-chain differences, and the embedded software might have been written somewhat differently, but these are all secondary to the fact that the two or more competing designs will be all roughly the same size, have the same cost to produce, and have the same power, all assuming they have the same features. The companies are then forced to compete in the market on price margin alone.

Now from the external fabrication house's point of view, it is advantageous to force common design practices across the ASIC design industry. These houses have developed business models that allow them to operate profitably by accepting ASIC designs that are produced by design centers that used their DDKs. These business models basically say that if the design customer used the digital design kit "as is" and the EDA tools in conjunction with these digital design kit files (assuming the design works), then the design will come out of the external fabrication house working or the external fabrication house will pay for it. Conversely, if the design customer decides to *not* use the digital design kit "as prescribed by the external fabrication house" but instead uses it with some changes and the resulting design does not work, then it is the design customer's fault and the external fabrication house has no fiduciary responsibility. Clearly, it is the external fabrication house's best interest to have solid DDKs. How can they do this? The easiest way is to allow margin in the various views of the kit, in effect making those views "bulletproof." Worst-case timing views are slightly slower, and best-case timing views are slightly better than expected. Worst-case power and best-case power views are guard banded. Similarly, for logical views, the external fabrication houses do not allow proper simulation of some actual although nonstandard functionality. In the end, view after view is a little worse performance or a little less functionality per model than would otherwise be expected. They are all "margined" to one degree or another.

However, this added margin is precisely what the design-center customer wants to have removed or, at a minimum, transformed. This removal would allow the design-center customer to compete in the market not on price margin alone, but on power, performance, or size, the first two of which can be viewed as feature additions and the last of which enables profitability even in a price-margin-alone environment. The transformation of the margin from something that is decreed by the fabrication house into something that can be controlled and potentially proportioned by the design-center manager allows judgment calls to be made. If there is no performance or area penalty, for instance, then such margin additions could be added back into the calculation without harming its competitive position in the marketplace. Contrarily, if the penalty is too large, it can be at least partially minimized to the extent required to allow sufficient cost-margin savings, again allowing the resulting finished device to be cost competitive in that same marketplace.

This book describes areas where the margins in the views can be found and how to determine their extent and how extensively these margins can be safely removed.

Wait, you might be thinking. I just said that actually doing this "margin removal" (or any "margin transformation") could cause the fiduciary burden to be shifted to the design-center customer. Yes, that is true. However, in the perspective of the business model of the design-center customer, the extent of the risk may be worth it. This is not a technical decision, but rather a financial decision. Control of that financial decision, as determined by the amount of margin removal that is being considered, is thus moved to the realm of the design-center manager. This book should give the designer-center manager that knowledge needed to make those decisions.

One additional comment: just because a little medicine might be good for you, more is not necessarily better. That dictum is also true in engineering. Use this book at your own risk. Many of the techniques on margin reduction listed herein are as dangerous as they may sound, and proper caution needs to be taken when partaking in many of them, otherwise the design center that you support may start to produce nothing more than expensive sand.

Now, before going to much further, I do have one apology to the reader. As I review this book, I find that I often use "negative language" (for instance, "if one doesn't do something, and as a result, something doesn't happen, then...."). I apologize. Nearly three decades in developing CMOS logic libraries, where the common function always involved an inversion (or NOT function) has trained my mind to think in that manner. I hope, however, that the average reader involved in the industry does not find this "negative" language any more awkward than I do.

1.1   ADDING PROJECT-SPECIFIC FUNCTIONS, DRIVE STRENGTHS, VIEWS, AND CORNERS

Beyond what the previous section says about margin removal (or margin adjustment), this book has a second goal: to allow adding items to a digital design kit in order to personalize it.

A specific digital design kit probably comes from the external fabrication house with three sets of timing and power views (abbreviated as PVT for "process corner, voltage corner, temperature corner"). These three PVT probably correspond to one worst-case processing corner, one best-case processing corner, and one typical processing corner. If the design center has a different market niche that requires a different set of voltages of operation or environmental temperatures, or if it is willing to accept slightly less or slightly more processing variation, then it has to pay the external fabrication house perhaps several hundred thousand dollars to generate and support such new PVT timing or power views. This book should allow the design center the knowledge of what it takes to make such a PVT characterization on its own. With that knowledge, the design center should be better equipped to make the decision whether to buy a new PVT from the external fabrication house, "make" it internally (with the business benefits but support cost of doing so), or "reengineer" the market decision such that it becomes unnecessary.

A specific digital design kit will have a limited number of logical functions, although that number might be very extensive. If the design center has identified some special function that is not easily built by the logical functions present, it might choose to add that function by means of a custom design. The external fabrication house may have the ability, assuming that the design center goes forward with the design, to say that the fiduciary responsibility for the design moves to the design center if this new function does not work but remains if the fault is the result of some other part not being modified by the design center. This may seem bad for the design center, but it is the model of many of them. These centers have specific "special functions or logic" that they have used repeatedly in the past. These special functions or logic are considered to be a competitive advantage for those design centers with such logic. They are in business specifically because they have such functional or logic blocks. This book tells the design center how they can better add such functionality within the external fabrication house's DDK framework.

A design center may determine that it can build a smaller, faster, or less-power-consuming design if it had access to the functions supported by the digital design kit, but it requires a slightly larger or slightly smaller version of a function (in terms of area of performance or power). If this slightly adjusted cell is a larger version of one that is already present, then the design center might be tempted to parallel up multiple copies of the function. If the slight adjustment is to make it smaller, actual trimming of transistor widths might become needed. This book will discuss the possible means of doing so safely.

A design center might use an EDA tool that is not supported by views delivered from the external fabrication house. Although this book will not explain how to develop such views, it will give the design center the knowledge of how to ensure that design-center–generated views are consistent with the views that are delivered by the external fabrication house. It will also describe the efforts needed to deliver and support such views across multiple designs and view and tool updates.

The bottom line is that even though some people see the DDK as cast in stone, the design center that is willing to extend or adjust such kits can deliver better designs (i.e., cheaper to produce, test, or support). These designs will allow those design centers to compete in their chosen markets on more than just commodity (i.e., price-margin) techniques. This book will allow the engineers and managers of those design centers additional degrees of freedom in making the decisions to extend and adjust such digital design kits.

1.2   WHAT IS A DDK?

A digital design kit is a collection of physical views of certain functions that a fabrication house is capable of processing at a given technology node. Coupled to this are the other physical (read place and route) views as well as the logical, temporal, power, noise, and test views that represent these various functions for (usually some, maybe most) of the various engineering design automation tools available on the market. Moreover, the typical DDK release will be for what has become known as a "Frankenstein" flow, with Cadence logical and LEF (library exchange format) views and Synopsys timing, power, and noise views and supported in the physical design kit (PDK), by Mentor verification decks, although there are exceptions to each instance, with Cadence, Synopsys, and Mentor being "the big three" EDA companies. These views are supplied by a fabrication house where any resultant design that uses these supplied cells and views will be processed. Some of these views can be adjusted, assuming knowledgeable guidance, with a fair amount of impunity from the fabrication house, while other views are touched with less impunity, even by the design-center experts. These views differ from and need to remain consistent with the decks and views in the PDK that also comes from the fabrication house.

More specifically, at the very minimum, a DDK will contain the following.

- Graphical Database System Stream-file format (GDS) (physical), which are views of one or more standard cell (stdcell) libraries, usually including IO-specific libraries, which can be used on designs that are destined to be processed in a particular fabrication house. These views are a binary format, viewable in any one of several GDS viewers. ASCII based human readable formats of GDS can be created through various EDA tools.
- Gate-level SPICE (simulation program with integrated circuit emphasis) netlist of each cell in the various libraries, SPICE being a cross between a physical view (which it represents) and a simulatable view, which is one of its two major purposes, and a verification source, which is its other major purpose. SPICE is ASCII based and easily human readable.
- Some further physical representation (possibly a physical LEF, an ASCII-based and human-readable description of the place and router pertinent parts of the GDS). This further physical representation is used by automatic place and route tools, in conjunction with a logical LEF, usually part of the PDK that contains information on layer definitions that the router can use during design place and route.
- Timing, power, and noise views, which are typically but not necessarily referred to as *liberty files*, a name that comes from the extension that is most often used ("lib") on versions of these ASCII files that were originally purely for Synopsys tools. Being ASCII, they are easily readable. More so, most liberty writers tend to output highly structured versions of the format, which further aids human understanding on reading. The Backus-Naur Format (BNF) for these files have since been "open sourced"—that is, they can usually be read and understood by most EDA timing tools. One aside: Synopsys has not frozen the BNF for its liberty files and thus care must be taken with every new Synopsys release in order to keep previous versions of liberty files from becoming antiquated.
- Logical and test views, typically being slightly adjusted versions of the same logical format, but with one geared toward aiding logical simulations and the other geared to representing actual internal nodes of a stdcell (or IO) function that allows for accurate tracking of "reachability and observability" for internal fault tracking. These views may come in one or both of two major logical tool formats (Cadence's open-source VERILOG file or IEEE's open-source VHDL format). They are both ASCII based and human readable.

- There may be other views, especially white papers, verification or validation documentation, and release-tracking documents, all in some form of human-readable format.
- Sometimes other, more EDA-tool-specific views can also be included.

Along with the preceding major views, support for special tools may be added to a DDK release by a fabrication house. These tools can include RAM/ROM memory compilers. Once installed on a system, RAM/ROM memory compilers, at the press of a few buttons, automatically generate all of the preceding formats for a specific RAM/ROM configuration supported by the fabrication house. The timing, power, and noise view is initially a machine-estimated version. Typical fabrication-house vendors, however, offer the option of characterizing the specific RAM/ROM in order to allow cleaner design closure of a design that uses that specific RAM/ROM, and these fabrication house usually offer this either free or for a small fee.

The reason that RAM/ROM compilers exist as opposed to a long list of usable RAM and ROM instances is that there is such a larger list of possible combined number of words, bits, and column-multiplexer (MUX) instances that could be chosen by design teams, all of which would therefore otherwise need to be supported.

Finally, there may be certain specific circuits (or analog subcircuit pieces, such as various layer resistors of multilayer capacitors or diode or PNP or NPN structures that can be used to build analog function in the particular technology node and in the specific fabrication house). Some of these, which are dependent on the technology node, will be custom and uneditable or only marginally editable structures (the more complex and deep the technology node, the more likely that this is so). Some will be just Boolean layer definitions. Most will have rather pared-down subsets of the previously mentioned views, just enough to allow some amount of simulation (usually in SPICE) and design-level verification (including netlist capability).

Now how extensive can any of the preceding be edited? It depends on the technology node and the desire of the design center. Deeper technology nodes will tend to have stricter requirements for the physical views than higher technology nodes. The real reason for this is that as geometries get smaller, they tend to go from drawn silicon shapes being transferred to the mask through lithographic treatments such as serifs at corners of polygons through optical-proximity correct techniques, through phase shifting, and to exact set widths and spaces. For the deepest technologies, what is drawn on the mask can bear little visible relation with what is desired on silicon. In those instances, the operative order is "Don't touch." Trust that the fabrication house understands better than the design team what certain polygons need to be, where they need to be, and why they need to be. However, for the higher nodes, even those in the 90-nanometer range, which includes some phase-shifted mask and some strong optical-proximity correction, it is worth pursuing with the fabrication house when a question on a physical view arises. For higher technology nodes than that, there is a good chance that you can find optimizations that the fabrication house has left "on the table." In addition, it is highly advantageous at these higher technology nodes to attempt to do just that. The reasoning is that these nodes have been "in the marketplace" for such a long time that every design house that wants to compete in a market will have access to the same generic package. Assuming that all design centers make the same decision on technology node (which is a valid assumption given that they would be privilege to the same marketing information), they will produce roughly the same size product in the same rough timeline. This means that the design centers will be competing on price alone. Having a "pushed" technology might just be the edge needed to undercut the opposition in such an environment. As far as the logical and timing views are concerned, we will discuss them later in the book, but even at the deepest technology nodes the fabrication house has generated "loose" views that a design team can beat and thereby find additional margin over the competition.

CHAPTER 2

STDCELL LIBRARIES

2.1 LESSON FROM THE REAL WORLD: MANAGER'S PERSPECTIVE AND ENGINEER'S PERSPECTIVE

Back when all logic design was accomplished by generating hand-designed schematics, it used to be said that once you became familiar enough with a certain logic designer's style, you could tell which schematics where done by that person versus which were done by other logic designers. In effect, just as is the case with classical painters and their paintings, each designer had a brush stroke that was all his or her own. Each logic designer was an artist in the field of logic design, admittedly some more than others.

Although the ability to see this artistry in an Register Transfer Language (RTL) netlist might be lessened because of the structural requirements of the RTL combined with the design-center restrictions on techniques, it is still sometimes observable in the resultant synthesized netlist that came from the original RTL. A netlist from some designers will have a different ratio of inverters to flip-flops or a different mix of Boolean functions or any of a number of other telltale fingerprints that would allow the seasoned observer to say one netlist was produced by one logic designer while somebody else produced another—that is, if the stdcell library is rich enough to allow such subtleties to arise.

A couple of years ago, a design center asked why I produced such large stdcell libraries. Typically, that design center would purchase a 350–450 cell library and be relatively happy. My usual deliveries were 650–850 cells. There were so many that the design team was lost in the details and could not figure out which cells they would remove. Because they had experience with successful using previous and much smaller

---

Engineering the CMOS Library: Enhancing Digital Design Kits for Competitive Silicon,
First Edition. David Doman.
© 2012 John Wiley & Sons, Inc. Published 2012 by John Wiley & Sons, Inc.

libraries, they knew that my offering was redundant overkill. They specifically asked me, as their supplier, if I could pare the offering down to just those typical stdcells that would be commonly needed. Because I had previously guaranteed that no complex stdcell function in my offering could be done in a smaller, faster, or less-power-consuming fashion by combining two or more other functions (with the possible exceptions of the full-adder and half-adder functions that I mention later in this chapter), I told them that the stdcell set was minimal. It was just more exhaustive in terms of functions than most commercial offerings. Because they had previous experience with a "more compact" library, I was told to prove it. We took a dozen or so functional RTL blocks that the design center had used over several previous technology nodes and proposed to test my hypothesis by synthesizing with a subset of the library picked by the design-center manager and again using a full set of my functions. We would look at synthesis time and resultant gate area as a function of performance for each RTL block. For block after block and frequency target after frequency target, the two versions of the library produced netlist that where within a percent or two of each other, sometimes with the pared-down library producing the better and sometimes the full library producing the better of the two. In most cases, the pared-down library would reach timing closure in minutes whereas the full library took two or three times as long. As the performance target kept getting greater, the sizes of the resultant netlist would hit asymptotic performance limits as illustrated in Figure 2.1.

Typically, these asymptotes were within a percent or two of each other for a given block for the two versions of the library. It looked as if the pared-down library was just as

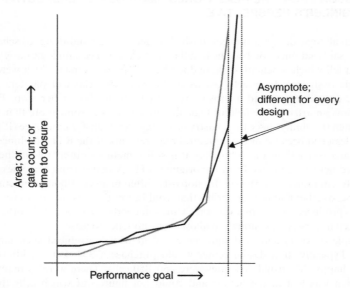

FIGURE 2.1   Showing the Asymptotic Nature of Performance Limits of Libraries and Technologies. For many applications that do not stretch the limits of a library's abilities, the ability of the library to achieve closure (in terms of time or density or power) is not significantly different from any other library. Only at the extreme edges of the ability of libraries do they show dramatic differences during design closure. Typically, such extreme limits are not nearly the same limits of other libraries designed for other design goals, although they are shown here to be nearly coincident.

effective as the larger full version, and the synthesis time was half or less for the pared-down version as opposed to the full version. It should be noted that as these asymptotes were approached, the "time to closure gap" tended to close, but this was faint praise for the full library if there was no real benefit until the very edge of the performance envelop was hit. I held my ground, but the design center was about to force me to give them what I knew to be an inferior stdcell library offering. Then we started the exercise of synthesizing a complex communication channel. It had both complex state machines and some significant arithmetic logic units in it, but these were not overly complicated as compared with many modern designs. The full library synthesized the netlist across the performance targets chosen in anywhere from 25% to 40% the time that the pared-down library did. The full-library asymptote was about 15% higher than the pared-down version. In addition, the netlists for the full library were on average 85% the area, at the same performance, as the pared-down library results. We had met the RTL writer whose output needed the full-library offering.

The bottom line—you will not need much of the following function list for your designs—that is, unless you do need them. In addition, you will not know a priori when you will run up against that need.

2.2   WHAT IS A STDCELL?

Standard cells, or *stdcells* for short, are the building blocks of the modern ASIC world. They can be thought of as the alphabet that allows the ASIC designer (or ASIC design team) to write the novel (that is, design the ASIC). They consist of several simple (in general) functions, usually in various strengths, that can be (potentially repeatedly) connected together in a network and then patterned on a piece of silicon, such that the result is a desired functional integrated circuit. Digital design kit (DDK) vendors, most often representing fabrication houses, usually offer them in a family of several hundred of the various functions and strengths. A sample of a synthesized circuit and a corresponding layout representation of a few stdcell are shown in Figure 2.2. Note that the relative areas of the various elements in the placed row of are valid, NANDs and NORs are larger than INV (the NOT in the example), and FLIP_FLOPs (relative sizes DFF in example) are larger still.

The actual choosing of which stdcells in a family to use in a design and the actual placing of these at various locations on a piece of silicon, both of which can be done manually, is usually done by software provided by various engineering design automation (EDA) vendors.* These various software tools, as mentioned in the previous chapter, work with representative aspects of the various stdcell that are pertinent to the particular software's requirements in order to perform properly their specific function.

Another type of placeable family structures that can be used to design and build integrated circuits (gate-array cells) exists and there is a continued market for them (field-programmable gate arrays), but they remain a niche market and will not be covered in this text.

Before we investigate the subject any deeper, it will be beneficial to understand that the actual stdcell layouts, as represented by the various rectangles in Figure 2.2, are DDK view. Although most DDK views are ASCII based (English readable), the layout is

---

*Sometimes called electrical or electronic design automation.

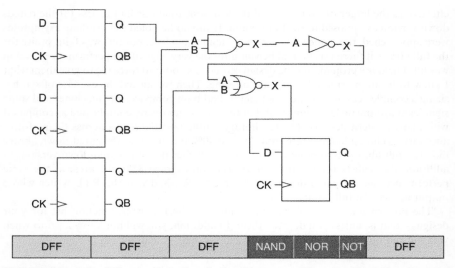

FIGURE 2.2 A Small Circuit of Stdcells and a Possible Layout of Same in a Single Row. Note that the relative sizes of the rectangles that represent the actual physical views of the various elements in the circuit are fairly close to actual ratios for those functions. Also note that the cells are not necessarily placed in any given order similar to the logic circuit.

graphical. As an example of what a typical layout may look like, the inverter (the NOT in the figure) might resemble something like that shown in Figure 2.3. The clear rectangle surrounding the shapes in the figure can be described as the stdcell outline. Although there may be shapes associated with the stdcell that extend beyond the rectangle (such as the NWELL), that rectangle represents the regular placeable structure of the stdcell. Such rectangles (or cell outlines or boundaries) are of consistent height (or multiples thereof) and consistent width (or multiples thereof). Other, more complicated stdcells will have more P-channel and N-channel transistors within their cell boundaries, but those cell boundaries will always be of given multiples of the basic stdcell boundary height and width. This feature allows them to "stack" in both the horizontal and vertical grid on a chip during place and route.

2.2.1 Combinational Functions

2.2.1.1 Single Stage The leading EDA synthesis tool over the last two decades, Synopsys's Design Compiler, requires at least one storage element, one inverting element, one combining element, and one tri-stating element (even if the RTL does not need a tristate function) to be in the stdcell library so that the tool can properly synthesize the netlist. The combining and inverting element could be combined as a NAND function or a NOR function. Therefore, the smallest stdcell offering, as illustrated in Figure 2.4, must contain a flip-flop, a tri-state, and a NAND or NOR.

Such stdcell offerings can be artificially built either by removing other elements in a more extant library offering or by adding the tokens "Don't touch" and "Don't use" to them. These are Synopsys constructs that prevent stdcells that are present in a Synopsys liberty file from being modified after the fact (the "Don't touch" construct), or instantiated before the

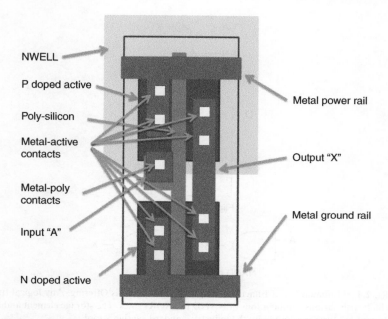

NWELL

P doped active

Poly-silicon

Metal-active
contacts

Metal-poly
contacts

Input "A"

N doped active

Metal power rail

Output "X"

Metal ground rail

FIGURE 2.3 Inverter (NOT) Circuit. The INV circuit given in Figure 3.6 might be arranged as above. The P-channel transistor at the top of the circuit is produced by the polysilicon crossing the P-doped active at the top of the cartoon. The N-channel transistor is produced by the polysilicon crossing the N-doped active at the bottom of the cartoon. It should be noted that the NWELL in which the P-channel transistor sits requires a connection to the power rail (or to some other voltage in one of the low-power applications mentioned later in the book), and the substrate in which the N-channel transistor sits requires a connection to the ground rail. Some technology nodes allow for such connections within the stdcell, and some technology nodes allow them outside of the stdcell.

fact (the "Don't use" construct), in a synthesized netlist, and then to watch Design Compiler synthesize some representative (for your design environment) circuits. The results of an experiment, doing just as described previously for a four-bit adder circuit, has been added in Appendix I. However, no viable stdcell library on the market contains just these three elements.

Instead, most stdcell offerings on the market today contain 300–400 elements representing 3–4 drive strengths for somewhere in the realm of 70–80 different combinational logical functions (and a set of sequential storage elements). These numbers, however, can be widely variable in some offerings. Some libraries offer more than 1,000 elements, a half-dozen drive strengths, or 120 or more logical functions singly or in combination. In addition, many current stdcell library offerings contain high-performance or low-power options that either are their own separate sublibraries or are part of the offering as a whole. These can be "high-Threshold Voltage (Vt)" (slower, less-power-hungry versions of the library elements) and "low-Threshold Voltage (Vt)" (hotter but faster versions). Some of these versions are designed so that they can be used intermingled with cells of the same basic architecture but of different Vt. Further discussion of this is offered later in the book.

In addition, most currently offered stdcell libraries contain circuits that are designed to lower dynamic and static power loss. Examples of the first are clock-gating cells. These are

14 STDCELL LIBRARIES

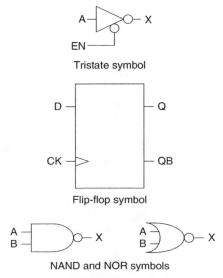

FIGURE 2.4 Minimum Set of Functions Required in a Synthesis Offering. Any logical function can be built with various combinations of NAND (or NOR) gates. The storage element and the tristate element are further required by the industry standard synthesis tool.

cells that a designer can insert into a clock tree in order to gate a clock that drives some of the flip-flops of a design to an off position. Doing so allows the flip-flops to hold their data without repeatedly needing to burn power when an unneeded clock pulse occurs. In addition, the clock tree itself will not burn dynamic power when turned inactive. These clock-gating cells come in four basic families, and Synopsys's Design Compiler has been able to handle them correctly. Examples of stdcell library elements that are designed to minimize static power, along with the high-Vt cells already mentioned, include some more esoteric elements such as various versions "state retention" flip-flops. Such flop-flops have small "master latch" sections, either in the place of or in addition to, the normal master-latch section (or small "slave latch" sections either in place of or addition to the normal slave-latch section, dependent on where the particular company plans on doing the low-power state retention) of the flip-flop master–slave circuit. These smaller master-latch or slave-latch flip-flops have some manner of an alternate power or ground supply driving these parts of the circuit. Such cells allow the power or ground to be reduced or removed from the vast portion of a circuit while the power remains active to the small state-retention master latch. This, in turn, allows power to be greatly diminished in designs that have sections that do not get used often but require instant recovery of their previous state when needed. These cells will be further described later as well. Note that some companies have invested seriously in such designs at a high level as previously described. That is why I have kept the description at such a level. At any rate, please protect yourself if you decide to produce such cells by doing a patent search on your particular version. By the way, some companies adjust the oxide thickness or the Vt of one or the other of the latches in order to get some of the benefits that a normal master-latch or slave-latch state-retention flip-flop without impinging on designs patented by other companies. In addition, voltage or frequency stdcell elements will be discussed later. These stdcells are designed to function over larger lower ranges of voltage of the power rail, albeit slower. Because power is a function of the square of the voltage, lowering the voltage rail in parts of the circuit that do not need to perform "as fast as possible" allows reduced power without harmful performance impact. All of these cells will be discussed shortly.

First, however, a discussion of combinational functions is in order. These various functions can be split into several different classes of logic. In CMOS technology, the basic logical function is an inversion. All single-combining operations always involve an inversion. One can logically combine two or more signals by AND (all inputs to such a function need to be high for the output to be high) or OR (any of the inputs to such a logic function can be high and the output is high). Slightly more complexly, one can use Boolean AND/OR or OR/AND signals as they combine in the CMOS logical function, but in any case this combination signal gets inverted as it leaves that logic function. To prevent that from happening, invert the resultant inverted signal (giving a double inversion, which is not similar to a double negative). Hence, the most basic functions in CMOS, as shown in Figure 2.5, are the INVERTER, the NAND (inverted AND), and the NOR (inverted OR). Note that the NAND and NOR symbols and circuits in the figure combine two incoming signals. The actual number of such inputs can be greater than two, subject only to the following discussion on stack height.

These three functions, viewed as P-channel transistor and N-channel transistor circuits, are shown in Figure 2.6. Note that many stdcell libraries, for convention, have signal names that occur earlier in the alphabet and represent signals that connect to transistors that are *stackwise* closer to the output and further from the power Voltage Drain Drain (VDD) and ground Voltage Source Source (GND) rails. This was of value when hand-drawn schematics were the prevailing practice because of the assumption that transistors closer to the output would switch faster. This is problematic in more-complex Boolean functions because P-channel transistors may be electronically nearer to the output whereas N-channel transistors are electronically farther away or vice versa. In addition, this practice has been rendered archaic by modern synthesis techniques that allow automated choosing of which functionally equivalent stdcell input to connect to which signal.

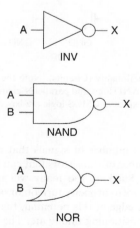

FIGURE 2.5   Basic CMOS Functionality (Symbols). Note that each is an inverting function, which is a primary characteristic of basic CMOS logic.

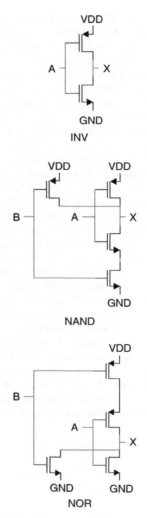

FIGURE 2.6 Basic CMOS Functionality (Circuits). Note the pattern of parallel P-channel and serial N-channel transistors for NAND logic and parallel N-channel and serial P-channel transistors in NOR logic. All further complexities in CMOS logic are based on various combinations of these patterns.

As already alluded to, the number of signals that are being combined in a logic function is limited by the number of transistors that can be stacked between a power or ground rail and the output. Stacking too many will not allow sufficient gate drain potential to allow the transistors to properly turn on or off completely (or efficiently). In addition, as stack reaches the edge of this potential, the last transistor out will tend to become an extremely slow transitioning timing arc. The CMOS physics showing this is not complicated and is covered extensively in any basic CMOS design text. For modern deep submicron technologies, a three stack (as many as three transistors between a rail and an output) is typically the most that can be handled. This limits the number of functions that can be handled. Note that for the lowest power designs, this becomes two-stack, further limiting the logic list. This will be further discussed in the low power support section of this book. Based on the preceding discussion, the number of basic one-stage (one combining logic circuit from any input to any output), two-stack, or three-stack functions can be limited to those shown in Table 2.1.

Note the severe reduction that would result in available functions in Table 2.1 if three stacks were removed for the lowest power designs.

TABLE 2.1   Basic One-, Two-, and Three-Stack CMOS Functions. These can be accomplished using single-stage implementation of either the basic NAND or basic NOR pattern.

Name	Function	Note
INV	A_	
NAND2	(AB)_	
NAND3	(ABC)_	3 stack
NOR2	(A + B)_	
NOR3	(A + B + C)_	3 stack
AOI21	(AB + C)_	
AOI22	(AB + CD)_	Can be used as a decoded MUX.
AOI211	(AB + C + D)_	3 stack
AOI221	(AB + CD + E)_	3 stack
AOI222	(AB + CD + EF)_	3 stack
AOI31	(ABC + D)_	3 stack
AOI32	(ABC + DE)_	3 stack
AOI33	(ABC + DEF)_	3 stack
AOI311	(ABC + D + E)_	3 stack
AOI321	(ABC + DE + F)_	3 stack
AOI322	(ABC + DE + FG)_	3 stack
AOI331	(ABC + DEF + G)_	3 stack
AOI332	(ABC + DEF + GH)_	3 stack
AOI333	(ABC + DEF + GHI)_	3 stack
OAI21	((A + B)C)_	
OAI22	((A + B)(C + D))_	
OAI211	((A + B)CD)_	3 stack
OAI221	((A + B)(C + D)E)_	3 stack
OAI222	((A + B)(C + D)(E + F))_	3 stack
OAI31	((A + B + C)D)_	3 stack
OAI32	((A + B + C)(D + E))_	3 stack
OAI33	((A + B + C)(D + E + F))_	3 stack
OAI311	((A + B + C)DE)_	3 stack
OAI321	((A + B + C)(D + E)F)_	3 stack
OAI322	((A + B + C)(D + E)(F + G))_	3 stack
OAI331	((A + B + C)(D + E + F)G)_	3 stack
OAI332	((A + B + C)(D + E + F)(G + H))_	3 stack
OAI333	((A + B + C)(D + E + F)(G + H + J))_	3 stack
TBUFE	A_ if EN=1 and EN_=0, tri-state if EN=0 and EN_=1, undefined otherwise	Admittedly, can be used as an open-source or an open-drain function as well, depending on other combinations of en and en_; rarely seen as stand alone in a commercially available library.

TABLE 2.2 **Two Special Yet Simple Two- and Three-Stack CMOS Functions That Demonstrate the Boolean Flexibility of the Technology. These are special cross combinations of both the basic NAND pattern and basic NOR pattern as explained in the text.**

Name	Function	Note
MAJ3	$(AB + BC + AC)\_$	Multiple implementations
MAJ5	$(ABC + ABD + ABE + ACD + ACE$ $+ ADE + BCD + BCE + BDE + CDE)\_$	Only one valid implementation in 3 stack.

In addition, two simple functions are sometimes added to a list such as the previous one. These are the two-of-three majority vote function, sometimes known as the MAJ3 function; and the three-of-five majority vote function, sometimes known as the MAJ5 function. The first can be implemented using an AOI222 or an OAI222 (see following analysis) or a cross of the two functions, but the second can only be done using the combination. These are listed in Table 2.2.

A demonstration might be required for the cited MAJ5. First, consider a simpler example: the MAJ3 previously mentioned. As given, the equation for the MAJ3 is $(AB + AC + BC)\_$, and it can be implemented with an AOI222. However, $(A + B)(A + C)(B + C)\_$ can be reduced to $(AB + AC + BC)\_$ through simple Boolean equivalents:

$$
\begin{aligned}
(A + B)(A + C)(B + C)\_ &= (AA + AB + AC + BC)(B + C)\_ \\
&= (A + AB + AC + BC)(B + C)\_ \\
&= (A + AC + BC)(B + C)\_ \\
&= (A + BC)(B + C)\_ \\
&= (AB + BBC + AC + BCC)\_ \\
&= (AB + BC + AC + BCC)\_ \\
&= (AB + BC + AC + BC)\_ \\
&= (AB + AC + BC)\_
\end{aligned}
$$

Hence, the MAJ3 can be done with an OAI222 just as well as the mentioned AOI222. These two implementations are given in Figure 2.7.

However, they are the same function. Therefore, for all input combinations that cause the P-channel stack of the AOI222 to pull the output up or the N-channel stack to

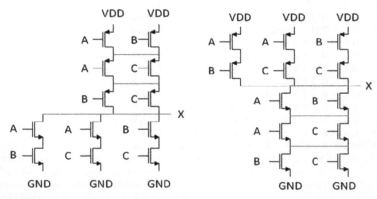

FIGURE 2.7 AOI222 (Left) and OAI222 (Right) Implementations of MAJ2 Function. Note that the one circuit has a three-stack P-channel and a two-stack N-channel pattern whereas the other circuit has the exact opposite.

pull the output down, the same input combinations will cause the same for the OAI222. As a result, the three P-channel stack of the AOI222 implementation of the MAJ3 can be replaced with the two P-channel stack of the OAI222 implementation of the MAJ3, and the three N-channel stack of the OAI222 can be replaced with the two N-channel stack of the AOI222 MAJ3 implementation. Hence, the MAJ3 can be done as a P-channel two-stack, N-channel two-stack as shown in Figure 2.8.

Finally, a similar analysis left to the reader gives the rather massive but still one-stage, three-stack MAJ5 as shown in Figure 2.9.

Some creative design usages for MAJ5 cells—for instance, in data-communication error detection and correction techniques and possibly in some low-power bus-inversion circuits and techniques—but typical stdcell libraries do not have MAJ5 functions. These cells require 60 transistors that further require 30 polyfingers in order to implement. Unless they are added for specific majority vote or data-stream validation reasons, their addition should be questioned.

Now assuming that three-stack functions remain viable and that two particular functions, the AOI333 and the OAI333 (both of which are in the preceding list), are viably added to a stdcell library, then a significantly larger family of functions can be added to this basic list. A typical AOI333 can be implemented with the circuit as illustrated in Figure 2.10.

Similarly, a typical OAI333 can be implemented with the circuit illustrated in Figure 2.11.

Note that there is significant similarity between the two structures. The only real difference between the two stdcells is that the horizontal connections between the P-channel transistors in the AOI333 are moved to the N-channel transistors in the OAI333. There is some slight reordering of which input signals go to which pairs of P-channel and N-channel transistors, with that ordering flipped across 45 degrees in the P-channel stack and again in the N-channel stack, which are done for consistency of pin ordering. Because this demonstrates that all horizontal connections can be moved, certainly some can be independently moved, and specifically one or two (as opposed to all) horizontal connections can be moved between the P-channel and N-channel stacks as well. Doing so, plus allowing the removal in some instances of transistors (perhaps by shorting across the source or drain connections to the particular transistors), allows a

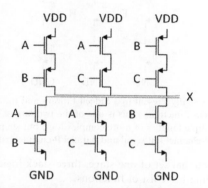

FIGURE 2.8 AOI222 N-Channel and OAI222 P-Channel Combined MAJ2 Function (Two-Stack). Replacing the three stack with the similar function two stack in only one or the other of the P-channel or N-channel while leaving the opposite two-stack side alone.

20   STDCELL LIBRARIES

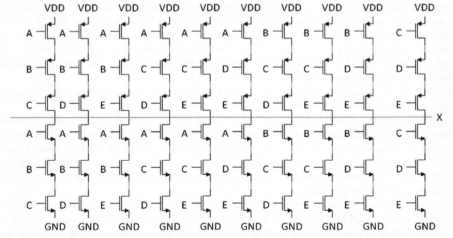

FIGURE 2.9   The AOI N-Channel and OAI P-Channel Final Version of the MAJ5 Function (Three-Stack). The similar but larger function follows the same reasoning for development as the text MAJ3 cell.

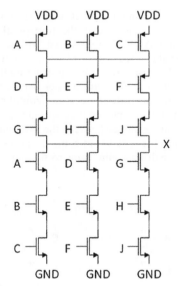

FIGURE 2.10   AOI333: Usually Present in a Stdcell Library and the Basis of a Large Family of Potential Additional CMOS Functions. This is one of the two largest "safe" three-stack Boolean functions from which multiple families of more complex logic but simpler circuit functions can be derived. Note the serial N-channel but serial and parallel P-channel pattern.

significantly larger additional set of one-stage, three-stack logical Boolean functions to be added to the preceding basic list of functions.

Figure 2.12 indicates the logical family of functions that can be added by moving one horizontal connection.

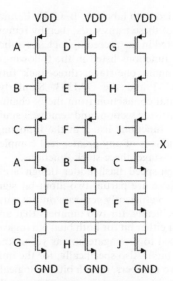

FIGURE 2.11 OAI333: Usually Present in a Stdcell Library and the Second Basis of a Large Family of Potential Additional CMOS Functions. This is the second of the two largest "safe" three-stack Boolean functions from which multiple families of more complex logic but simpler circuit functions can be derived. Note the serial P-channel but serial and parallel N-channel pattern.

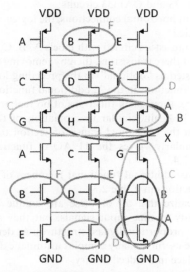

FIGURE 2.12 A First Group of Extensions from the Basic AOI333/OAI333. This graphically represents the logical extensions and circuit simplifications that lead to multiple families of stdcells.

Note the transistors circled with labels A, B, C, D, E, and F. By shorting across the source or drains of several of these transistors, thereby removing them from the circuit, and removing the signals listed in the preceding logical symbol of this family, this allows the 22 AOAOI and OAOI functions listed in the following 44 additional functions in Table 2.3, all of which remain one-stage, three-stack functions of no greater real complexity than the AOI333 and the OAI333. Similarly, one can start with an OAI333, move one horizontal connection from the N-channel transistor section of the circuit to the P-channel transistor section, and remove a similar set of transistors to get the 22 OAOAI and AOAI functions listed in the following 44 additional functions. Again, none of these functions is of any greater real complexity than the AOI333 and OAI333, and all remain one-stage, three-stack functions.

Those familiar with high-speed flash adder design are aware of *propagates* and *generates* (P's and G's). These are partial two-input-bit signals, independent of input carry bits, that can be used to efficiently generate output carry signals capable of use in full-adder calculations. Specifically, for two numbers that are to be added for matched bits of those two numbers, if either bit (or both bits) is a logical active high (1), then a P is generated (which will be used to propagate any as yet unknown incoming carry signal past those matched pair of bits). Also specifically, for the numbers that are to be added for matched bits of those two numbers, if both bits are logical active high (1), then a G is generated (which will be used to generate a carry out past those matched pair of bits). A flash carry, then, becomes the Boolean function of "generate any new or propagate the earlier carry." The function equation is $C_n = G_n + P_n C_{n-1}$. This can be extended to earlier bits as well: $C_n = G_n + P_n G_{n-1} + P_n P_{n-1} C_{n-2}$, and this can continue to significantly deeper stages as well. This style of equation, $A + BC + BDE$ and the related $A(B + C)$ $(B + D + E)$, are just examples as to why just some of the preceding functions (specifically those noted as "Useful in flash adder") in Table 2.3 are so useful. Additional examples using state machines as well as other data path constructs are plentiful.

However, we are not yet done with derivative circuits, all of which are based on slight modifications to the AOI333 and OAI333 circuits.

By moving two adjacent horizontal connections, the logical family of functions shown in Figure 2.13 can be added.

Here again, note the transistors circled with labels A, B, C, D, E, and F. By shorting across the source drains of these transistors, thereby removing them from the circuit, one can remove the signals listed as "optional" in the preceding logical symbol. This family thus allows the 11 OAOI functions in the 22 additional functions shown in Table 2.4, all of which remain one-stage, three-stack functions of no real greater complexity than the AOI333 and the OAI333. Again, similar manipulation of the OAI333—moving two horizontal connections in the N-channel transistor section to the P-channel transistor section and shorting similarly—gives the 11 AOAI functions in the 22 additional functions listed in Table 2.4.

There are actually two additional functional members of this family, but they are duplicates of the two functions in Table 2.3.

It is important to realize that all of these are unique combinational functions, admittedly across as many as nine input signals, but they are all potentially useful, proportionally small, and proportionally fast structures. As demonstrated with the flash-carry analysis, they can be useful in many complex arithmetic data-path or state-machine synthesis. They have a place in a stdcell library.

I should give a final word on some of the more common one-stage functions that can be seen in many stdcell libraries, full adders, and half adders. Functionally, a half adder is

TABLE 2.3 **Resulting First Set of Functions from Previous Extensions from the AOI333/OAI333 (Previous Two Pages). The table is an illustration of the USUAL but not ONLY one-stage stdcell Boolean additions to a typical stdcell library offering.**

Type	Function	Note
AOAOI	$(((AB + CD)(E + F)) + GHJ)\_$	
AOAOI	$(((AB + CD)(E + F)) + GH)\_$	
AOAOI	$(((AB + CD)(E + F)) + G)\_$	
AOAOI	$((AB + CD)(E + F))\_$	
AOAOI	$(((AB + CD)E) + FGH)\_$	
AOAOI	$(((AB + CD)E) + FG)\_$	
AOAOI	$(((AB + CD)E) + F)\_$	
AOAOI	$((AB + CD)E)\_$	
AOAOI	$(((AB + C)(D + E)) + FGH)\_$	
AOAOI	$(((AB + C)(D + E)) + FG)\_$	
AOAOI	$(((AB + C)(D + E)) + F)\_$	
AOAOI	$((AB + C)(D + E))\_$	
AOAOI	$(((AB + C)D) + EFG)\_$	
AOAOI	$(((AB + C)D) + EF)\_$	
AOAOI	$(((AB + C)D) + E)\_$	
AOAOI	$((AB + C)D)\_$	Useful in flash adders.
OAOI	$(((A + B)(C + D)) + EFG)\_$	
OAOI	$(((A + B)(C + D)) + EF)\_$	
OAOI	$(((A + B)(C + D)) + E)\_$	
OAOI	$(((A + B)C) + DEF)\_$	
OAOI	$(((A + B)C) + DE)\_$	
OAOI	$(((A + B)C) + D)\_$	Useful in flash adders.
OAOAI	$((((A + B)(C + D)) + EF)(G + H + J))\_$	
OAOAI	$((((A + B)(C + D)) + EF)(G + H))\_$	
OAOAI	$((((A + B)(C + D)) + EF)G)\_$	
OAOAI	$(((A + B)(C + D)) + EF)\_$	
OAOAI	$((((A + B)(C + D)) + E)(F + G + H))\_$	
OAOAI	$((((A + B)(C + D)) + E)(F + G))\_$	
OAOAI	$(((A + B)(C + D)) + E)F)\_$	
OAOAI	$(((A + B)(C + D)) + E)\_$	
OAOAI	$((((A + B)C) + DE)(F + G + H))\_$	
OAOAI	$((((A + B)C) + DE)(F + G))\_$	
OAOAI	$((((A + B)C) + DE)F)\_$	
OAOAI	$(((A + B)C) + DE)\_$	
OAOAI	$((((A + B)C) + D)(E + F + G))\_$	
OAOAI	$((((A + B)C) + D)(E + F))\_$	
OAOAI	$((((A + B)C) + D)E)\_$	
OAOAI	$(((A + B)C) + D)\_$	Useful in flash adders.
AOAI	$((AB + CD)(E + F + G))\_$	
AOAI	$((AB + CD)(E + F))\_$	
AOAI	$((AB + CD)E)\_$	
AOAI	$((AB + C)(D + E + F))\_$	
AOAI	$((AB + C)(D + E))\_$	
AOAI	$((AB + C)D)\_$	Useful in flash adders.

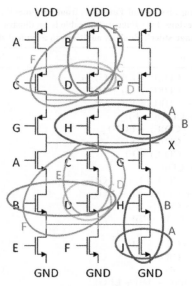

FIGURE 2.13 A Second Group of Extensions from the Basic AOI333/OAI333. This graphic representation shows additional logical extensions and circuit simplifications that lead to multiple and more esoteric families of stdcells.

TABLE 2.4 Resulting Second Set of Functions from Above Extensions from the AOI333/OAI333. The useful additional Boolean one-stage cells that can be added to a library release.

Type	Function	Note
OAOI	(((A + B)(C + D)(E + F)) + GHJ)_	
OAOI	(((A + B)(C + D)(E + F)) + GH)_	
OAOI	(((A + B)(C + D)(E + F)) + G)_	
OAOI	(((A + B)(C + D)E) + FGH)_	
OAOI	(((A + B)(C + D)E) + FG)_	
OAOI	(((A + B)(C + D)E) + F)_	
OAOI	((A + B)(C + D) + EFG)_	
OAOI	((A + B)(C + D) + EF)_	
OAOI	((A + B)(C + D) + E)_	
OAOI	((A + B)C + DEF)_	
OAOI	((A + B)C + DE)_	
AOAI	((AB + CD + EF)(G + H + J))_	
AOAI	((AB + CD + EF)(G + H))_	
AOAI	((AB + CD + EF)G)_	
AOAI	((AB + CD + E)(F + G + H))_	
AOAI	((AB + CD + E)(F + G))_	
AOAI	((AB + CD + E)F)_	
AOAI	((AB + CD)(E + F + G))_	
AOAI	((AB + CD)(E + F))_	
AOAI	((AB + CD)E)_	
AOAI	((AB + C)(D + E + F))_	
AOAI	((AB + C)(D + E))_	

comprised of just an EXOR and an AOI22. Similarly, the full adder is comprised of a three-bit EXOR and an AOI222. Personally, it appears to me that it would be better to build these functions by means of wrapping the two appropriate functions within a soft wrapper. This would then let the automatic place and route tools place them separately "where they are needed" as opposed to combining them in a hard function that needs to be placed at a compromise location. This will be discussed further in a later section.

However, I have never gotten a customer to go along with my assertions. Hence, if your design team desires to keep these types of functions in your synthesis runs, then do so. They do not hurt; they are just redundant.

2.2.1.2 Multistage Several other functions exist that are customarily added to a typical stdcell listing. These are all additional multistage circuits. These are shown in Table 2.5.

TABLE 2.5 Basic Two-Stage CMOS Functions. These are the usual two-stage families on which the most popular synthesis engines rely.

Name	Function	Note
BUF	A	Double inversion
AND2	AB	
AND3	ABC	
OR2	A + B	
OR3	A + B + C	
XOR	AB_ + A_B	
XNR	AB + A_B_	
MUX	AS + BS_	
MXB	(AS + BS_)_	
NAND3i	(ABC_)_	Can be done with 3-stack NOR (or less) into a 3-stack NAND (or less).
NAND4ii	(ABC_D_)_	Can be done with 3-stack NOR (or less) into a 3-stack NAND (or less).
NAND5iii	(ABC_D_E_)_	Can be done with 3-stack NOR (or less) into a 3-stack NAND (or less).
NOR3i	(A + B + C_)_	Can be done with 3-stack NAND (or less) into a 3-stack NOR (or less).
NOR4ii	(A + B + C_ + D_)_	Can be done with 3-stack NAND (or less) into a 3-stack NOR (or less).
NOR5iii	(A + B + C_ + D_ + E_)_	Can be done with 3-stack NAND (or less) into a 3-stack NOR (or less).
NAND2i	(AB_)_	Can be done with 3-stack NOR (or less) into a 3-stack NAND (or less).
NAND3ii	(AB_C_)_	Can be done with 3-stack NOR (or less) into a 3-stack NAND (or less).
NAND4iii	(AB_C_D_)_	Can be done with 3-stack NOR (or less) into a 3-stack NAND (or less).
NOR2i	(A + B_)_	Can be done with 3-stack NAND (or less) into a 3-stack NOR (or less).
NOR3ii	(A + B_ + C_)_	Can be done with 3-stack NAND (or less) into a 3-stack NOR (or less).
NOR4iii	(A + B_ + B_ + D_)_	Can be done with 3-stack NAND (or less) into a 3-stack NOR (or less).

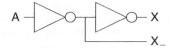

FIGURE 2.14 Once Common but Now Fairly Forbidden Two-Stage CMOS Stdcell Function. Because the delay to the last output is dependent on the loading of the intermittent stage, it is best to not have such cells in a non–"expert use only" library release. However, this can be eliminated by the general rule of only allowing multiple outputs on sequential cells.

Those functions in Table 2.5 with inverted inputs are especially appropriate for synthesis. Design Compiler searches for AND/NAND and OR/NOR functions with one or more inverted ENABLE signals. Note that this multistaging of NANDS and NORS can be extended to Booleans as well (AOI into OAI or vice versa), but the resultant functions (AOI or OAI with some inverted inputs) are less likely to be used because Design Compiler does not search for these more complicated constructs in netlist. In addition, because of the large number of Boolean functions already listed, plus the number of possible ways to stage them onto succeeding inputs, staging one after another would cause significant and cumbersome increases in the library functional offering. Hence, they are not included here.

As an appropriate additional, just as there are BUF functions, AND functions and OR function inversions of INV functions, and NAND functions and NOR functions, each Boolean listed in the previous tables in this chapter could be duplicated with appropriate negated output. Although these were not listed, they might certainly be added to a typical stdcell offering.

Also note that the preceding list does not have full-adder and half-adder circuits. They can be added, and most stdcell libraries do offer them, but they are just combinations of AOI222 and three-bit XOR. Design Compiler or other synthesis tools will just as easily synthesize adder circuits from these. In addition, without the need to have both functions combined, the separate functions can be placed by an automatic placement and routing tool at separate optimal locals as opposed to a compromise single local.

Older stdcell libraries sometimes have other multiple output stdcells such as a combined INV/BUF. Note that these elements are always either inefficient (because of the need for isolation of the two outputs from each other) or difficult to use (because of the loading on one of the nonisolated outputs, effectively causing added delay on the other output). Figure 2.14 illustrates such an element. These should be avoided at all times.

2.2.2 Sequential Functions

We have yet to describe the needed sequential cells (flip-flops and latches). For pure synthesis purposes, the flip-flop offering can be limited to D-type flip-flops only (Design Compiler does not synthesize JK flip-flops and uses D-type flip-flops for toggle flip-flops). However, every design team eventually comes across a legitimate need for a JK or toggle flip-flop and can add one. In addition, design teams usually require many integrated functions to be added to the otherwise limited flip-flop offering. These can include synchronous clears, synchronous sets, combined synchronous clear or set, asynchronous clears, asynchronous sets, combined asynchronous clear or set, clock enables, data enables, positive clocked, negative clocked, various prioritizations of some of these added functions over others of the list, and, probably most important, MUX-D scan inputs (for test purposes). With the preceding description in mind, Table 2.6 gives a useful but not necessarily function-priority-ordered complete set of flip-flop stdcells.

For typical design practice, where all flip-flops need to be scan capable, this list can be further paired down to just the scan flip-flops. However, legitimate reasons exist to have flip-flop circuits that do not contain the added MUX-D. A specific example would be the need for high-speed communication channel design. By removing the scan test requirement for the flip-flops in such communication channels, those channels can be made to run at faster speeds, allowing the reaching of performance goals with less special or

TABLE 2.6   Fairly Complete Set of Common Flip-Flops and Latches (Both Scan and Nonscan Versions), LSSD Functions Excluded. Although other sequential functions certainly do exist, these make up the more useful set.

Name	Function	Note
DFFP	Positive clock D flip-flop	
DFFRP	Positive clock D flip-flop with asynchronous reset	
DFFSP	Positive clock D flip-flop with asynchronous set	
DFFRSP	Positive clock D flip-flop with asynchronous set and reset	There is an inherent race condition between asynchronous set and asynchronous reset: designs need to be warned not to rely on coming out in a known state if they are both going inactive at the same time.
DFFCP	Positive clock D flip-flop with synchronous clear	Synchronous clear also works as a data enable.
DFFEP	Positive clock D flip-flop with clock enable	
SDFFP	MUX-D positive clock D flip-flop	
SDFFRP	MUX-D positive clock D flip-flop with asynchronous reset	
SDFFSP	MUX-D positive clock D flip-flop with asynchronous set	
SDFFRSP	MUX-D positive clock D flip-flop with asynchronous set and reset	There is an inherent race condition between asynchronous set and asynchronous reset: designs need to be warned not to rely on coming out in a known state if they are both going inactive at the same time.
SDFFCP	MUX-D positive clock D flip-flop with synchronous clear	Synchronous clear also works as a data enable.
SDFFEP	MUX-D positive clock D flip-flop with clock enable	
DFFN	Negative clock D flip-flip	
SDFFN	MUX-D negative clock D flip-flip	
JKFFP	JK positive clock flip-flop	
JKFFN	JK negative clock flip-flop	
LATP	Positive nonscan lat	
LATN	Negative nonscan lat	
SLATP	Positive scan lat	
SLATN	Negative scan lat	

custom circuitry. Test coverage of such channels can be had with a functional test as opposed to a scan test. With that said, however, most stdcell libraries offer both scan and nonscan versions of their flip-flop listings.

Positive and negative scan and nonscan latches are also listed. Most design teams would insist on the addition of these functions, which are useful in so many circuits.

For all of these sequential cells, several possible circuit designs exist. As an example, a flip-flop can be designed as a "transmission gate" flip-flop, with the MUX at the front of the master and slave latches being accomplished by some combination of transmission gates and tri-statable inverters, or it can be designed as a "Boolean" flip-flop, with the MUX at the front of the latches being accomplished with Boolean functions. The particular circuit design that is used should depend on the primary goal of the library of which the sequential cell is part. A transmission-gate flip-flop will tend to be faster (with minimal setup and hold constraints) but will take more space to layout (because of the extra routing required to accomplish the opposing P-channel and N-channel gating within the transmission gates). A Boolean flip-flop does not suffer from this need for the more complicated routing of the transmission gates, but it tends to be slower with longer setup and hold constraints. Hence, it may make sense to have examples of both types within the same library. The reason for this is that few of the various cones of logic tend to be true long paths in a typical ASIC, requiring the fastest setup constraints and propagation delays. As a result, if both types of flip-flops are available, then most cones of logic can use the denser Boolean flip-flops, whereas only the true long paths that truly need them will use the transmission-gate flip-flops.

2.2.2.1 *Metastability* A typical storage element such as a flip-flop or a latch has internal feedback loops (a flip-flop has two, a master feedback loop and a slave feedback loop; a latch has one such loop). In their simplest sense, these feedback loops are composed of two inverters feeding into each other. There is usually some sort of MUX circuitry in order to merge the incoming signal into the loop and probably some sort of isolation circuitry on the output as well, but the coupled inverter design is the basis of the loop. These loops and the mechanism that cause the storage element, be it a flip-flop, a latch, or a RAM bit cell, to actually store data. Either one side of the loop is active while the other is not (which is one of the two stable states) or the first is not while the other is active (which is the other stable state). The basic operation of the rest of the circuit is to merge the incoming signal appropriately and to interpret correctly which of the two stable states means what. However, in any such feedback loop with active circuitry, there is an inherent race condition. One or the other of the two inverters can start to change the state of the node connecting it to the other inverter fast enough that both nodes can be active or inactive at the same time. Then each node tries to resolve the other node to the opposite case. This is an unstable event, and when it happens it can usually resolve itself within not too many cycles. However, during that length of time, the storage element (flip-flop, latch, bit cell), is not functioning properly. This is called *metastability*. It is caused either by the incoming edge of the synchronous data signal happening too close to the clock edge, being removed too soon after the clock edge, or being settled for too short a time around the clock edge.

It is real, and it does happen. In addition, it is extremely difficult to simulate because you really cannot tell how close to the clock edge the data edge must be or how slow or fast the slope of either the clock or the data signal has to be. One could be a fraction of one picosecond away from causing metastability during a simulation and not be able to see any indication of it being close to occurring.

So, what can be done about it? Certain stdcell design techniques can make the flip-flops and latches less prone to having this occur. These include using significantly different strengths on one of the two inverters in the loop as opposed to the other or using significantly different P-channel versus N-channel lengths in the two transistors of at least one of the inverters, thus changing the voltage when it switches well away from the usual halfway to rail. However, metastability happens, and no storage element is immune to it. The best way to handle it is to ensure that when the cell is used in a given design, the signals feeding the data input of the storage element occur significantly far away from the clock edge and are stable for enough time that metastability does not occur. Methods to do this are described in the Chapter 8.

2.2.3   Clock Functions

The clock functions described here fall into two categories. First, there are the buffering cells. These are used to increase the drive strength of the clocking (or control) signals that are distributed across the various chip designs. Second, there are the clock-gating cells. These are used to cleanly enable and disable clock signals.

The nature of a clocking signal is that it needs to control several, perhaps thousands of storage elements, and it needs to do so in a synchronous manner. It thus needs to have fairly sharp signal edges, and those edges need to arrive at the various storage elements at as close to the same time as possible. The reason for the first of these is obvious. As an edge falls over, the point at which it causes the storage element to grab the data signal value becomes more difficult to define (because of voltage, process, and temperature fluctuation uncertainties and on-chip variations in how the transistors of the particular storage element will switch). A sharper clock edge passes through these threshold uncertainties more quickly, and the instant of data acquisition becomes better defined. The reason for the second is that in modern digital synchronous design, flip-flops, latches, or RAMs across a design, perhaps even physically located on opposite sides of a design tend to need to grab all the data "all at once." That is to say, the data are determined and registered within the smallest and most defined time possible after each clock cycle. Are there design techniques such as "useful skew" that can take advantage of slightly delaying or slightly speeding up some flip-flops or other storage elements clocking versus the remainder of the flip-flops on a particular clock? Yes, but these techniques are variations off the base principle of straight digital synchronous design, and all still require a known, although admittedly nonzero, amount of delay.

So how do design teams accomplish this? The primary technique is brute force. Figure 2.15 shows the first three levels of what is usually known as a *clock tree*. There is no set number of levels for such a clock tree. Sometimes, just one level, that of a large buffer (or inverter dependent on the flavor of the clock signal), suffices; sometimes the number could be four or five or more levels, although even for the largest designs, this number tends to be no more than six. In addition, the figure shows a regular fan out of more than three. There is no fixed form with this either. Typically, fan out is at the least two, causing a regular bifurcation of the clock signal network, but it could be larger, requiring fewer levels in order to generate a large number of leaf nodes. It is possible that there will even be different fan outs at each node in the clock tree. Beyond all of that, there are several variations of this. If the number of flip-flops or storage elements that need to be driven is small, then it could be as simple a clock tree as a signal buffer. If the requirements for synchronicity are tight enough in a particular technology, then it could require a messing together of various stages after some or all of the buffering stages in the tree. If certain

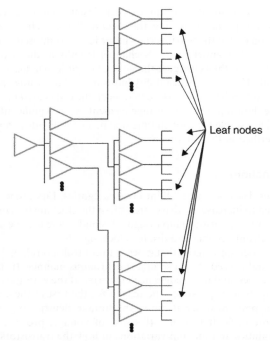

FIGURE 2.15   A Basic Clock Tree. Also a useful design technique for any high fan-out signal

synchronous skewing requirements exist, then some branches of the tree could be purposely delayed, for instance, by adding an additional one or two stage of buffering that the rest of the clock tree does not require. This book is not a treatise on clock-tree design but one on, among other things, designing elements that clock-tree designers would need in order to make efficient clock trees; hence we will leave further discussion of the various clock trees for now. However, in general, the image is a typical clock tree.

Each stage tends to drive larger stages, cumulatively (and sometimes absolutely) as the clock tree fans out. At the *leaf nodes* of the tree, the individual clock signal leaves drives out to a subset of the total storage element loading the clock signal. The brunt force comes into play at these nodes and at each of the internal levels of the clock tree. Each succeeding level drives large capacitive loads. At the leaf nodes, the edges of the signal have to be sharp. Hence, the solution: make the buffers (or inverters) used in the clock tree large. You will find that most stdcell libraries have a set of CLKBUF buffers and CLKINV inverters that are "reserved" for the clock tree (or other signal tree, typical of control signals such as RESET and TEST_MODE). The real reason they are reserved is that they tend to be the largest drive strengths of any buffer or inverter in the stdcell library. Hence, they take up larger footprints on silicon and consume more power during operation. These can be used in any signal in the design, but the stdcell vendor (or fabrication house) would like to restrict their use to just special cases such as clocks and control signals. Now, is there any other characteristic of these large buffers and inverters that they have in common? Yes. Most stdcells tend to have P-channel to N-channel ratios that are geared to "fastest possible performance" (or "lowest possible power consumption," or "smallest possible footprint"). None of those three "best possible" bear directly to another basic requirement of clock trees: the need for "equal propagation." Typically, along with these cells being larger, the typical P-channel to N-channel ratio will be adjusted in order to allow for equal rising-edge and falling-edge propagations and equal rising and falling edge rates. Realizing this and knowing that it might make sense to allow for NAND, NOR, and MUX usually leads to the addition of several additional functions in the "clock buffer and inverter" list. Specifically, a typical stdcell library has several drive strengths of "equal rise and fall and propagation," CLKNAND, CLKNOR, and CLKMUX2. These usually do not get the largest strengths that the CLKBUF and CLKINV do because they are intended to be used as early in the clock tree as possible (which makes sense because doing so later in the clock tree would require repeating the NAND or NOR function on the signal appropriately).

This discussion is pertinent to the stdcell library user because this is one of the few physical view issues that, if you have the need to change, you may not need to go back to the fabrication house. There are usually enough drive strengths of each of these types of cells that if a finer granularity of drive strength is needed, then it is possible to build a hard macro by combining two or more of the extant cells and then either connecting in router metal or "soft connecting" in verification. The result is a hard macro with the drive strength needed without adjusting any polygon in the stdcell library.

The second category of clock functions that are usually included in a stdcell library are the clock-gating functions. Clocks tend to be significant consumers of power on modern designs, partly because they tend to be made of large clock trees and must drive large amounts of capacitive load and toggle at the clock rate, which is the fastest toggling on a design (with the exception of internal glitches from asynchronous race conditions inherent to much design that is done, including RTL synthesis). Hence, there is a need, especially in low-power design, to minimize clocks when they are not in use. The solution is to somehow gate off a clock. However,, in digital synchronous design, there is an issue with the normal method of gating off a signal. Normally, a signal can be AND or OR with another signal; when applied, the result can be forced to a logical active or logic inactive as needed. But the application of that signal can occur asynchronously with respect to the original signal. This can and does cause more than occasional glitches to arise in a cone of logic as one signal goes active or inactive asynchronously with other signals in that cone. Usually, in modern RTL synthesis, this is okay as long as the signal settles appropriately at the next flip-flop input by the next clock edge. Allowing it on a clock itself, however, can cause "runt" clock pulses. Figure 2.16 illustrates this possibility.

Allowing such runt clock pulses can easily cause intractable operation on a chip because some flip-flops, latches, or bit cells recognize the runt pulse as a legitimate clock edge and some do not. Hence, it is desirable to find a way to gate the clock on or off only when the clock is in an appropriate phase (that is, when it is inactive). Clock-gating cells do this function. They usually come in four flavors:

1. those that set an active high clock signal off with an active high enable;
2. those that set an active high clock signal off with an active low enable;
3. those that set an active low clock signal off with an active high enable; and
4. those that set an active low clock signal off with an active low enable.

Figure 2.17 will illustrate how the first of these four classes of clock-gating functions operates. The operation of the remaining three, which amounts to just inverting either or

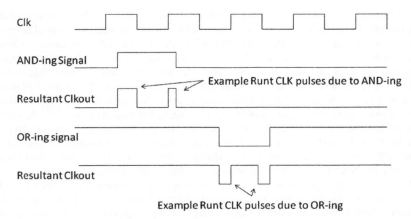

FIGURE 2.16 Runt Clock Glitches: Why Clock Gating is Important. Care needs to be taken in order to ensure that runt asynchronous signal pulses (such as on clocks and resets) do not propagate.

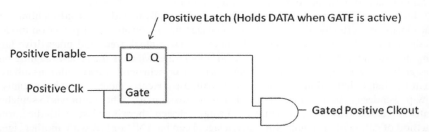

FIGURE 2.17 High-Level Representation of One Flavor (Positive Enable, Positive Clock) of a Clock-Gate Function. The four flavors are the keys to the power-reduction techniques used in the leading synthesis tool.

both of the incoming ENABLE signal and the polarity of the latch, is an operation left to the reader. In Figure 2.17, the latch will allow the ENABLE signal to pass through when the CLK is inactive but prevents the signal from going past the AND. When the CLK signal goes high, if the ENABLE signal is active, then the latch output is active, and the AND function passes the CLK signal as CLKOUT. On the other hand, if ENABLE is inactive when the CLK signal goes active, then the latch output is inactive, and the AND function prevents the CLK signal from propagating to CLKOUT. This amounts to the first of the four flavors of clock gating just listed.

2.3 EXTENDED LIBRARY OFFERINGS

2.3.1 Low-Power Support

Low-power design support can mean many things. The technology node that is being considered largely influences how extensive the low-power extensions need to be. For larger technology nodes, above 180 nanometers, design support tends to be just the previously described clock-gating cells. As the technology moves deeper, the support of multi-Vt designs might become needed.

By the way, if the stdcell library, along with the fabrication house that supports it, does have multi-Vt capabilities, then it is important to be able to place different Vt cells next to each other in an automatically place and routed design. This is because design centers, wanting to minimize power, will most likely synthesis with the highest Vt library (lowest power, least performance) available and then to accomplish timing closer, after automatic place and route, by substituting the cells in the failing long-paths with lower Vt cells (assuming that the same cell is available in each Vt). Figure 2.18 shows before and after views of automatically placed and routed stdcells having long-path cells substituted for by lower Vt stdcells.

Notice that the there are regions where the opposing Vt regions can come together corner to corner (as in the Four Corners geographical feature of the United States). These corner-to-corner contacts cause lithographical difficulties, and some fabrication houses will not support then. The lithographic issue is that the masking plate for the two large regions of antireflective chrome comes together at a point. In reality, these two regions either will have to be separated by a small gap (the regions having been pulled back from each other) or will have to be merged. During incoming mask inspection into the fabrication house (or outgoing inspection out of the mask house), these gaps or mergers can cause some to reject the mask. This is truly a lithographic issue and not an electrical issue. Any transistor that could be affected by the pullback or the merger must be already far enough back from the edge of the stdcell in case the cell is used at the edge

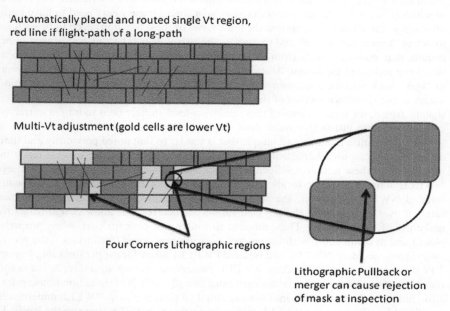

Automatically placed and routed single Vt region, red line if flight-path of a long-path

Multi-Vt adjustment (gold cells are lower Vt)

Four Corners Lithographic regions

Lithographic Pullback or merger can cause rejection of mask at inspection

FIGURE 2.18   "Per Cell" Automatic Placement of Different Vt Cells Leading to a "Four Corners" Lithographic Issue. In actuality, cells can be designed such that this is a nonissue (by ensuring that no transistor source or drain falls within the region where such region lithographic pullback can occur).

of an automatically placed and routed region. However, working with the fabrication house in order to allow this potential DRC pseudoviolation is imperative to allowing this sort of design-center effort.

However, for deeper submicron technologies, even multi-Vt design is not enough. What is needed is twofold. The first need is simple: allow some amount of voltage-performance scaling. The basic concept here is that only some parts of a circuit are required to meet critical performance goals. Most of a circuit can run at a lower voltage and hence at a lower level of performance and still meet the timing requirements for the circuit. A lower voltage directly affects power consumption. Hence, while the critical sections of a device run at one voltage, significant parts of a design run at a second lower voltage. Alternately, entire sections of a design can be shut down when not in use. To support these design techniques, however, certain functions need to be added to a library and these may not be present in a fabrication house's stdcell offering.

The first of these is a level-shifting function, but it is more than the typical threshold-adjusting kind that can be part of a stdcell library and is used to shift logic signal voltages between lower-voltage cores and higher-voltage IO structures. When just dependent on the different voltages used in a voltage-frequency design, this kind of level shifting may or may not be necessary. For designs that allow active shutting down of parts of the circuit, there needs to be a series of functional changes that need to be part of both the circuit and the simulation environment. The added feature of these internal level shifters is that of isolation. In the circuit design of these isolation-level shifters, detection of shutdown and the prevention of unknowns from propagating into the parts of the circuit that remain active need to be addressed. This can be done by either retaining the last known good signal logic level or by driving it to a known state when shutdown is detected. The choice of which is better is a design-center decision, so a stdcell library may need to have both offerings in the library. In addition, the logic views of these functions need to be such as to detect "logical unknowns" and to react as the circuit does. There also may need to be circuits that prevent signals from the remaining active side from propagating into a shutdown section of the design. Allowing such signals may allow some amount of power to "leak" back into the shutdown side by means of reverse biasing the P-channel transistors in the shutdown section of the design that happen to be attached to these active signals. Again, the logical views of these isolation-level shifters need to inhibit correctly such propagation just as the circuit does.

WELL isolation is another technique that is similar to that noted previously and that design centers can use. Admittedly, fabrication houses must have the process ready in order to allow these techniques. Such process is usually made of but not limited to deep NWELL support in order to allow isolated PWELL or no deep NWELL, allowing isolated NWELL. However, the stdcell library offerings of these fabrication houses rarely if ever directly offer the physical support structures that allow design teams to make use of the processes. These amount to router row endcaps that, when properly placed and in conjunction with the correct isolation processing (for instance, the previously mentioned deep NWELL and isolated PWELL). allow for proper isolation. Figure 2.19 is a rough example of a deep NWELL router row endcap structure, but it is for illustrative purposes only; actual endcaps must comply with the fabrication house rules. Also, note that this figure only makes sense strictly if there is a deep NWELL underneath the PWELL; otherwise, the PWELL within the resultant NWELL ring and the PWELL without the resultant NWELL ring are connected underneath the resultant NWELL ring. These physical support structures cannot be simulated, so design centers need to take care of the proper usage and library support engineers may need to include some

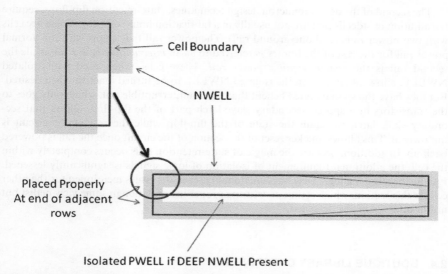

Cell Boundary

NWELL

Placed Properly
At end of adjacent
rows

Isolated PWELL if DEEP NWELL Present

FIGURE 2.19   Example of an NWELL Endcap Used with Deep NWELL to Provide Isolated PWELL for Low-Power Design. Alternately, with no deep NWELL and without the endcaps, isolated NWELL (allowing NWELL pumping) can occur.

sort of test that design managers can incorporate into the design-verification process. This is not a trivial task. The best that I have seen is an identification layer for the isolated WELL that is combined at extraction and verification with Boolean constructs that flag errors if the identified isolated WELL exist and touches nonisolated WELL. Fabrication house extraction decks, straight out of their physical design kit offerings, may not be critical enough in verification of this and may need to be adjusted.

With isolated WELLs, other functions become possible additions to a stdcell list. These are the WELL-pumping cells. Well-pumping is the act of shifting the voltage of isolated WELLs from the CMOS normal mode of NWELL driven to VDD and PWELL driven to GND. Shifting the voltages of the substrate WELL, isolated from other non-isolated WELL, can reduce the power that is consumed by the circuits placed in the isolated WELL. This is at the cost of performance (or conversely increase the perform-ance of circuits in the isolated WELL at the cost of power consumption, assuming that the pumping goes the opposite way—that is, above VDD for NWELL or below GND for PWELL). But just as the isolation requires the addition of WELL isolation endcaps, plus the process option previously described, WELL pumping can require active and large-area WELL-pumping circuits. The techniques to build a WELL pump are described in many lower-power design books and will not be repeated here, but support of these cells will need to be added to the library if the design team requires it. As already noted, these cells do not tend to have logical view simulation capability (although they do have a function in the SPICE environment). Because they do not add function in the logical view environment, though, there is a need to be able to add these cells in the physical view and in the verification view but not in the logical, timing, or test view. In addition, just as previously noted, this is not a trivial task, ensuring that these cells are properly added to some view netlist but not to others.

The second of the deep submicron design techniques, state-retention flip-flops, require the addition of stdcells that are not usually in a fabrication house offering. These are cells with two power rails (and one ground rail). The power rail that aligns with the normal power rail for the rest of the library is referred to as an *interruptible power rail* while the second rail is the *noninterruptible power rail*. These cells, when used with isolated NWELL, allow all circuits on the isolated NWELL to be turned inactive when desired. But they have the special added benefit that the noninterruptible rail, which only goes to the transistors in a special secondary slave-latch part of the flip-flop, allows that secondary slave latch to retain the state of the flip-flop while the rest of the circuit is unpowered. This allows quicker reset of the section of the circuit once the rail is powered back up. In addition, because the magic of state-retention cells occurs completely within the cell, the additional requirement of isolation of logic signals is significantly lessened. Finally, because the logic view has functionality, even the issues mentioned for the other cells in this section, which is that of assuring proper instantiation in some view netlist but not in others, are also lessened.

2.4   BOUTIQUE LIBRARY OFFERINGS

Much as the name implies, boutique libraries are stdcell libraries that exist for, and are used in, special and unique design environments. As such, they may not need much of the optimization that is being described throughout this book. They tend to be "optimized" in a different manner.

An example of a boutique library would be one for a technology shrink of an extant ASIC design (or design block) that is a known good (read functionally correct). The design center may not wish to resynthesis from a RTL, which would require some amount of formal validation of the resultant netlist, or the original RTL may have been lost, or too significant an amount of postsynthesis manual manipulation of the gate-level netlist may have taken place. Perhaps another company that had been using a different stdcell library supplier has been bought and the task is to transfer (known as *port*) a netlist into the correct design environment for use by other internal design centers in the same technology node. However, for whatever reason, the design center has a gate-level netlist that it generated for a different technology node in a different stdcell library and it needs the exact same netlist in a technology node and environment that is somehow foreign to the original design. One way to go about this is to create a set of library stdcells that match functionality and name to the original stdcell library. This could be done with soft macro wrappers, around some subset of the new stdcell library that represents the original stdcell library cells. Another is to use only those functions and cells in the new stdcell library that are used in the original library. A third is to port in the original stdcell library, assuming the technology node is the same, and use it as a hard-macro block of the original design. There are other methods as well. Which one is used is highly dependent of the level of difficulty in porting the design, the level of availability of the original stdcell library, and the allowed amount of access to the original stdcell library.

In all the preceding cases, however, the desire is to match functionality and possibly size and performance as opposed to optimizing the stdcell library offering. Does this mean that other views (specifically, logical and timing) need to match? For the first of thee, the answer is no. If adjustments to the logical views allow for better reachability or better observability during testing, then this needs to be addressed. For the second, the answer has to be no. Moving across a technology node will necessitate a new set of timing and power numbers, even moving between fabrication houses as the same technology node forces a new set of timing and power numbers.

2.5 CONCEPTS FOR FURTHER STUDY

1. Multistage, two-stack versions of the stdcells listed in this chapter:
   - As technology nodes continue to shrink, the need for reducing the number of transistors stacked between rails and outputs will increase. However, all of the functions presented in this chapter have been shown to be useful in multiple designs over many technology-node generations. It might be useful for the student or interested engineer to attempt to replicate as many of the given functions in two-stack logic with increased staging depth.

2. Transmission-gate–based MUX versions of the various cells:
   - All of the MUX logic presented in this chapter was implicitly designed with a Boolean-based MUX function in mind. However, there are times when a true transmission-gate–based (or tri-statable inverter) MUX would be beneficial (with significantly faster performance). The student or interested engineer might attempt to design such MUX (and explore pin-order benefits in such designs).

3. Boolean-based FLIP-FLOP versions of the various cells:
   - Almost all FLIP-FLOP circuits designed over several generations of technology nodes have used some amount of transmission-gate–based MUX circuitry inside the actual stdcell (usually, but not always, three tri-state inverters and one true transmission gate). However, as technology nodes head toward 20 nanometer, efficiently handling the significant amount of signal crossing within the cell that is required by such a circuit topology can become cumbersome, One alternative is to use Boolean constructs within the stdcell as opposed to transmission-gate constructs in order to MUX the appropriate signals together. The student or interested engineer might find it interesting to attempt to design a Boolean-based FLIP-FLOP for each FLIP-FLOP listed in the chapter. One caveat: Boolean-based FLIP-FLOPS tend to be slower in both output propagation and synchronous input constraint.

4. Pulsed versions of the various cells:
   - Several design centers have attempted to use pulsed logic where a logic function does not resolve until an active pulse is recorded on a transistor (or more than one transistor) within the cell. Such designs have been implemented and used, with varying degrees of success, over multiple technology nodes. The student or interested engineer might be interested in attempting to design such a set of functions for many if not all of the functions listed within the chapter.

5. Dual-flavor logic:
   - Many years ago, I was involved in a current mode logic (CML) design. CML technology inherently provides both the true and complement flavor of a function for every function capable within the technology. The resultant design that I produced had the smallest number of inverter functions (all used as buffers) of any design in which I have ever been involved. Since then, when I build CMOS MSI- and LSI-level logic blocks, I take great pride in attempting to

reduce the number of inverters required. I do so by building both flavors of signal values within every cone of logic (and then removing those that become unnecessary). To do so, I employ a special set of logic functions that supply both the true and complement outputs of each function output (and usually require the true and complement version of each function input). The student and interested engineer might find the design of such cells for many to most of the functions presented in this chapter to be of interest.

6. Majority vote usage:

   • As mentioned in the chapter, the two-of-three majority vote MAJ3 and the three-of-five majority vote MAJ5 stdcells are usually not included within a stdcell library offering. The student or interested engineer might be inclined to develop some RTL, perhaps for a communication channel error detection and correct circuit and see how the synthesized netlist work change as a function of the inclusion versus exclusion of these two stdcells.

7. Minimal versus maximal stdcell library synthesis:

   • Appendix I tells of synthesis of a four-bit full adder with a minimum number of stdcells versus a richer offering of the same. The student or interested engineer might be inclined to repeat this exercise using various other RTL in order to understand why a rich Boolean offering is important to allow smaller and faster synthesized netlist.

CHAPTER 3

IO LIBRARIES

3.1 LESSON FROM THE REAL WORLD: THE MANAGER'S PERSPECTIVE AND THE ENGINEER'S PERSPECTIVE

Several years ago, I was asked to build all of the possible offerings out of a company's Input–Output (IO) compiler. The compiler was old and unsupportable. The engineer who had originally built it had long since retired or moved on, and upper management wished to get rid of it. The tool had some interesting features. Designers could choose various input and output options but could also get various strength pull-up and pull-down resistors that would be integrated within the placement boundary of the resulting IOs. Although from a usefulness point of view this was a "nice to have" feature, it allowed several small issues to creep into the resulting design. Specifically, the IO produced, whether specified as a pure input or a pure output, was actually a full tri-statable IO. Also, every IO, whether or not it had a pull-up resistor, a pull-down resistor, or a bus-hold circuit specified as needing to be attached to the pad, actually had all of these, including various strength versions of each, attached to the pad. In addition, the cells were huge. All of the circuitry for all of the various drive strengths was present in all of the IOs, and because the IOs were the same width, the IOs were nothing more than metal connection gate-array structures as opposed to true stdcell IO structures. Management had told me that it thought that there were more than 1,400 different IOs that could be built from the compiler and asked me how long it would take me to produce each and then productize each so that supporting this compiler would no longer be necessary. When I realized what the compiler was doing, in about one day I produced a global cell

Engineering the CMOS Library: Enhancing Digital Design Kits for Competitive Silicon,
First Edition. David Doman.
© 2012 John Wiley & Sons, Inc. Published 2012 by John Wiley & Sons, Inc.

(it had to be all-encompassing because all of the circuitry was present and would therefore be tested during chip integration).

When I told the managers what I had done and why, they had me produce a list of the more than 1,400 possible IOs. We took the list to the design teams and asked which ones that they would ever use (asking for proof of usage on their earlier designs). Approximately 100 IOs were then produced with separate layouts. In addition, the design teams had to go back and retrofit new production tests for all of their old designs that had used the gate-array IO versions in order to test the functions of the IO that, although they were not intended, were still present and needed to be "test validated" that they were assuredly turned off.

By the way, in the new pared-down IO, I removed the pull-up and pull-down resistors from inside of the IO boundary. These were now placeable structures alongside the IO and could be added or changed in the netlist as opposed to having to replace the IO cell. This limited the versions of the IO that were needed. Admittedly, the design teams now could not be assured that the pads on the design would be exactly so far apart. Sometimes, room for a pull-up or pull-down resistor had to be added to an IO ring. Only one design team complained and not too loudly.

3.2 EXTENSION CAPABLE ARCHITECTURES VERSUS FUNCTION COMPLETE ARCHITECTURES

As just shown, there are many different ways to implement a set of IO functions. Why should this be so? Are not IOs just larger versions of stdcells? Yes, in fact, in many ways, IOs certainly can be considered just large awkward stdcells. However, that very awkwardness allows for this myriad of ways to implement IOs. Because of the large number of combinations of the various functions, drive strengths, pull-up and pull-down resistors versus bus hold circuits, pure input versus pure output versus tri-stated bidirectional capabilities, IOs can be viewed as populating a middle ground between the stdcell realm and the memory compiler realm. The realm of the stdcell is one of every cell being an extant member of a proffered set. The realm of the memory compiler is one where every instance is created only when and if needed. As stated, the realm of the IO offering is between those two extremes. This is further confirmed because even characterization thresholds lead to hundreds or even thousands of variations, many of which will never be used in any design exist. If the design center's environment is such that it can truly define a hundred specific IOs, then it is beneficial to optimize these IOs just as you would the most used stdcells. However, if this is not the case, and it usually is not, then building an extensible set of pieces makes sense. Note that an extensible set of pieces does not imply the need to build or maintain an IO compiler. The pieces can be instantiated appropriately by the design center during chip integration.

For instance, Table 3.1 shows possible functions and options that can be the basis of a workable extensible set of IOs capable of being combined in the gate-level netlist of a design while being completely maintainable by the typical IP organization supporting said design center.

The list alone could mean that more than 7,500 IOs might need to be developed—and it is not an extensive or complete list. Several additional options are certainly possible, such as which layers the resistors are to be built (with the implied different characteristics of those layers across temperature and process and voltage) or the number of metal layers out of which each IO is built. For the design center that wishes to add on to a fabrication

TABLE 3.1 **List of IO Function Options and Elements. This is a small sampling; nearly as many IO functions can be defined as are typically in a stdcell library release.**

Channel:		Type:
CMOS		Pure I
TTL	(CMOS characterized to TTL levels)	Pure O
USB	(differential)	Bidir
PCI		
DDR	(1,2,3)	Attachments:
I2C		Bus Hold
Other private licensed structures		1k Pull-up
Analog		1k Pull-down
IO Power and Ground Rails		5k Pull-up
Core Power and Ground Rails		5k Pull-down
Combined Power and Ground Rails		10k Pull-up
		10k Pull-down
Drive Strengths		50k Pull-up
1 mA		50k Pull-down
2 mA		100k Pull-up
4 mA		100k Pull-down
8 mA		
16 mA		
32 mA		
64 mA		
128 mA (rare)		

house's supplied IO offering or even to replace it, one way to pare this effort down is to separate the pull-up and pull-down resistors and bus hold circuit from the IO and make them the separate structures as previously mentioned. By ensuring that the pad of the IO has a direct connection to a pin on the core side of the IO and by making these pull-up and pull-down (and bus hold) structures separate netlist-capable elements, the list can be quickly pared down to a more manageable 250 or so elements. In addition, implementing the resistors as netlist capable allows for the later addition of currently undefined versions of these elements without having to redevelop a large number of extant IOs. Figure 3.1 shows how such elements would be placeable in an IO row in a design. By the way, notice that the resistor in the example is placed adjacent to the IO to which it is connected. This does not have to be the case in a netlist connection such as this. If there is space in an IO row—for instance, near a corner where radial bonding might force IO placement further apart—then the resistors might be placeable there and be autorouted back to the appropriate IO located somewhere else in the IO ring.

Beyond the preceding issue, there is the need to properly supply power to the IOs. Because they are the largest transistors (widthwise) on most designs and they need to drive external traces and sourcing and to sink significant amounts of current, IOs tend to need wide metal buses to supply or remove the current used by these structures. IO designers realize this and usually supply two or more layers of wide metal that over lay the entire IO structure and that can be connected by abutment with adjacent IOs, providing wide paths to the nearest IO power or ground pins, with the frequency of placement of those pins being defined by the fabrication house. However, if the design center has an IO-limited design (that is to say, the amount of logic on the design is placeable and routable in a region smaller than is available by close packing of the IO ring), then there is a strong tendency to skimp on this fabrication-house–supplied

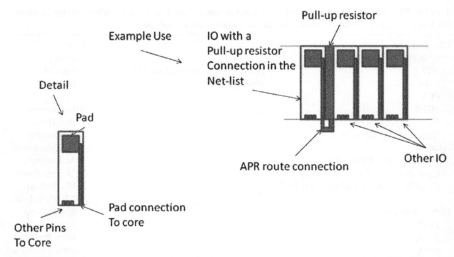

FIGURE 3.1 IO and Resistors Placed in an IO Row. This is not construction of complex IO "on the fly" but chip-level integration of any given pull-up and pull-down resistors (or even bus-hold clamping). Such an effort allows design engineering the flexibility of adding such features without relying on the a priori existence of such additions having been built into the IO. It amounts to an IO feature set multiplier because any given IO that has the ability to accept additional connections to the pad can have the pull-up or pull-down resistor (or bus-hold clamp) added.

frequency of placement. The librarian supporting that design center might be able to help the situation here. Many IOs have either extra space on higher levels of metal over the IO structure. Some have not completely populated the entire metal stack with metal rails. The widening or the addition of metal for the entire allowed stack can supply the extra width needed to supply the IO power demands at a less-frequent placement of the IO power and ground pads. Conversely, some IO structures really do have a full stack of metals, but the IO-limited design might be routable in fewer levels than with this maximum. It would be nice to reduce the cost of the design by reducing the number of metal layers in the IO to match the number of metal layers used in the route. Care must be taken here because the IO pad will have stacked layers up through each metal to the pad, so removing a layer might mean the processed design might have missing connections from the pad to the remainder of the circuit. This is especially true if the fabrication house is unaware of the design center having removed such higher metal layers from the IO. However, of course, the fabrication house needs to be made aware because it will need to understand that the part will not require this now missing level of processing.

One final way to reduce the number of IOs is to design only true bidirectional IOs. Obviously, these can be used as pure inputs or as pure outputs, when controlled properly as well as used as bidirectional devices. The penalty here, as mentioned in the opening section, is that the circuitry for the input remains on pure outputs. Similarly, the circuitry for outputs remains for pure inputs. Hence, proper production-test vectors must be maintained for the validation of these redundant circuits (in order to assure that they do not interfere with the proper operation of the design). However, there is a benefit to using bidirectional IOs in this manner. Specifically, the output side of bidirectional IOs used as pure inputs and the input side of bidirectional IO used as pure outputs can be used to cheaply and efficiently implement additional access points for various scan chains and other alternative testing structures and strategies, a feature which many design centers would find appealing.

3.3 ELECTROSTATIC DISCHARGE CONSIDERATIONS

Electrostatic discharge (ESD) considerations should be addressed across all aspects of a design: core, ground and power rails, memories, analog, and IOs. However, because it is common for ESD structures to be added to IO circuits (or, at the least, in the IO rings), the discussion has been placed similarly in this chapter.

Several structures in a fabrication house stdcell and IO library offering can be adjusted and added to in order to reduce the manufacturing margins that the fabrication house desires and that design-center users abhor. The ESD structures should not be considered one of these. A fabrication house has long-standing experience in how the material that is produced in that fabrication house responds to ESD events. ESD experts exist in the design world outside of fabrication houses, but no amount of outside learning is as directly applicable to the internal to the particular learning developed in a particular fabrication house. That being said, the ESD experts inside the fabrication house have developed the structures and the "use rules" that they know will offer the level of ESD protection on the designs run on their processes guaranteed by the fabrication house. Trust them and use them "as is." Even the addition of "extra" ESD structures created by the design centers could be detrimental, interfering with the operation of the extant fabrication-house–supplied ESD structures by providing inadvertent paths for the ESD events to travel to harmful areas of the design.

However, there are going to be design centers that will not take this advice. Hence, the following is a very preliminary primer on ESD events. It is not meant to be a complete education on the physics of ESD or the engineering of ESD-tolerant circuitry. It is meant as a listing of items to be thought about if the design center that is being supported requires its own version of ESD circuitry.

An ESD event can be defined as an excessive charge being applied across two or more pins of a device. The pins can be a signal input or output to any other signal input or output or to any power rail or ground rail. It can also be between any power rail to any other power rail or any ground rail, or it even can be between ground rails. It is the job of the ESD circuit to dissipate this excessive charge without causing serious or permanent damage to the rest of the design or to itself. ESD events can happen in many ways, but they are usually tested in one or more of three manners.

First, there is human body model (HBM), which is meant to model what happens when somebody that may have a static charge on his or her skin touches a structure such as an electrical circuit and imparts charge to that circuit (actually it dates from the 19th century when people were worried about static electricity causing dynamite explosions and methane gas explosions). A typical test circuit for this is shown in Figure 3.2.

The basic operation of the circuit in Figure 3.2 is to build a large charge on the capacitor (the model specifies this capacitor as 100 pico-farad) and then to switch the dipole and allow the charge to dissipate through a 1.5 kilo-ohm resistor to the device under test. Once the ESD event has occurred, the IV curve of the tested signal is checked for differences from what it was before the test. The level of the charge and the number of times that the test is performed before IV-curve degradation occurs determines the level

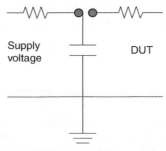

FIGURE 3.2   Typical Human Body Model ESD Testing Circuit. This is the most common and oldest defined ESD testing circuit.

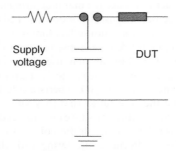

FIGURE 3.3   Typical Machine Model ESD Testing Circuit. The alternative can be viewed as a variant of the HBM as defined by various Japanese semiconductor companies.

of HBM capability to which the design is capable, ranging from class 0 (devices that fail at 250 V) to class 3B (devices that pass to greater than 8,000 V).

Second, the ESD test mode known as the machine model (MM) may be used. This is meant to be a worst-case version of the HBM. It was developed over the past few decades by Japanese semiconductor companies in order to emulate what happens as devices run through automated circuit-board–handling machines. The test circuit, which is shown in Figure 3.3, does not look too different from the HBM version.

The basic operation of this circuit is to build a static charge on a 200-pico-farad capacitor, two times the size of the HBM version capacitor, and then discharge it across the device under test through a 5-micro-henries inductor (as opposed to the HBM 1.5-kilo-ohm resistor). Just as before, after each application of the charge through the device under test, the IV curve is inspected for degradation. As with the HBM, there are levels of success defined. Class M1 is for devices that show degradation after events of 100 V or less, class M4 are for devices that pass for voltages beyond 400 V, with class M2 and class M3 in between.

Third, one common ESD test mode is known as the charged device model (CDM). It also is for testing what can happen to a device as it is handled in automated handling machines. Specifically, the model deals with charge that builds on the device and then discharges to the handling machine. By the way, many consider this method to be the best emulation of real-world ESD events that typical occurs. The test structure for it is rather different that the first two, in that the device under test is more integrated into the test circuit as opposed to the periphery of it. Figure 3.4 shows a typical CDM test circuit.

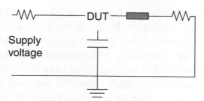

FIGURE 3.4 Typical Charged Device Model ESD Testing Circuit. This is the only "through device" ESD test model.

The basic operation of the CDM test circuit is to build a large charge inside the device under test and have it discharge through a pin of the device and then through a parasitic inductor and a 1-ohm current-measuring resistor to ground. Because of the extremely short duration of the discharge (typically about 1 pico-second), the charge flux can be extremely destructive and devices tend to not pass for very high voltages (although the testing organizations allow for passage for voltages up to or more than 2,000 V). The levels for this test are class C1 for those devices failing below 125 V and up to class C7 for those devices that pass beyond 2,000 V.

One caveat on CDM should be mentioned. Although HBM is more than 100 years old and MM is decades old, CDM is relatively new and still evolving. The number and duration of the CDM test as well as the test module is still evolving.

The bottom line, however, is this: any "increased ESD tolerance" design enhancement circuit that a typical design center wishes to add to the extant fabrication house ESD circuitry must take into account that the three described test will be run for every pin (or pair of pins) on the device. Also, realize that any result failures may or may not be limited to the IO regions of the design. At a microprocessor company where I used to work, the four large clock buffers deep within the circuit constituted the recurrent point of ESD failure for a custom-designed microprocessor unit, be it used stand-alone or as an embedded model. After several rounds of redesign of the ESD structures, the ultimate solution was the removal and redesign of the clocking circuitry.

Ultimately, using the fabrication house ESD circuitry gives the design-center user the luxury of not having to worry of such concerns, by replacing or enhancing the design house "user knows best" circuits, which could easily be risk adders to a design since the design house is potentially inexperienced in the fabrication house's processing knowledge. Hence, feel free not to take the fabrication house supplied ESD structure if you wish, but do so at your own risk.

3.3.1   Footprints

3.3.2   Custom Design Versus Standard IO Design Comparison

Many custom design houses, even to the present day, believe that there is value in a design center maintaining a full custom-design flow. This may, in fact, be true for many reasons. One reason in particular, however, needs reevaluation.

Specifically, custom-design houses may believe that full custom-designed IO structures can fit more closely together than standard automatically placeable IOs such that they maintain a marketing edge (more closely packed IOs on IO-limited designs means smaller devices, which means cheaper-to-produce devices). That is to say, any amount of automatically placeable standard IO structure will allow a less-dense placement of IOs around an IO ring. This seems logical because, in all cases, this is true for full custom circuit design in the core of the device. If that is the case in the core, should it not also be the case in the IO ring? However, the layout engineer at the design center may not view the standard IO structure appropriately. Although it is true that a typical placeable IO structure is a rectangular element, it does not need to be so. A rectilinear IO structure, as long as it is constructed in a manner that allows for adjacent placement of other rectilinear IO elements, is just as efficient as a regular rectangular IO structure. Figure 3.5 represents this idea.

With this in mind, there is little if anything that can be gained by forcing a full custom IO placement methodology. Note that there is one possible argument that could, at first glance, still be made in support of a full custom IO methodology, that of placing additional ESD or rail noise canceling capacitors in the "holes" located that a replicatable rectilinear structure still does not support. But on reflections one can still realize, as in Figure 3.6, that

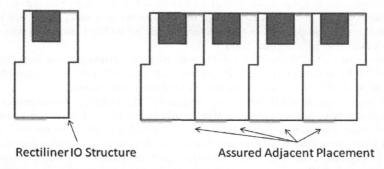

Rectiliner IO Structure   Assured Adjacent Placement

FIGURE 3.5 Rectangular and Rectilinear Repeatable Automatic Placeable IO Structures: Rectilinear shapes can be difficult for stdcell APR, but not so much so in terms of IO placement.

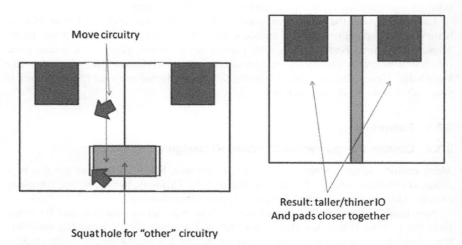

Move circuitry

Result: taller/thiner IO
And pads closer together

Squat hole for "other" circuitry

FIGURE 3.6 Why IO "Holes" Still Need to Be Tall and Thin. Squat holes force squat IOs and are more likely to allow IO-limited design.

a 'hole' of a set area is still smaller in total width if it is as tall as the remaining IO. Hence, making these structures tall and narrow allows for smaller running IO total width.

The areas of the IOs in both the left and right sides of Figure 3.6 (white rectilinear shapes on the left and white rectangles on the right) are equivalent. In addition, the areas of the red "other circuitry" rectangles on the left and right sides of the figure above are equivalent (actually, the red rectangle on the right side is slightly larger in area). The tall and thin "other structure" on the right side may be larger in area than the squat version on the left, above, but in terms of width, the right side is significantly smaller. This is partially because of the rearrangement of circuitry within the IO, which then allows the denser placement of IO pads. Hence, even with the individual components being potentially smaller with an arrangement as on the left, the combined arrangement on the right becomes smaller. Therefore, it is always possible to match, and sometimes beat, the density of full custom IO rings with standard automatically placeable IO. In addition, the use of these standard IOs can greatly reduce the amount of layout engineering effort required for the design of a device.

In addition, Figure 3.6 inadvertently points out an interesting feature of IOs. Rarely does a "squat" IO make sense. In almost all cases, a tall, thin IO will allow for smaller die. The only case where this is not true is an excessively core-limited design. In such instances, it may be possible to embed the IO within the core (actually around the periphery of the core, but each IO surrounded on three sides by circuitry). In all other cases, a proper chip floor plan will have an equal or smaller area than a tall, thin IO. Hence, it is always better to have a main offering of tall, thin IOs (thin enough to allow close packing of the pads to the limit of the chip-assembly capabilities) with a potential special IO offering of squat IOs.

3.3.3 The Need for Maintaining Multiple IO Footprint Regions on an IC

One final IO topic affects the design and support of a proper IO library offering: the need for the maintenance of multiple IO libraries, perhaps from different vendors, perhaps including special IOs designed previously from the internal design center. In many if not most of these cases, the aspect ratio of the IO could be different. Also, the ordering of the power and ground rails overrunning the IOs could be different. Having such maintenance requirements can and will eventually cause the need for the inclusion of these various IO, from various sources, on the same design. Figure 3.7 gives an example of the

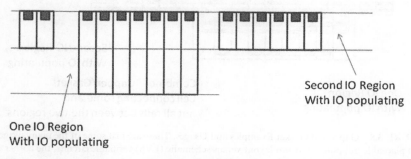

One IO Region
With IO populating

Second IO Region
With IO populating

FIGURE 3.7 Example of Two IO Regions of Different Sizes on a Side of a Design. The LEF placement decks, of course, need to comprehend the need for multiple IO sites of varying heights.

side of a design with IOs from two different sources. This can happen easily enough even if the IOs are from the same vendor. For instance, it could be that the shorter of the IO are the normal general purpose IOs (GPIO) from a vendor and the taller ones are the DDR1/2/GDDR3 IO with their much larger circuitries required to support three DDR standards from the same vendor. Things can get a little problematical if the division of the regions is too near a corner of the design. This is because the taller IO coming down into the core can severally limit access to the shorter IO on the immediate far side of the corner. Aside from that, Figure 3.7 is representative of what can actually happen on a design. Because each of the vendors of the multiple IO libraries will have no incentive to support any of the other competing vendors, it will become the responsibility of the internal (design team) library support engineer to figure out how to allow such multiple IO on the same design.

Luckily, the requirements for such support are easily implemented. First, we need to combine the regions for the two (or more) types of IO in the "logical LEF." Each IO library offering will have a logical LEF and, assuming that the layers are consistent (or can be made consistent), with an easy parsing of these LEF files identifying the few lines and cut and pasting them from one of the LEF into the combined LEF is readily accomplished. However, the second item is a little harder. These are the required *gapers* and *gaskets* that will be used to physically stitch together or physically separate the various IO regions appropriately. Figure 3.8 gives a rather high-level example of how such items could be used by expanding on the examples in the preceding figures. The extent of these gapers and gaskets depends on how different the underlying structure of the various IO are and how complicated the design's requirement is for stitching various power and ground rails together. Notice, for instance, in the example that not all rails have been connected between the shorter GPIO region and the taller DDR1/2/GDDR3 regions. Obviously, in those cases when a rail is not connected to an adjacent region, those rails need to be connected to one or more appropriate rail IO pad(s) in the isolated section.

As can be seen, gasket structures tie various IO rail (and well structures) appropriately together, and gaper structures provide for enough room between structures that are not to be tied together so that all appropriate design rule check (DRC) rules based on the physical design kit are maintained.

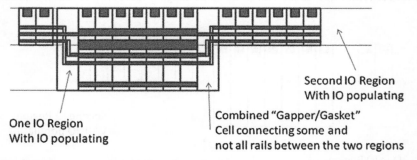

Second IO Region
With IO populating

One IO Region
With IO populating

Combined "Gapper/Gasket"
Cell connecting some and
not all rails between the two regions

FIGURE 3.8   Gaper and Gasket Examples and Usages. These are just some of the many usages of such placeable but potentially not layout versus schematic (LVS) verifiable circuits, the corner cells being another example.

Usually, there is no extractable electrical circuitry in these various gapers and gaskets. Hence, there is no need to add them to the design netlist—and there is no need to support them with much beyond the appropriate physical views of the graphical display standard (GDS) and physical LEF and the always-present documentation views. However, this might not always be the case. It could be, for instance, that to isolate noise it is desirable that a series of layout versus schematic (LVS) extractable decoupling caps or maybe a bidirectional double diode-isolation structure is to be part of the circuit. If this is the case, then they will need to be added to the netlist by the design team and their views be expanded beyond the physical to include Pre-extracted Circuit Netlist (CIR) and maybe extracted SPICE and power or performance views.

In the end, most design centers should resist the desire to massage the IO offering (because of the ESD issues mentioned previously, if nothing else). However, items such as these gapers and gaskets will always be required, and few IO vendors will supply them. This is where the design center's librarian will do most of the work on IOs.

3.3.4 Circuit Under Pad

Circuit under pad (CUP), sometimes known as *bond over active* (BOA) is a commonly used method of IO design that is often implemented to reduce total chip area. As is obvious from the name of this method, the actual bond pad for an IO is physically located over the active circuitry incorporated within the IO structure.

The one real restriction for the implementation of this concept is that the physical structure of the circuitry underneath the bond pad must be robust. This is because during the bonding process the downward-directed pressure applied to the bond pad by the bonding process has the potential of crushing any nonrobust circuitry. For instance, the circuitry under the pad should be as regular as possible and as "larger than minimal" as possible. Because most logical circuitry for a typical design does not meet either of these criteria (random logic tends to be nonregular, and typical design performance requirements mean that circuitry tends to be as small as possible), the typical circuitry under a pad has to be special. Fortunately, for IO structures in general the need for inclusion of ESD within the cell, together with the regularity of those ESD structures (and the robustness of those structures), meet the previously mentioned requirement. As a result, almost all CUP implementations involve ESD circuitry.

However, this is not the entire requirement for implementation of CUP. The reason for this is that advanced processing technologies use low-conductivity (low-K) dielectrics. Low-K dielectrics offer multiple benefits to modern design including but not limited to lower interlayer and intralayer capacitances), allowing faster circuit performance. However, one issue with low-K dielectrics is that they have little structural support capability. Hence, even the most robust and regular ESD circuit could still be damaged or even destroyed by the bonding process. Therefore, it is important that the IO design that incorporates circuit under pad must actually build some form of physically sound structure from the solid silicon substrate up through the metal structure that is being cantilevered over the regular ESD structure. Figure 3.9 graphically represents this requirement. In all such cases, it is important to note that these support structures are electrically conductive, so silicon contacting must be to electrically isolated regions (for instance, common P substrate for ground and isolated N-well for Voltage Drain-Drain (VDD) or signal pads).

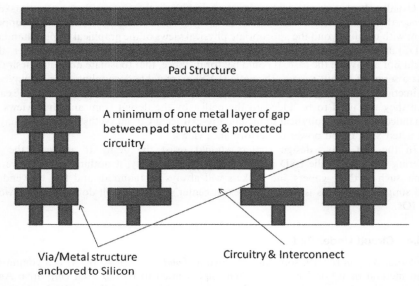

FIGURE 3.9   Protecting Circuitry Under a Bond Pad. Structurally weak low-K dielectrics mean that circuitry placed underneath a bond pad should be protected by otherwise structurally sound construction. Depending on the width and length of the actual pad structure, further vertical VIA or METAL pillars can sometimes be needed within the actual CUP area.

3.4   CONCEPTS FOR FURTHER STUDY

1. Parts are parts:
   - As mentioned early in this chapter, one can design a potentially large set of function and drive strength (and pull-up and pull-down) specific IOs. Alternately, one can build a series of subcircuits determined by analyzing the desired list of all possible IOs and finding the set of parts of the various IOs that allow all of the IOs to be built. The student or interested engineer might find it beneficial to determine the number of such subcircuits and build an IO compiler tool that not only places the proper subcircuits together to accomplish a particular desired function but also builds the proper temporal views.
2. Rail-to-rail ESD coupling circuits:
   - Most modern IOs contain multiple power and ground rails. All of these rails at some point must either connect through the IO ring to a pad or actually be part of the IO ring, including a pad. Such separated rails can cause ESD issues. One possible way to reduce ESD susceptibility is to provide some amount of interrail path for ESD currents to dissipate. A student or interested engineer might find development of such an interrail ESD protection circuit to be a practical effort.
3. Placeable decoupling capacitors:
   - In core-limited designs, it is beneficial to add decoupling capacitors as often as possible in the IO ring attached to the various power and ground rails within the ring. A student or interested engineer might design a placeable decoupling

capacitor that can be automatically added to a design and automatically connected to any pair of IO rails, together with a tool to automatically add such cells.

4. Radial-placement algorithm:
   - The capability to design dense IO often far outpaces the ability of bonding tools to accomplish the actual bonding. As a result, it sometimes is a binding requirement that IOs placed near the corners of designs have to be placed at a radial increasing pitch. A student or interested engineer might find it useful to build a software tool that reads a package-bonding grid and automatically determines the increased grid that is required for IO placement and actually builds the automated place and route (APR) placement file that accomplishes it.

5. Multitiered IO:
   - One means to push IO pitch in an IO-limited design is to use multitiered IOs (assuming that the bonding tool used by the design center allows this). A beneficial design task for a student or interested engineer is to develop an IO architecture that allows for such tiered-design tools, together with the software, needed to automatically place and route them.

6. Proof that nonsquare placement in IO-limited designs yields smaller devices:
   - Many designs tend to have the same number of IOs on each of the four sides. In an IO-limited design, such an arrangement will produce a larger chip size than some alternative arrangements with uneven amounts of IOs on each side. A student or interested engineer might find a mathematical proof of this to be interesting. (*Hint:* Think about dependent variables from your calculus class.)

7. Inter-IO stdcell APR regions:
   - In core-limited designs, there are often large regions between banks of IO that can be used for APR of stdcells. However, use of these regions tends to be inefficient in many APR tools. A student or interested engineer might be able to develop a partitioning tool that will determine the amount of stdcells that can fit within such a region, find such a subset of cells, and partition them so they can be readily placed in such a separated region.

8. Multiple-function IO:
   - The exact opposite of the parts-are-parts concept is to attempt to combine the totem-pole transistor stacks for various IO functions into a multifunction IO (with various subsets of the totem-pole stack being enabled on or off in order to convert the particular IO from one such function to another). A student or interested engineer might attempt to design single IOs that could be used as alternative IO types.

CHAPTER 4

MEMORY COMPILERS

4.1 LESSON FROM THE REAL WORLD: THE MANAGER'S PERSPECTIVE AND THE ENGINEER'S PERSPECTIVE

The very first technology that I was involved with (in the industry as opposed to at a university electrical engineering department) was at 1.25-micron very-high-speed integrated circuit (VHSIC) level design. VHSIC was a U.S.-government sponsored program in the early 1980s that had among other goals kick-starting what was then viewed as a foundering U.S. high-technology industry that was apparently falling behind the technical innovations of the rest of world. The need for such a program back then, the lasting effects of this program on the future trends of the integrated circuit design world, and the success of the of the designs that came out of that environment either directly or indirectly in many aspects of modern society over the last 30 years can be debated. However, it did allow for my earliest employment in the industry.

One of the integrated circuit designs that the VHSIC contractor that I worked for was developing was an early version of a microprocessor. That microprocessor was a simple von Neumann architecture that involved a data path (and a controller) that interfaced with an off-chip memory. However, the data path had two sets of "internal-to-the-device" memory connected to it. The first was a large memory array (for the time) of half-word addressable, 32 bits by 32 words, that had contents that had to be guaranteed stable between microprocessor operations (except for those that actually wrote to one of those internal memory locations). The second was a smaller, 12 word by 32-bit "scratch-pad" memory, again half-word addressable. The contents of these locations were meant to hold

Engineering the CMOS Library: Enhancing Digital Design Kits for Competitive Silicon,
First Edition. David Doman.
© 2012 John Wiley & Sons, Inc. Published 2012 by John Wiley & Sons, Inc.

partial results of microprocessor operations; as such, they were not guaranteed to be stable at the microprocessor operation boundary and were not even addressable through any allowable (to the outside user) memory operation. The classic six-transistor bit cells (see Figure 4.1 for an example of such a structure) used in these two types of internal-to-the-data-path memories actually differed in terms of the sizes of the six transistors of which they were constituted. The bit cell for the larger "operation boundary stable" memories had wider transistor channels, a larger voltage difference between the two sides of the latch, and a larger signal-to-noise ratio, all compared to that of the bit cell in the scratch-pad memory. Both types of memory met specification requirements; it is just that the specification for the scratch-pad memory was less stringent. While this always amazed me, I could understand the reasoning. The microprocessor operation stable memory had to be stable for as long as needed. Any instantaneous noise on the chip had to be absorbable by the memory such that they could remain stable for as long as needed, perhaps for years of operation between memory accesses. The scratch-pad memory had to be stable for a few hundred clock cycles at most and usually significantly less, depending on the microprocessor operation. In addition, because I was also involved in the microcoding of the various operations, I could go back and verify how long any given piece of information had to remain stable. The less-stringent specification requirements for the scratch-pad memory allowed for a smaller designed bit cell that, at 1.25-micron technology even though it was used just 384 times (i.e., 12 words by 32 bits), saved significant space as a result.

These are important because they are the only two entire memory designs I ever worked on that did not involve development of or at least use some kind of memory-compiler effort. Are custom memories needed in the modern design environment? Most certainly, and they are discussed later, but the compiler constitutes the most-frequent memory support that the typical Intellectual Property (IP) design and support engineer will be involved with.

Although a compiler can be thought of as "just" an automatic physical stacking tool for previously built physical representations of extant bit cells, they are actually much more than this. A good compiler will also deliver the entire preliminary view set for the various engineering design automation (EDA) tools (including estimated liberty files at whatever desired process, voltage, or temperature corner is requested) for the design center's chosen methodology for the completed memory. Further, it will build, extract, and instigate true characterization of the actual requested memory such that a final set of instance accurate design views will be releasable in some reasonably short time in the future. The compiler might be smart enough to not just assume some arbitrary multiplex factor and memory blocking ratio, or to blinding accept the customer supplied request for these such factors. Since it is aware of the arrangement and interaction of the underlying bitcells, it could, as an alternative, suggest different arrangements of these factors which could better match the customer's input design goals for the output memory (in terms of density, performance and power). Fabrication houses and other IP vendors usually have such compilers consistent with their other IP ready for use. In that case, the level of support for such a tool for a design center might be as simple as installing it in some release area and allowing the design-center engineer free usage (and unencumbered access to the IP vendor's application and support staff). Conversely, it might mean the actual pressing of the button and monitoring the generation of the deliveries for that given instance. If a development and support engineer is lucky, neither eventuality will occur and the design center will need the IP development organization to actually develop such a compiler.

By the way, in most conditions, a read-only memory (ROM) is usually constructed by use of a variation of the compiler. In the case of ROM, in addition to the other above-named features, the compiler will convert some input file representation of the data that are to be present in the ROM into a placeable "program deck" graphical display standard (GDS) that is delivered along with the rest of the ROM structure and views.

In addition, register files, which amount to arrays of stdcell (or stdcell-like) stackable sequential cells (sometimes flip-flops, sometimes latches dependent on the desired function), are constructed by use of a variation of the compiler.

Beyond the automatic generation of memories that a compiler offers, there are at least two other types of arrayed sequential cell designs that the library support organization will encounter over the course of supporting design teams. The first of these is the one-time programmable (OTP) memory or the nonvolatile memory (NVM), which are in some manner or another versions of programmable read-only memory (PROM) technology. In other words, these devices in some "beyond normal, typical operating conditions" can be written to such that the data stored in the memory are stable even if the power for the device is shut off. Such "beyond normal, typical operating conditions" usually but not always can be done only once and usually but not always involve some level of blowing away a fuse structure within the memory. OTP and NVM memories are typically technology specific. OTP and NVM are notoriously difficult to design without extensive testing (if for no other than to understand how much voltage should be applied to the fuse in order to ensure that it is properly and stably blown with no chance of self-healing and how much voltage can be applied to the fuse without harming the surrounding regular CMOS logic). As a result, in most if not all technologies, the typical structure for such logic is set array (usually a small set of bits in a small set of words) in a self-contained complete block with known and published specifications. There is no way to either expand or reduce the number of bits per word or words per structure. The design that requires more bits or words is reduced to accepting the inefficiency of using a second such structure. The design that requires fewer bits or words is reduced to accepting the inefficiency of having unused circuitry. Contrary to the above case for the development of compiler technology internal to the IP support organization being a value-added function, the need for development of an OTP or NVM capability internal to the IP support organization should be avoided. The mentioned need for extensive development and testing and the chance of "getting it correct" should persuade against underaking such an effort.

The other type of arrayed sequential structure that an IP development and support organization may encounter—custom memory array—is how all memories, including the two types of arrays on the VHSIC part on which I worked, were once done. The need for such as this comes down to the lack of need for the full development of a compiler for a given type of array. The ability to allow some features into a compiler—for instance, byte writeability or readability—might be sufficient to offset the extra effort of actually ensuring that the feature is correctly implemented within that compiler. But other features—for instance, an N-read, M-write multiport RAM (say, three read ports by two write ports)—may come out of a design center request once in a lifetime, let alone once in a technology. As a result, building that capability into a full compiler is not worth the effort, and such requests—if and when they do happen and cannot be "otherwise designed out" by the requesting design center—become a custom-array development. There is a large number of possible interactions between the various ports in a multiport RAM. A given design house may need to design such a multiport RAM only once or twice at a given technology node. Hence, it may be suboptimal for a design team to commit staff to such a critical but low occurance function and the IP support organization may not want to do the actual development.

This chapter will explore these three areas of memory design.

4.2 SINGLE PORTS, DUAL PORTS, AND ROM: THE COMPILER

Figure 4.1 shows a classic six-transistor bit cell. Such a cell is usually physically drawn in some manner of repeatable structure that is capable of being arrayed along both the x and y directions. This is done so that, together with the bit-line signals and the write signal (sometimes known as *word-line*) and the power and ground signals, they can efficiently fill a two-dimensional space with little to no wasted space either between rows or columns of the constructs. The result is an array that can be connected to address decodes, row and column drivers, and sense amps in a way that produces a working single-port memory.

The actual construction of the bit cell is a manual process. Indeed, the bit cell itself is usually drawn with "pushed" design rule check (DRC) rules such as to make the cell as small as possible without causing it to fail (beyond some exceedingly small failure rate, which will be discussed more shortly). Pushing DRC rules is done because the possible arrays of memory bits can be quite large in both number of bits per word and number of words per array. Every small decrease in bit-cell area can reap exceptionally large amounts of density improvement and area decrease at the chip level. Such pushing of the DRC rules works because the bit cells are always arranged in set patterns, possibly with sacrificial rows and columns either around or in among the working bits in a known pattern. Lithographic analysis is run on those patterns, and the optics are set such that the "pushed" DRC-rule bit cells remain functional. By the way, this is why the library development and support engineer should never attempt to "improve" a bit cell unless

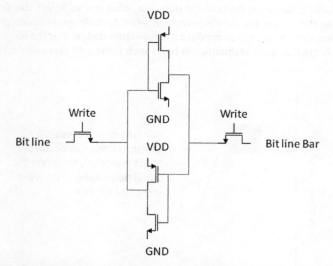

FIGURE 4.1 Along with generating the various temporary "cookie cutter" EDA views and scheduleing the development of the real EDA views, most but now all singleport static RAM compilers array such bitcells along with the various required address decode circuits, row and column driver circuits and sense amp circuits that combine to make up a memory instance. Dual-port, ROM, and register-file compilers operate in a similar manner. In addition, the bit cell for such arrays as dual-port SRAM look remarkably similar to the above but with additional bit lines connected through additional pairs of N-channel transistors controlled by additional write lines.

the fabrication house has asked for assistance in developing a special case bit cell specifically for the design center's described purpose. There are many places in a digital design kit (DDK) that a library developer can find useful margin. The memory-array bit cell is not one of them.

Although this book is not a treatise on memory bit-cell design, a quick review of the function of a bit cell and how it is measured is in order, especially in regard to noise and the need for tighter standard deviation performance for memories. As such, Figure 4.2 shows a typical bit-cell signal-to-noise eye diagram. It consists of the voltage swings of the two cross-coupled inverters that make up the middle section of the bit cell. As one inverter's output is pulled "high" (toward the power rail), the opposite side is pulled "low" (toward the ground rail). Because the two inverters are almost universally designed as mirror reflections of each other and they are cross coupled, which one is high and which is low depends only on how the bit-line voltages were set the last time the write signal was active. As a result, there are two transfer curves in the diagram, each a mirror reflection of the other. When the latch, which is made from the two cross-coupled inverters, is set one way, the circuit is operating in the opening between the two curves in the upper left of the diagram; when the latch is in the opposite sense, the circuit is operating in the opening between the two curves in the lower right of the diagram. The pushing of the "inside curve" of each opening from the "outside curve" downward and to the left (in the graph) is what determines the signal-to-noise ratio (SNR) of the bit cell. The larger the SNR, the more likely the bit cell will not fail because of some noise upset event during operation. Unfortunately, this is accomplished by growing the transistors in the bit cell. Doing so causes the bit cell itself to grow. A larger bit cell goes against the original goal of saving chip space by making the bit cell as small as possible.

The preceding reasoning leads to the question, what is a sufficient size for the SNR? This leads to the reason for the discussion. Most stdcell design is accomplished using views and files that are geared to produce a three-sigma design. For the most part, this is an admirable goal, and the techniques to build such views and files and decks have been

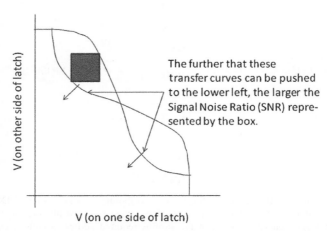

FIGURE 4.2 The Bit-Cell Eye Diagram. By increasing transistor sizes for the cross-coupled inverters inside the bit cell in width and length, the more robust the signal-to-noise ratio becomes and the less likely that a single bit error is to occur. However, increasing the bit-cell transistors goes against the other goal of maximizing the density of a structure.

largely successful at this. It can be argued, as this book does, that in some cases a full three-sigma view (for instance, see Chapter 9) is unneeded and wasteful margining. The reason for this is that if a few hundred transistors on a design operate in a high power range, they will be swamped by the typical operating range of the hundreds of thousands or millions of transistors on the same device. If a few hundred transistors are slower than, say, some three-sigma liberty suggest, and if those transistor are in a long path on the chip, then there is still a significant probability that the device will function correctly. This is because of the other transistors in that particular long path operating not at best case, or even typical case, but just at better than worst case.

This is not the case for a memory, however. Every individual bit cell is completely operated independently by six transistors. If any one of those six transistors is too far out of margin, then the bit cell is broken. If too many bit cells are broken such that they swamp the redundancy measures that may be built into a typical array on a typical design, then the design is broken. Literally, then, a few hundred transistors, or even just one transistor, being bad in a memory array can cause an otherwise working design to fail. Hence, memory design does not operate on the same three-sigma design range on which the rest of the chips operate. Three-sigma design techniques imply that approximately one in every 741 transistors operates outside of the design margin that the designer expects. That means for a six-transistor bit cell that one out of every 120 or so bits in a memory array falls outside of the design range. If a memory array has millions of bits, then it has to be designed to sigma values of around five and a half or more. As such, special proof of concept designs in terms of electrical performance and process lithography has to occur and be understood, and bit cells must be modified to allow the resultant lessons learned. Although this is a further argument for not touching a vendor's bit-cell IP, it is also an argument for adding redundancy to an array wherever and however possible.

Regardless, a memory compiler will take the bit cell and array it in two dimensions consistent with the bit and word count requested and in combination with the other associated structures already mentioned in such a way as to make a memory compliant to the specification of the customer's inputs. It will generate the preliminary views, including documentation, of the generated arrays such that the design team requesting that array can continue designing and floor planning the blocks of circuitry that will contain such arrays. It will instigate the actual final characterizations of the arrays and later regeneration of views, again including documentation, for those arrays such that they will be ready for the original requesting design team as timely a manner as possible.

Why should such a tool be needed? Well, the actual two-dimensional stacking of the bit cells and the surrounding drivers, sense amps, and the rest is tedious, time consuming, and error prone if done by humans. In addition, the generation of the various views, either the original short-term use temporary views or the final accurate views, is likewise tedious, time consuming, and error prone. Although this is certainly true for the same operations for stdcells, the "final" stdcell views are done once (or, at worst, a few iterative times) per library release. Because there are so many possible arrangements of bits per word, words per array, MUX factors, access modes, performance requirements, power requirements, and the like for memory arrays, there are far too many to generate every permutation of possible memory arrays all at once. Hence, if done by hand, memory arrays would need to be generated one instance at a time, usually after a request for that specific arrangement is made. If it is done by hand, then the person who did the job correctly on the previous occasion might not be available to repeat the job on a different array, so the quality of the hand-generated array is questionable. All of these reasons lead to the need for a repeatable process for a proven method of generating all of the views for the sparsely populated list of requested arrays within a given technology node. Hence, for all of these reasons, the memory compiler becomes a viable addition to the library support and development effort.

Why limit the effort to just certain memory-array possibilities? Typically, a compiler will be limited to single-port memories, dual-port memories (either true dual port, or one read, one write port), and ROM (and possibly register files). The effort usually stops at this list because usually the demand for memories with port selections beyond that becomes relatively rare. There is an effort to develop the bit cell for a multiport bit cell; beyond that, there is an effort to add the generation of such multiport memories into a compiler and testing the feature after that. If all of this effort does not "pay off" with easy generation of several instances of such multiport memory, is just not worth it. If the design center to which your library organization supports does, in fact, regularly need triple-port memories of various sizes or other multiport memory in various sizes, then do not hesitate to develop the compiler that will support such generation, but this will be the rather rare case. Typically, a multiport memory beyond a dual port occurs once or twice in a technology node. These will be addressed later in this chapter.

One more item should be addressed before we leave the topic. ROM and the register files are usually generated by compilers. As for ROM, the only requirement beyond those in the typical memory compiler is that the input of a program file that converts the desired code that is to be loaded into the ROM into actual ROM fuse location features (by whatever fusing mode the ROM is made—for instance, active fuse or VIA connection). As for the register file, consider the "bit" to be either a regular stdcell of a specially architected version of one such that it can be fit together in an area-efficient manner. Aside from these two special features, the ROM and the register file easily fit into the same concept of compiler as the single- and dual-port memory. In addition, the generation of the preliminary data views and the scheduling of the final data views of each are just as possible using compiler technology. Finally, the need for such generation is the same as for memories. If the design center requires ROM or register files, then certainly the addition of such capability is desired.

4.3   NONVOLATILE MEMORIES: THE BLOCK

Another type of memory that is typically required by design center devices is that of some sort of nonvolatile memory or one-time programmable memory. The basic heritage of such devices comes from U.S. military requirements in the 1950s: a way to efficiently store program data inside the original Atlas missiles. As part of that development program, the or programmable read-only memory was developed. A PROM could have data programmed into it by means of some special functional modes in combination with higher-than-expected voltages. This combination would permanently blow fuses inside of the device so that when that particular memory location was accessed, the resulting output would be determined by which fuses in that particular byte had or had not been blown. The programming methodology for the device was somewhat of a one-time event. Although a fuse could not be later "unblown," the PROM could be further run through the programming sequence again, with further previously nonprogrammed fuses blown. By the way, with the cost of silicon devices at the time, a minor industry developed to determine efficient coding of data such that when reprogramming a PROM was required, the likelihood of having to blow additional fuses within a previously used PROM was better than the likelihood of requiring an entirely new PROM device.

Eventually, ROM technology progresses to the modern OTP or NVM device. Most of these devices remain fuse based, although some fuses can now be healed enough to allow complete reprogramming at least a few times.

At any rate, OTP or NVM is a physically specific geometry. In other words, the physical structure that makes up the circuit is specific for the functionality of the circuit. Whereas one can add extra rows and columns to a regular memory array until either some Current Resistance (IR) drop or drive limit is reached in order to grow the number of bits per address of addresses per block of an array, this is not the case with a typical OPT or NVM. The structure will be designed in a single format (or a small selection of formats at most) with a given number of bits per address and a given number of addresses. As such, if a customer requires more than the given number of bits or addresses, that customer must have more than one such OTP or NVM device on the design, each individually addressable. There is no cheating in adding just another row or column of bits. The structure is a hard block of circuitry.

Typically, such structures are supplied by vendors, the fabrication house itself, or a third-party vendor that has prequalified its IP inside that fabrication house. As such, these versions of IP are not compiler ready. The IP provider supplies the various required views, and the structure is left on some IP design shelf within the library design and support organization. If there are design center methodology-specific views that the vendor does not provide, then the library design and support organization can supply them, but it probably behooves the organization at least to prevalidate these views with the original IP vendor.

An example of what can happen if such precaution is not taken is the case of a 0.35-micron NVM that I was recently marginally involved with. The IP vendor supplied the netlist for the device with three terminal transistor models (source, gate, and drain but no bulk connections). This particular vendor had previously supplied this NVM structure to other companies and each of those other companies had not required four terminal netlists. Unfortunately, the design center that was interested in this structure had developed a methodology that required bulk connections on the transistors in the netlist. This was fine; after all, it seems easy to add such a bulk connection to the SPICE netlist, assuming that the circuitry is basic CMOS. The only two things that are needed is to add a VDD to the regular bulk location in the M cards representing the P-channel transistors (which would be the insinuated fifth field of the M-card call) representing the normal N-well connection to the power rail. Similarly a GND to the regular bulk location in the M cards representing the N-channel transistors (which, again, would be the insinuated fifth field of the M-card call) representing the normal P-sub (or P-well) connection to the ground rail. The M-card calls for these would look something like the following:

- XMNNNN SOURCE_CONNECTION GATE_CONNECTION DRAIN_CONNECTION BULK_CONNECTION PFET L=...
- XMNNNN SOURCE_CONNECTION GATE_CONNECTION DRAIN_CONNECTION BULK_CONNECTION NFET L=...

where fourth connection, the fifth field (that of BULK), was missing in the delivered SPICE netlist from the NVM IP vendor.

However, these were two complicating features in this particular instance of usage.

- The NVM IP vendor neglected to inform the company that there were several isolated N-well areas inside the NVM structure. Because these were included, when the NVM circuitry had to endure high voltage during NVM programming, it actually could endure it. These N-well were not directly tied to the power rail. The fact that the netlist did not have a bulk connection for the transistors did not seem to be a problem for the vendor because they had verified the structure internally to their design facilities on multiple occasions, and the uncommon rail N-well transistor were buried well within the structure of the layout. No user, it was assumed, could attempt to connect to any signal buried that deeply within the design. In addition, as I have already mentioned, they had successfully deployed the NVM IP to multiple other users who had successfully used it on multiple designs.

- Unfortunately, the second complicating factor was that the design center in which this NVM structure was to be used was on a deep N-well design (for noise-isolation purposes, the overchip design was a noise-sensitive mixed analog–digital device). Deep N-well effectively ties all N-wells placed over it together. Hence, no N-well placed over DEEP N-well is isolated. It is true that there were deep N-well "keep out" rules that could have been brought to bear on the situation, and these would have allowed the DEEP N-well to be not placed under any N-well identified as requiring isolation. However, because there was no need in the IP vendor's mind to point out the isolated N-well deep within the NVM layout, these keep-out rules were never used (nor apparently required). As a result, the silicon came out with the isolated N-well in the NVM IP connected to the N-well across most of the rest of the chip. It was not considered a first-pass functional design.

This is just one of a seemingly innumerable list of possible miscommunication issues that can and often do arise during the acceptance of third-party IP. The bottom line here is that such IP is usually supplied by outside vendors, either the fabrication house or the third-party design house. They are the experts in the design of the circuit and layout. However, the internal user is the expert on the desired use of that externally supplied IP. Significant effort is required to uncover and methodically answer such simple questions as, does this have any isolated N-well and does itd use require or assume the tying together of all N-wells?" The function of the library support organization in such instances is to facilitate such in-depth exploration.

4.4   SPECIAL-PURPOSE MEMORIES: THE CUSTOM

Along with the previously mentioned OTP or NVM third-party IP, one additional array IP can possibly and even probably come from outside sources. When no other general-purpose compiler output fits the problems, there is a need to develop custom solutions. As previously mentioned, although these instances can and do happen they do so only *rarely*. Because of this, it is unlikely that design-center engineering or library development and support engineering will be completely at ease with the intricacies of such design. As a result, it might be better for the design-center organization and the library development and support organization to have a third-party vendor that is more familiar with such custom development to do the actual design and then port it into the design center (or onto the IP shelf). The reason for this is that the outside third-party vendor of such custom memory arrays is much more familiar with the pitfalls of such customer memory design.

Some of the possible issues that could arise are the interaction of the several of the multiple bit-line pairs with which the bit cell is connected. In a single-port memory, the bit cell shown in Figure 4.1 is the sum total of the circuit (aside from the row and column drivers and the sense amps). The two N-channel pass gates, both controlled by the same write control (or word-line) signal are the only interaction with the interior four-transistor toggle latch. Set the bit lines in one direction, pull the write-line signal active, and the latch is set one way. Set the bit lines in the other direction, pull the write-line signal active, and the latch is set the other way. Allow the bit lines to float and pull the write-line signal active and the state of the latch is transferred into a marginal change of potential between the two otherwise floating bit lines (hopefully, without destroying the state of the bit cell in the process). These cause the attached sense amps to convert that minimum voltage difference into a true full-swing logic signal that can be grabbed by an output of the memory's full logic-level swing sequential cell.

Things get a little more interesting on a dual-port memory. In that case, there are four N-channel pass-gate transistors surrounding the four-transistor toggle latch. These four pass gates, operating again in pairs with each pair controlled by its own write-line, connect two pairs of separate bit lines to the latch. Each pair of bit lines operates separately, as the single pair operates in the single-port bit cell. However, there can be complications when the two separate write-lines attempt to cause either two reads or two writes (especially if the writes are to opposite logic values but possibly even if both writes are of the same logic value) or a read and a write simultaneously to the same bit cell. How the bit cell reacts to such concurrent access depends on many issues, some of which might not be very controllable. For instance, bit-cell transistor geometries are usually so small and bit-cell DRCs are usually so pushed that improper residual charge on the combination of the four floating bit lines might be enough in some process corners to actually flip the bit cell, destroying the intended data. As a result, most dual-port memories restrict or even prohibit such access (of any read-and-write combination). This may extend to adjacent clock-cycle accesses as well as same-cycle accesses. A well-defined dual-port compiler can build the logic to enforce this prohibition, protecting the data in the cell array while simultaneously storing and scheduling such attempted dual accesses to safe, nearly adjacent clock cycles.

However, much beyond the single-port and the dual-port case, such interactions can quickly become nearly intractable. A local to the design center or library group memory designer, no matter how adept at memory design, might easily miss such an interaction, allowing a faulty (or at the least marginal) multiport array to be released. Again, recall that these memories are not made by a compiler (because of the usual lack of demand for such memory arrays); they are made by hand. Generally speaking, those designers who implement such arrays two or three times in a career are not capable of the task, no matter how capable they are of developing more common memory-array arrangements. Hence, as I mentioned earlier, it behooves the library design organization to farm this effort out to the third-party IP provider who can do this repeatedly for many customers and at every technology node.

Another form of complication to the "normal" memory is the "tag" or content addressable memory (CAM). Such memory arrays are used extensively in general-purpose microprocessors (holding "currently common data and instructions" that are more likely to be needed shortly by the compiled code being run at that instant in the von Neumann machine) and in communication processors (holding pertinent communication channel addresses—for instance, in a TCP-IP Internet protocol channel). The concept here is that the newest data that come across a data path or a communication channel replaces, if it does not yet exist, the least most recently accessed data within an array, no matter where that data exist within that array. To do so efficiently, a structure that allows for simultaneous (or near simultaneous) reading of multiple addresses within an array has been developed. Such CAM operation can be easily defined, but such implementation has to be accomplished with a custom design to be anywhere near optimal. Again, unless a custom CAM designer exists within the design center or the library organization, it is best to farm such development out to the third-party IP vendor who is better capable of implementing it correctly.

4.5 CONCEPTS FOR FURTHER STUDY

1. Bit-error rate (BER) calculations:
   - Although not given in this book, BER equations are fairly well known and can be found in many locations. The student or interested engineer might find it useful, however, to attempt to construct such equations together with or without some simplifying assumptions.

2. Redundancy and spare bits:
   - Using the just constructed BER, the student or interested engineer might choose to calculate the expected failure rate for given BERs for various sizes of RAM in various technology nodes and determine when extra rows and columns (and how many of each) are required to ensure higher RAM integrity.

3. Encoded ROM arrangements:
   - Although not always possible in all applications, one way of making smaller ROM is to attempt to encode two or more adjacent columns of bits into a smaller set that then requires proper decoding. The student or interested engineer might be inclined to develop such algorithms and then to use open-source ROM compilers, together with Register Transfer Language (RTL) for the decoding, to compare the size advantage with the performance disadvantage.

4. Bit-cell design and comparison:
   - The chapter gave an admittedly perfunctory description of the analysis useful during the design of a bit-cell circuit. However, several good books on RAM design have better descriptions. The student or interested engineer might benefit by attempting to design a given bit cell in a given technology node.

5. Bit-cell physical layout:
   - As a follow-on to the previous concept for further study, the student or interested engineer might attempt to physically lay out the just designed bit cell such that it can be stitched together in an arrayable structure.

6. Custom RAM design:
   - As a second follow-on to the preceding concepts for further study, the student or interested engineer might find it important to develop sense amps (with correct column MUX counts) and row and column drivers that align properly with the customer bit-cell layout such that a custom RAM can be constructed for a given technology node.

CHAPTER 5

OTHER FUNCTIONS

5.1 LESSON FROM THE REAL WORLD: THE MANAGER'S PERSPECTIVE AND THE ENGINEER'S PERSPECTIVE

Designs are timing (and power and noise) closed to some set of operating conditions. Those operating conditions often include a nominal voltage (plus or minus some prescribed range) and some range of temperatures. In addition, for manufacturing purposes, a range of processing is also a factor in the timing, power, and noise closure process. Indeed, Chapter 8 will point out that the timing views are usually described as at a given "corner" representing a given process, voltage, and temperature (PVT).

On one communication channel design on which I worked while in the data-communications design center in a large microprocessor company, it was important to determine as soon after start-up at what voltage would the dev ice be operating. This is not a difficult issue to resolve. One could develop a solution circuit in several different manners. One possible solution is to use a voltage controller oscillator (VCO) that has a well-defined relationship between the operating voltage and the oscillator frequency with periodic reading based on some external clocking system of the number of toggles of the output indicating the voltage. Some complications involve the effect on the frequency by the process and temperature at which the circuit operates during the reading. Such complications can usually be overcome, however, mostly through semisignificant levels of analog circuit design.

For the particular data-communications design to which I refer, there was a further complicating factor, however. Not only did the determination of operating voltage have

Engineering the CMOS Library: Enhancing Digital Design Kits for Competitive Silicon,
First Edition. David Doman.
© 2012 John Wiley & Sons, Inc. Published 2012 by John Wiley & Sons, Inc.

to occur as soon after start-up as possible and with a fine degree of granularity, but also the circuit that would be used to make the determination had to consume no power after the determination was made; in fact, it had to be shut down. These three requirements forced the circuit design away from that of a VCO and toward something far less complicated. The final circuit consisted of multiple voltage dividers, each with separate pairs of impedances, a band-gap voltage generator, a comparator for each voltage divider, a power on reset, and a free-running linear feedback shift register (LFSR) that could be enabled. The output of the LFSR both latched the results of the comparators and, one clock edge later, disabled the power connection of the rest of the circuit and then further disabled itself one clock cycle after that. Figure 5.1 gives a general graphic description of the circuit.

Another design that I worked on required us to determine the process corner at which the circuit had been made along with the voltage and temperature at which it was operating. The circuit needed to generate, in a testable and deterministic manner, a practically random signature that could then be broadcast across a network of similar devices so that the network would soon be completely defined by which unique device was connected to which unique device. This was for an early self-determining, self-aligning, distributed computing network. Although it was important to be able to test the practically random feature—at least to see that the voltage and temperature coefficients were inexact enough that they would combine with the fabrication range to produce something close to a completely random number—it was also important that the

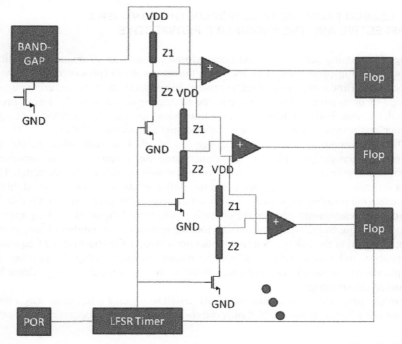

FIGURE 5.1 A Start-Up Operating Voltage Detection Circuit. It was important, in the device in which this was implemented, that the design that determined the operating voltage determined that voltage as soon after start-up as possible and then consume no power afterwards.

combined real-life effect of the three was nondeterministic. The circuit that was chosen was a series of three otherwise poorly controlled but otherwise free-running phase-locked loops (PLLs) that could be enabled, each with a different central frequency, none of which was a harmonic of either of the others (thereby maximizing the chances of non-deterministic interference between the three PLLs). Further, placement of the three was made in a common and otherwise isolated well and rail in the corner of the design. One of the three PLL was the generator signal that fed another linear feedback shift register for a given length of time (as determined by the isolated system clock). Once a system signal, the LFSR would be enabled for a set length of time. The resultant signature became the device's signature. The LFSR was run once and only once for any given network. The reason for this was the likelihood of the LFSR signature being radically different if run a second time, thus disabling the distributed computing network while a new signature for that particular node was communicated across the given network. Figure 5.2 represents the basic circuit.

Another design on which I worked included a previously designed microprocessor that was intended as an embedded core. It just so happened that the microprocessor had been designed with an internal transmission-gate–based bus. Although this was sufficient for the microprocessor onto itself, several customers required the devices into which the embedded microprocessor was to be placed to be tested at start-up against integrated circuit quiescent current (IDDQ) specifications. Unfortunately, the transmission-gate controls were undefined at start-up because most of the logic was in a random unknown state at that moment. Everything resolved after an asynchronous reset was taken active, but the various customers required IDDQ testing before such a reset event. The solution that was implemented was twofold. First, an extra term was added to each asynchronous control to the transmission-gate bus that forced the transmission gates inactive at power up (these extra minterms being controlled by a power-on reset, or POR). (By the way, because of the already limited performance requirement of the embedded-core version of the microprocessor, these added minterms did not otherwise hinder such performance.)

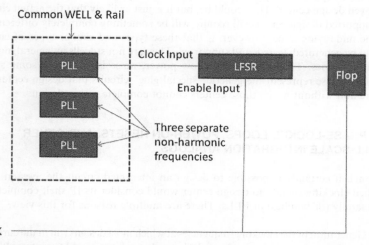

FIGURE 5.2 A PVT-Based Randomizing Circuit. The goal here was to make the three PLLs as susceptible to process, voltage, and temperature changes as possible. Hence, all of the normal noise-minimization techniques that one would normally find in a PLL circuit were missing from these.

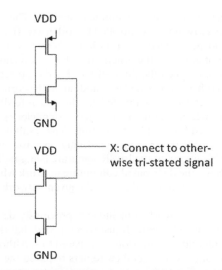

VDD

GND

VDD — X: Connect to otherwise tri-stated signal

GND

FIGURE 5.3   A Weak Bus-Hold Circuit. The transistor widths were minimal. The goal was not to "hold" the signal in one state or the other once it had been forced to that state but to assist the signal to just get to a rail state and not to float at some power-draining middle voltage. If a noise-glitch forced the toggle of the state, fine. The bus-hold helped force it to the new state.

Second, a bus-hold function, a very weak one-pin latch, was connected to each signal in the bus. The weak bus-hold held the disabled signals at POR without causing any significant delay in actual operation. Figure 5.3 shows the bus-hold circuit.

These three circuits represent design-center specific "added" circuits to an IP library offering that a library development and release (and support) organization is likely to be required to support (and possibly develop). Is it likely that these three will be required by any given design center? They could be, but it is just as likely that the actual circuits that the supported design center will require will be something completely different.

The underlying issue, however, is that these types and similar types of circuits will need to be generated, tested, and supported. They are not stdcells in general, nor are they IO or memories. They might be for low power, noise isolation, or some such other function. These represent the generic "special glue" circuits that design centers seem to require and without which those centers cannot complete a chip design.

5.2   PHASE-LOCKED LOOPS, POWER-ON RESETS, AND OTHER SMALL-SCALE INTEGRATION ANALOGS

Although it certainly is possible to design an integrated circuit the uses an externally supplied clocking signal, no design center would consider its IP shelf complete without some family (or families) of PLLs. There are multiple reasons for this view.

- Generation of the clocking frequencies at which most designs run in the current design world is many orders of magnitude faster than what external (or internal) oscillators (OSCs) can cleanly supply. Therefore, it is usually necessary to have a frequency-multiplying function. That multiplying function can be generated by a PLL.

- Another reason for a PLL is for data-capture techniques. Specifically, many communication channels (for instance, fiber-distributed data interface, or FDDI) embed the clock into the data format in order to save the need for a second clocking signal. FDDI is a synchronous communication system, but the determination of the synchronicity that is generated by the broadcasting device is left to the receiving device. The only clues to that synchronicity are the data waveform. "Oversampling" the waveform at multiples of the expected clock frequency can detect when the received signal changes. Repeated application of this technique can allow, in a predictable manner, an estimate of when the broadcasting devices clock edge occurs.

- A third reason a PLL is needed is "clock dithering." Many modern devices have to pass some amount of electromagnetic radiance testing, especially medical devices. In such cases, the design is placed in a testing chamber and allowed to operate "as it normally would" while the radiant frequencies that it generates are evaluated. If there is too much broadcasting in a specific critical range, the part will fail. One means of ensuring that more parts pass is to allow (actually force) the clock frequency to drift in a prescribed and repeated manner over time. Effectively, the part may be radiating some of the time in a critical range, but the length of time that the part does so remains below critical limits. The clock adjusting that allows such an operation can be accomplished by adjusting the multiplication factors in a multiplying PLL. By the way, companies that require such dither-capable PLLs have patented many different dithering circuits. The IP support organization that is asked for such a PLL would do well to run a patent search and ensure that it can find some alternate yet unpatented method.

There are many other reasons to add PLLs to an IP delivery. However, the preceding will probably be required by at least some of the design centers that any given IP support organization is supporting. Table 5.1 breaks the list down to a manageable range of probably required "typical" PLL IPs.

How these various pieces of PLL IP are acquired (externally purchased or internally developed either inside the IP organization or within the design center) depends only on the abilities of the designers employed within the design center and the IP organization (as compared against the design specification of the various required PLLs) versus the purchasing abilities of the company.

TABLE 5.1 Listing of Typical PLLs That Can Be Found in an IP Offering. The listing is not complete by any means. The design center or centers that the IP organization supports should have the final vote as to what PLL functions and frequencies are desired.

Function	Frequency	Comment
Frequency Multiply	sub MHz	May not be required as IP, could be developed on an "as needed" basis
Frequency Multiply	low-mid MHz	Typical for video applications; consider low jitter techniques (divide output by N)
Frequency Multiply	high MHz	
Frequency Multiply	GHz	
Data Capture	sub MHz	
Data Capture	low MHz	
Dithering	low MHz	
Dithering	mid-high MHz	

Another piece of "slightly larger than small-scale integration (SSI)" level IP that is typically required by a design center is a series of POR structures. The uses of such devices might be questioned after all with a sufficiently broad stdcell library offering in terms of functions; it should be sufficient to build nearly any circuit that will lock into a known state within a small number of clock cycles. Surprisingly, this is not necessary the case, even with extremely broad stdcell libraries.

A classic case is a nonreturn-to-zero (NRZ) communication pattern. NRZ communication protocols are synchronous protocols that represent one of the logical states as a steady-state voltage and the opposite logical state as a change in voltage. Figure 5.4 demonstrates such a communication protocol together with two possible voltage waveforms that could represent such a protocol. The fact that there are two such waveforms, either of which is optimally proficient in conveying the data, is the basic benefit of the protocol and the basic problem with building such a circuit either to generate or to read such a data stream. Figure 5.5 gives a perfectly viable circuit for such a protocol generation circuit using "toggle flops." However, dependency on the initial state of the toggle flop at the outset of a testing of the circuit (i.e., at power-up) is ultimately critical to the resultant test vector. There is a basic polarity issue here. Assuming the toggle flop comes up in one state, at no time can a test vector ever cross to become a test vector that is produced when the toggle flop originally comes up in the opposite state. However, each resultant vector is perfectly valid. As can be imagined, test engineering will have difficulty with such an

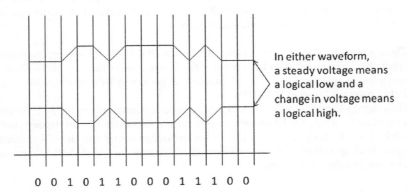

FIGURE 5.4 Nonreturn to Zero. NRZ protocols allow for exact opposite waveforms that never coverage into a single pattern to both be correct and a legitimate representations of a data stream.

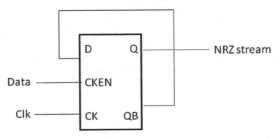

FIGURE 5.5 A Viable NRZ Generation Circuit. Without a reset, the toggle FLIP-FLOP in the circuit will produce a legitimate but otherwise nontestable data stream.

arrangement. Somehow, the tester has to understand that if the device under test exactly matches the test vector or exactly does not match the test vector, then the part is performing correctly. The part is only broken if it sometimes matches and sometimes does not match the vector. The solution here is obvious. Set the state of the toggle flop at reset. That is why toggle flops should always have resets, even though the reset will not change to functionality (in such usage as NRZ coding) one bit. Now, because many or most design centers are reticent to use resets, given that the engineers have been trained on the lack of need for such signals, the reset that goes to such a toggle flop may have to come from somewhere. Hence, this is a good reason for having a POR circuit. It allows for the reset of otherwise nonlocking sequential circuits without forcing a reset input to the chip.

One aside: there are multiple versions of POR cells. One important feature that not all POR circuit design has in common is brownout resistance. Specifically, in many instances, power rails can drop significantly during nominal chip operation. A POR circuit that is connected to and triggered by changes on the voltage on that power rail should not trigger inadvertently when such temporary brownout occasions occur. A POR design should be tested against the voltage level and length of time at that level on the attached power rail that will trigger the POR to generate a reset signal. The design center being supported should be advised of such parameters.

Another special need for a design center is some level of decoupling capacitance (DECAP) structure. Every design-center engineer is afraid of noise induced on his or her designs. Further, they are afraid of how that noise can cause intermittent and generally untestable failures of those designs. As such, they usually require the addition of stdcell DECAP that are capable of automated place and route (APR), which should be part of the stdcell library offering, with all the required views associated with it [graphical display standard (GDS), library exchange format (LEF), carrier-to-interference ratio (CIR)]. But they also require post-APR, postverification additional circuitry that can be added wherever otherwise possible, typically in a manual process. These DECAPs tend to be larger structures and are placed in arrays between any exposed voltage rails and substrate. They also tend to be large polysilicon to active transistors. Alternately, if the structure is capable, with both sides of the DECAP capacitor structure accessible, then the DECAP capacitator can be used between any two exposed voltage rails. The goal of the design-center engineer is to place these as often as possible in various locations around the device in the hope of suppressing noise from being injected on the rails of the device. I cannot personally vouch for the effectiveness of these structures, but I have yet to find a design-center engineer who has not requested them.

Finally, every design center has also asked for P-cells (or skill code) for each of the resistor layers, capacitor structures, and inductor structures, together with all of the monitored and measured N-channel, P-channel, NPN, and PNP structures that the fabrication house can produce on the technology node to be built and supplied with the IP.

Bottom line, all of these structures can and will be useful to the design center during chip-building operations and should be supplied by the library support organization.

5.3   LOW-POWER SUPPORT STRUCTURES

Most "deep submicron" bulk-technology nodes have distance-to-well-ties and substrate-ties DRC rules that allow that such ties can be removed from each individual cell and be placed separately at a much-reduced frequency. This will be covered a little more extensively in the next section, but it is mentioned here because of the added benefit that it affords in terms of usability for a typical stdcell IP offering. Specifically, stdcell IP that does not have embedded well ties, and substrate ties can be used without modification in various low-power implementation schemes that otherwise manipulate the (potentially isolated) well or substrate.

As transistor features shrink through deeper submicron levels of technology nodes, leakage (static power) tends to grow. This is because of several factors, the physics of which are outside of the realm of this book, but include thinner transistor channel lengths and thinner transistor oxide thicknesses. Figure 5.6 gives a good representation of this explosion in static power.

So what can be done to reduce such static waste of power that otherwise occurs at these deep submicron-technology nodes? Already mentioned in the library chapter are multi-Vt stdcells. As mentioned in that chapter, the physics of the situation are beyond the scope of this book. Suffice it to say that transistors show less leakage at higher Vt but at the cost of slower performance. This can be resolved by being able to mix stdcells of different Vt. This is because such stdcells can be mixed in the same logic to help power consumption in the cones of logic that do not limit performance (non–long-path cones) without forcing the long-path cones of logic that retain the lower Vt stdcells to lose performance. Hence, it is desirable to a design center for the library support organization to provide such multi-Vt stdcells, and, as described in the architecture section, in such a manner as to allow side-by-side automated placement of stdcells of such divergent Vt. However, a multi-Vt stdcell library does not directly take advantage of the sparse well-tie, sparse substrate-tie feature mentioned at the beginning of this section, and low-power support structures are not the issue. Specifically, well ties essentially "pump" the well to the appropriate rail voltage. Because that rail is voltage drain drain (VDD) (for N-well) or ground or voltage source source (VSS) (for substrate), the pumping action is actually direct tying of the wells to the rails. However, just as they can be used to do this pumping

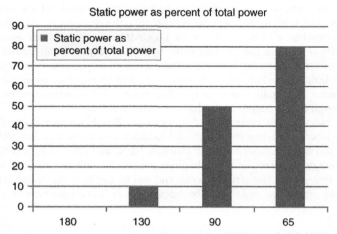

FIGURE 5.6   Static (Leakage) Power as a Function of Technology Node. As a design center moves from one technology node to the next, engineers are often shocked at the increased amount of leakage of a transistor. When multiplied by the increased number of transistors on a design at that technology node compared to the number that could be placed on a device at the next higher technology node, the combined increase often forces increased power-management techniques. Those techniques will force additional SSI-level IP offerings by the library support organization.

or tying, similar structures (without direct connections to the rails but with the connections to the wells, can be used in combination with active voltage pumping (or dividing circuits) to pump or tie the wells to different voltages (for at least part of the time). This allows the bulk connection of the transistors to be forced off the rails, lowering static power consumption (for at least part of the time). Hence, the following classes of added items need to be included in a low-power IP offering:

- "floating" well ties, where the actual well tie or substrate tie in separate cells or in the same tie cell are connected to a routable signal pin as opposed to being directly connected to the appropriate rail (N-well to VDD, substrate or P-well to GND or VSS);

- well-pumping circuits, typically implemented as standalone embeddable hard blocks that can either continuously pump the appropriately isolated well (which implies that the stdcell library that will sit in the well-pumped APR region will need to be characterized with that bulk voltage) or will pump the appropriately isolated well to a disabled state when activated;

- appropriate well or substrate isolation endcap structures (see Figure 2.18); and

- one additional logic-isolation device described as follows.

Design-center engineering should be involved in any well-pump or well-isolation design (or at least specification). In addition, as can be imagined, the inclusion of these structures will affect the views of the rest of the stdcell library offering. Not only will the aforementioned "pumped bulk" characterization liberty file need to be delivered, but also the function of the cells during an "off" bulk state will need to be described in a logical view. Finally, the test views of the stdcells will need to be modified to allow for reachability and observability testing of the synthesized netlist in situ.

As to the additional logic isolation device, it is needed to:

- ensure that inactive undriven signals emanating from the appropriately disabled well regions do not propagate unknowns (either in logical simulations as X's or in actual operation as transient changes in the voltage of these undriven nets) into the other regions of the device that remain active; and

- ensure that active signals emanating from the regions that remain active do not drive into the otherwise isolated and disabled well.

As such, the structure consists of a pair of isolation-inversion (or double-inversion buffering) circuits or pass gates that effectively cause the previously mentioned isolation. Attention must be given to the effect of floating input signals from the disabled side and the effect of voltage drop from a driven side of the structure across to a P channel or N channel on the disabled side.

These devices, together with their required logical views are described further in the Chapter 3 of this book.

5.4 STITCHING STRUCTURES

No device is ever completely composed of pure circuit structures. Invariably, various blocks of logic will need to be somehow attached with each other. The same will be true for the stdcells and the IO. It is a practical impossibility for a device of sufficient complexity to implement the device function from 100% stdcells, memories, IOs, and analog without any "white space" between the various structures. As a result, there is a need for additional "stitching" structures, which are typically but not necessarily without any netlist-verifiable circuit additions, to be added to a complete IP offering.

The typical set of such stitching structures include core-fill cells, IO-fill cells, DECAP cells, chemical or mechanical polishing (CMP) fill cells, spare-logic cells, probe-point cells, antenna diodes, and test-debug diodes, among others.

5.4.1 Core-Fill Cells

Core-fill cells are a small set of spacer cells that can be included between the automatically placed stdcells within a placed block, allowing for the continuous attachment of wells and rails across a given row. Some current APR tools based on engineering design automation will automatically extend these geometries and actually advertize this feature, but using such geometric well and rail extension can be error prone if the cell architecture is not along the lines of that presupposed by the APR tool. Hence, it is always prudent to include such core-fill spacer cells within an offering. The absolute minimum number of such cells is one, that being the width of the minimum placement grid. This will allow for automated filling of open spaces of any feasible width between the placed cells. However, it will also fill the resulting GDS with the largest number of such core-fill spacer cells. This could become overwhelming to the various tools that otherwise handle, manipulate, and verify the device GDS. A better approach is to build a small set of such core-fill spacers, each larger than the previous, such that they can be used from largest to smallest to successively fill smaller openings between the placed cells. A typical set of such core-fill cells widths get defined by the Fibonacci sequence up to some usually impossibly large width. For instance, CORE-FILL1, CORE-FILL2, CORE-FILL3, CORE-FILL5, CORE-FILL8, CORE-FILL13, CORE-FILL21 would be sufficient to efficiently fill any intercell spaces up to 33 placement grids in width. Again, the algorithm for use would be to fill the areas with CORE-FILL21 first, then CORE-FILL13, and so on down to CORE-FILL1. This will allow the same amount of closing of the aforementioned open space between APR cells without the heavy extended handling of an exceedingly large GDS database. Care should be given, by the way, to using these appropriately. If part of a row is to be isolated from the rest of the row, for whatever reason, such core-fills could accidentally break that assumption and cause the unwitting connection. Also, if there are instances where wells are to be connected but rails are not connected, or vice versa, then special core fills allowing the appropriate such connection while disallowing the other must be developed.

5.4.2 IO-Fill Cells

IO-fill cells amount to the same function as core-fill cells. However, they are specifically for inter-IO spacer filling and well and rail extension. These are placed in the IO ring between IOs of similar footprint. For a IO-limited device, why would such a device ever be used, forcing two otherwise adjacent IO apart, growing the already area-limiting IO ring? Many fabrication houses have radial bonding rules. That is to say, as IOs are placed near the middle of any of the four sides of a device, the bond wires will tend to run nearly perpendicular to the side of the chip, and the bonding tools allow for close packing of IOs. But as the IOs get placed nearer to the corners of the chip, the bond wires connected to them tend to have to be swept further from the ideal perpendicular, which causes the bonding tool to not be able to effectively bond with such close-packed IO pitch. Hence, IOs placed near a corner of a chip need to be spaced farther and farther apart. By the way, this added spacing, although it is an ever-increasing amount as the IOs get placed closer to the corners, that amount remains small, being first one extra placement grid beyond minimum, then two such grids, then three, then more, but never much beyond five or six such grids. As a result, where there existed a legitimate need for a set of core-fill cells, each of increasing width, there is usually no need in terms of the explosion of the size of the database from extensive usage of IO-fill cells. Hence, usually one size, a minimal placement grid in width, is sufficient. Now, however, the other axis, just as in the core-fill cell case, does still hold. Care should be taken to supply enough variety of IO fillers, each allowing a given rail or well connection but limiting some others as the design center is likely to require. By the way, one special IO-fill cell is the "pin 1 indicator" cell. It is usually a minimum-width IO-fill cell with some such indication, usually in top metal, and it is used in the design centers to indicate pin 1 of the device being placed next to that particular pin.

5.4.3 DECAP Cells

APR versions of the DECAP cell were described earlier in this chapter. The larger non-APR versions are usually added at chip integration. These are small enough to be used within the stdcell APR regions and usually are placed as often as possible in order to provide some amount of reduction of local rail noise by forcing some capacitance between power and ground rails within that APR region. A DECAP is usually some sort of wide and long N-channel or P-channel device, with the polysilicon gate connected to one rail and the source or drain connect to the opposite rail. The gate capacitance in the wide and long transistor supplies the rail noise localized suppressing capacitance. As in the core-fill cells, also APR capable, these can come in a variety of sizes, the larger of which are usually placed first, followed successively by the successively smaller versions in an attempt to fill the open areas between cells in a placement without causing too much bloat in the size of the GDS database. The difference here is that the smallest DECAP cell, because of the area restrictions of the capacitive structure itself, cannot be just a single or even just a few placement gridwide. The smallest APR-capable DECAP is probably approximately five or six such grids in width. This means that the general Fibonacci sequence of sizes become something along the lines of DECAP5, DECAP6, DECAP11, DECAP17, and maybe DECAP28. Here the fact that these are not harmonics of each other will give the added benefit of the localize capacitance between the rails is not "regular" enough to pass some frequency of rail noise while hindering others. On the other hand, because the sizes of the DECAP do not go all the way down to a single placement grid, if core filling with DECAPs is desired, a set of minimal sized core fill will still be required after that DECAP filling.

5.4.4 CMP-Fill Cells

Chemical or mechanical polishing fill structures amount to additional types of core-fill cells. Most modern fabrication processes require some range of polygon density on critical layers in order to facilitate the chemical–mechanical planarization of the device being processed as it goes through the series of steps in fabrication of that particular layer. For many of these critical layers, tiling engineering design automation (EDA) programs exist that allow for the addition of enough "floating" tiles of any particular layer into the final integration GDS database. Although these EDA-based tiling pro-grams work well for the routing layers, they can easily break down when attempting to "tile fill" on layers that are either extensively defined or exclusively defined within the cell, such as active. The reason for this is that the typical stdcell is significantly smaller than the typical "measurement rectangle" used by the EDA tiling tools in order to calculate layer-density numbers. Although it is prudent to ensure some amount of measuring of layer-by-layer density checking, adjusted for this smaller measurement rectangle within the stdcell, thereby minimizing the likelihood of chip level tiling concerns, there will still remain the possibility of large sparsely populated APR regions, fill otherwise by core-fill cells. This is where the CMP-fill structures arise. As such, they have some amount of "small floating tiles" within them that, when used as core fills, ensure easier compliance with chip-level CMP density checks. Again, because these small tiles do take some space, the CMP-fill cells cannot be made as small as single placement grids. Hence, although they probably do not need as large a set of various-sized cells (making too many can cause significant variance in the density of the particular layer in question), they still will not be able to fill completely open spaces between APR cells, and regular core-fill cells will still need to be added.

5.4.5   Spare Logic Cells

Spare logic cells are groupings of the more common stdcells—for instance, one or two INV, a NAND2, a NOR2, and a scan flop, with their inputs tied inactive and usually clumped together in a large placeable stdcell, which is preplaced multiple times around a design APR region. These spare logic cells are there to allow the potential for a fix, in metal layers only, of any detected logic errors. Some EDA-based APR tools allow for the a priori shooting of individual representative cells, as opposed to these clumped together set of cells, but this might not always be optimal. The reason is that the results of indi-vidual scattering shoot of individual functions might mean that any logical fix might require a combination of cells to accomplish might require routing the "fixing nets" over significant distances just to reach appropriate scatter shoot cells. Scatter shooting these clumps of cells minimizes these risks because there will exist a representative set of functions adjacent to each other within the clump. By the way, note that this ability to fix all such logical errors cannot be ensured because the more specialized the fix, the more the likelihood that it will require added functionality within a clump or within scattershot individual cells that was not anticipated beforehand. Design centers sometimes have a preferred mix of specific functions and drive strengths that are added to such clumping of spare cells. Hence, it is often the case that the library IP offering will require multiple versions of these structures. By the way, depending on the verification methodology, these spare cells, either individually scattershot by the EDA APR tools or done so in clumps, may or may not require CIR views. If the methodology requires any transistor on a device to be present in a layout-versus-schematic netlist, even those that are turned completely off, then these cells will require the appropriate netlist. If the methodology allows for these to be "not there," then obviously they do not need such CIR.

5.4.6   Probe-Point Cells

Depending on the technology node and the possibility of actually successfully placing probe needles down onto signals, probe-point structures can be added. These can be used to route a priori pertinent signals such as clocks and resets and state-machine controls to the topmost layer of the chip in the hope of measuring or detecting glitches could be useful in the debug of any detected improper operation of the device. Depending on the test devices in the companies' debug labs, some tools of which can detect the voltage potential of pieces of metal without physical contact, such probe points might be extremely useful even for technology nodes below the threshold of such successful needle probing.

5.4.7 Antenna Diodes

Antennae diodes release built-up charges on a signal as successive layers or routes get processed, one on top of another, before that charge can damage the connected source or drain (or gate) of the driving (or loading) transistors.

5.4.8 Test-Debug Diodes

For technology nodes where probe points no longer occur, such structures can actually glow in the substrate as a function of the voltage potential of the attached signal because of voltage discharging. The use of these devices requires significant back grinding of the substrate in order to allow detection of the glowing diode through the substrate, but such detection can easily be mapped back to front in order to properly identify the signal-attached diode or the ones that are discharging.

5.4.9 Others

Additional nonlogical library elements include the following.

- Well ties and substrate ties as described in the previous section on low power.
- Rail decoupling diodes that allow for connection of some or all of the rails within a device through stacked diodes such that rail-to-rail ESD events are more easily handled while simultaneously allowing for significant isolation between the various power planes of the device. Such structures tend to be custom or semicustom because of the a priori unknown number of voltage differences between the various power rails. Still, such structures need to be built so that they allow proper automated handling during the APR development of the device.
- All such required gapers and gaskets between the various IO ring regions, including corner cells, as described in the IO section of this chapter.

5.5 HARD, FIRM, AND SOFT BOXES

About 20 years ago, I was working in a data-communications design organization. One design that we built used a media access controller (MAC) device, part of a communications channel architecture that fulfilled part of the function of the data-link level, the second tier from the bottom physical-link layer, of the seven-tiered open systems interconnection (OSI) model. The design was built in conjunction with an external customer who, as part of the business arrangement between the two companies, agreed to give the register-transfer language (RTL) of the controller to our company so that we could use the design as we saw fit as long as we also produced a design for that company's specific use based on that RTL. The one caveat that the customer insisted on was that it would not give up the timing constraints for the design nor agree to verify the resultant timing of the MAC design, complete to timing closure, more than once. The problem was that the customer was only interested in a single instance of the MAC in its device, whereas my company was interested in producing a QUADMAC, with four instances of the same circuit capable of separately and independently controlling four such communication channels (as in a hub).

The solution that was implemented was to build the customer version of the MAC as a hard-core one hierarchical layer deep in the synthesized netlist, together with the surrounding logic and IO rings one layer above in the same hierarchy. This allowed for the single timing closure called for in the customer contract but allowed for the identification of an embeddable core of the design, thence known to be timing clean and closed, that could be placed and used four separate times together with some surrounding logic for the internal QUADMAC design. One further complication was that both the customer and the internal data-communications organization were interested in making the two respective designs as small as possible, which meant that they would both need to be IO limited. Unfortunately, the form factor forced on the single instance design for the external customer by its IO ring meant that the core could not be beyond a certain width, whereas the form factor forced on the fourfold implementation for the internal division by that IO ring meant that the four cores could not stack vertically beyond a certain height. As a result, even though both designs were demonstrably IO limited exhibited significant empty space, the core had to be automatically placed and routed at an exceptionally high density (more than 90%). The empty space in the external customer design was vertically above, but below the dense core, which stretched to the sides of the available core area, and the empty space in the QUADMAC was horizontally left and right of the four copies of the dense core, which stretched to the top and bottom of the design. This was effectively a development of one of the first embedded cores, together with the methodology for such implementation and usage. By the way, the company was not willing to pursue a patent on the concept or the methodology for such a core.

The above description illustrates the development of what is sometimes referred to as a *hard core*. Hard cores represent IP that exist in an explicit polygon-based format. IP vendors supply such versions of their IP for many reasons. It might be that they have a strong desire to protect their IP from the risk of reverse engineering. This could more easily occur if the IP was just a netlist or even just an RTL description. It might be that the IP supplier is interested in preventing poor functionality in a not truly digital synchronous design by forcing usage of a known good physical implementation. It might be that the IP supplier is not interested in spending the application engineer (AE) resources that would be required to support repeated timing closure by every customer interested in the supplier's product and instead had implemented it only once and would not reopen the design to satisfy a customer's whim.

In many instances, an IP support organization will find that it needs to support a development or delivery of such a structure to a design center. What is involved in such an effort? On the good side, hard cores can be thought of as large-scale integration (LSI) sized stdcells. As such, they can be no more troublesome to support than any other piece of stdcell IP. Just as a stdcell will have physical views (GDS, LEF, CIR), logical views (Verilog or VHDL), and temporal views (liberty), the hard core also will. Because of the size of core and the potential need to protect the IP from reverse engineering, there may be restrictions and simplifications to the various views. For instance, the physical view GDS may not actually be delivered from a third-party vendor but only be resident at the design-fracturing locale. A simplified LEF, containing only pin and blockage data, would be delivered to the design-center customer and used for APR with the rest of the chip, and the actual GDS would be added once the otherwise completed design was delivered back to the fracturing facility. Similarly, the logical views of such IP may be somewhat different from that of a normal (and smaller) stdcell. Again, for protection of the IP from reverse engineering, the IP vendor might deliver a transaction-based model of the logical functionality of the hard core. Such a logical model will be able to accept input signals and to diagnose issues with those signals, informing the user where certain typical events occur such as stack overflows or underflows or even incorrect ordering of data and control signals. In addition, such a simplified core will be able to issue pseudo-information back, typically on a cycle-accurate manner. Such activity would allow a design center to accurately judge if its connection to and software driving of the actual core, if present in its netlist, was correct. At the same time, such a simplified core both limits the exposure of the IP vendor from risk and limits the otherwise cumbersome size of the design environment that would occur if that IP were actually present.

One alternative to these cycle-accurate models (or in addition to them because many vendors supply multiple models of their IP that can be used in different developmental environments) is the encoded model. Specifically, EDA-based logical simulators allow the inclusion of encryption-encoded versions of accurate models that are theoretically unreadable by humans. Many vendors deliver such products. The problem is that such encoded models are delivered to encryption standards sufficient for the time of delivery of the model to the customer (and to the level of encryption that the EDA logical simulator tool can handle). However, that does not mean that shortly thereafter that the particular encryption could not be broken by a competent engineer, thus exposing the IP vendor to reverse engineering of the no-longer-protected hard core. One further complication is the need to align the testing of the vendor core with the testing of the rest of the design. If a vendor's IP has, for instance, certain numbers of scan test channels running through it, then it behooves the design center to accurately connect the leading and trailing ends of those chains either directly to IO pins or to other scan chains internal to the rest of the design. However, that means there is the need to understand how those scan chains that are internal to the core test that core. Thus, to develop accurate scan test vectors, the IP vendor is exposed to a reverse-engineering risk. Many such IP vendors get around this by requiring their scan chains to be directly brought out, in some scan mode, to the pins of the chip, and to further prohibit these scan chains from being connected to the other internal-to-the-chip scan chains.

A second, less-restrictive version of IP that a support organization may have to deliver and support is a *firm core*. The difference here is that the firm core, although defined to the synthesized netlist level, is not delivered as a single piece of placed and routed GDS. A netlisted block can be further instantiated within the customer design and be placed and routed along with the rest of the customer's netlist. Depending on the reason for the IP vendor's delivery of such a core, the delivered netlist may be hierarchical or flat. If flat, then there is probably some idea of IP protection in the thoughts of the IP vendor. The problem with this is that such a netlist has little to no information relevant to the placement of the various cells used in that netlist. As a result, although the placement tool can attempt to keep as much of the netlist within the delivered block together, it may not place the cells of the firm core optimally among themselves. This is worse than the remainder of the design-center netlist because that netlist will be hierarchical and the EDA placement tool can take cues from the hierarchy during placement. Hence, APR of such firm cores is, by definition, problematic. They will require more that the usual amount of AE help from the external IP vendor and from the internal IP support organization. Aside from this problem with flat netlist, the rest of the firm-core delivery is simplified. Because it is not physical, there are no physical views beyond the netlist, assuming that the stdcells in which it was synthesized are available to the design center. The netlist suffices for the logical and test views. The only temporal views are the vendor-supplied timing-constraint files.

The easiest version of an IP-delivered core is that of the *soft core*. Here an IP vendor delivers a synthesizable RTL to the design center, either directly or through the internal design support organization. That piece of IP will have timing constraints and have been successfully synthesized within the technology node (otherwise, the design center would not have purchased the otherwise broken IP). Here the IP support organization's support level is that of facilitator. Bugs and request pass from the design center to the IP vendor, and updates, fixes, and answers pass back to the design center form the IP vendor. Internal IP support organizations enjoy this kind of support.

5.6 CONCEPTS FOR FURTHER STUDY

1. Optimized placement of core well and substrate tie cells:
   - It used to be a common stdcell design practice to add well and substrate ties within each standard cell. This usually meant that each cell was up to 10 or 12% larger than the rest of the circuit that actually accomplished the specific function would require. Developments in the processing of integrated circuits, starting at about the 130-nanometer technology node, allowed for a significant decrease in the required frequency of placement of such well-tie and substrate-tie structures. As a result, as mentioned in the chapter, such ties were removed from the stdcells. They became placeable elements unto themselves. The student or interested engineer might find it useful to develop an algorithm that allows for such tie placement at one-fourth of the required DRC frequency (four times *larger* than the maximum DRC distance) within a given row over the majority of APR region. Further, this algorithm should still allow for any given placed stdcell to be compliant and within one DRC maximum distance from a well and a substrate tie.

2. Diamond script:
   - In the chapter (as well as in the earlier chapter on stdcells), it was mentioned that many low-power design techniques require that stdcells be designed such that functions of a given Vt can be automatically placed adjacent to functions of a different Vt. As mentioned in Chapter 2, this means that sometimes a "four corners" placement can arise where two cells of one Vt are placed at the lower left and upper right of a intersection of their corners with two cells of a different Vt placed at the lower right and upper left of the same intersection. This means that the implant mask for the two Vt will have lithographic issues. The student or interested engineer might like to develop a "diamond" placement algorithm that finds all such four corners and places a diamond structure of one of the Vt (on its own GDS layer that can be Boolean merged in during design fracture) at an appropriate size to pass all DRC at these particular locations.

3. Hard digital delay lines:

   • Along with the need for such structures as PLL, many communication channel designs require digital delay lines (DDLs). These are usually inverter chains, with an added N-channel transistor, with gates controlled by a voltage determined by a charge pump. That charge pump is then controlled by the relationship between arrivals of the input and the output of the inverter chain. It is designed to deterministically delay a given regular clock signal a known amount. The student or interested engineer might be inclined to develop such a circuit such that it is expandable to a given clock frequency and a given delay.

4. VDD region isolation cells:

   • In low-power design, it is often the case that a particular region of a chip is required to be powered off while the remainder of the chip is still active. Such design requires circuitry that prevents active signals from the active regions of the design from propagating within the shutoff region (possibly partially powering the region by back biasing the N-well and hence the region power rail) and that prevents inactive signals from propagating unknowns into the active regions of the chip. The student or interested engineer might be inclined to design such a circuit, including the logical views that would be required.

5. Charge pumps:

   • The fact that the well and substrate ties are usually not connected to the power and ground rails within stdcells in most modern technology node libraries allows for the low-power technique of pumping a well so some other potential than would normally be found in a digital APR region. The student or interested engineer might find it useful to determine the drive requirements of such a structure as a function of the number of transistors expected within a given APR region and then to build an appropriate circuit to supply this at a given technology node.

CHAPTER 6

PHYSICAL VIEWS

6.1 LESSON FROM THE REAL WORLD: THE MANAGER'S PERSPECTIVE AND THE ENGINEER'S PERSPECTIVE

I was taught to program in the 1970s, an age when the cost of silicon-based memory was relatively expensive. As an example, a program that I submitted for a project was marked downward because I had used a byte instead of a bit in order to store a true–false value. For comparison, UNIX core kernels were compiled in less than 64 kilobytes of storage space during the same period. I argued with the teacher that there was no other true–false variable in the program and that making the variable a bit would take up just as much room in the computer memory, but I was told that if this program later became a sub-routine in a larger program, then my argument might not be valid. I sometimes think back on that programming class when I see how large the current software packages are for the home PC. At any rate, after the lesson the particular software teacher gave to me, I have always translated into my hardware and, specifically, library efforts.

Several years ago, I was tasked to design a 90-nanometer stdcell library for a large microprocessor company, emphasizing density over performance or power. The company had previous used an outside library vendor for designing, developing, and supplying the stdcell library at the 130-nanometer technology node. Several key members of the division management team felt that had gotten their money's worth at the 130-nanometer node. They felt that we should use the same vendor at 90 nanometers. Some design rule check (DRC) rules limited the ability to route as many signals in several layers within the cell, and the design-center customer wanted to limit the amount of metal 2

Engineering the CMOS Library: Enhancing Digital Design Kits for Competitive Silicon,
First Edition. David Doman.
© 2012 John Wiley & Sons, Inc. Published 2012 by John Wiley & Sons, Inc.

within the cell, so I proposed the construction of an architecture with nine tracks (number of horizontal router rows, which equates to height) at 90 nanometers. The 130-nanometer solution was an eight track. When the vendor heard that I was promoting a nine-track architect, they knew that they had me beat and that they could manipulate the division management into purchasing the vendor's eight-track solution by convincing management that the key metric was cell density. They came in with a presentation of several NAND and NOR layouts, all matching my layouts in width and all being 89% (8/9) of my cell's height.

NAND gates (and NOR gates), especially two-input varieties, are notoriously non-dense. The reason for this is abundantly clear. A two-input NAND (or two-input NOR) consists of two P-channel and two N-channel transistors, adding up to what is usually defined as a one-gate equivalent (one gate being equal to four transistors). In addition, there are three pins for the cell: two inputs and one output. Because it is beneficial from an automated place and route (APR) point of view to allow each pin a separate vertical track, the minimum size for such a cell is four tracks wide by the cell height (which can be given in terms of number of tracks tall). The four-track width is because of one track per pin, plus the equivalent of a half-track on either side, in order to allow not having to deal with intracell issues. The four transistors can easily fit inside the four-track width. As a result, library vendors can usually show dramatic raw gate densities for their stdcell libraries by producing a squat cell architecture, laying out a two-input NAND in that footprint, and extrapolating the size of the rest of the library. For instance, an eight-track stdcell library will appear denser than a nine track, and a five track will appear denser than a six track.

The inconsistency in the thought process to such an approach is that not all functions can be produced in just four transistors. Routing together larger numbers of transistors and connecting the various inputs to perhaps multiple regions of a stdcell soon swamps the number of cell internal routing channels. Because of this, FLIP-FLOPs are notoriously limiting to shorter (in terms of tracks) cell architectures, but they affect many, or even most, stdcells beyond simple NAND and NOR. Hence, the true density of a taller (i.e., more tracks) may be better if the taller cell height allows for a sufficiently large numbers of stdcells such as FLIP-FLOPs to be decreased in width.

With that knowledge, I asked if the vendor had any other functions completed. I knew that it had won the account for another technology node and that it wanted to show magnanimity, so it shared the sizes for all of its cells, even giving an electronic file over in an Excel spreadsheet format during the meeting.

I had earlier gone back to the five largest integrated circuits that had been done with the 130-nanometer library and had build a histogram of all the cells that had been used across those five designs, including how frequently each cell was used. Because it was the same vendor, the cell names had remained consistent with its new 90-nanometer offering. At the meeting, in front of the vendor and in front of the division management, I plugged the vendor's sizes into the spreadsheet that I had prepared earlier and compared the resulting weighted average for designs that had actually used the earlier library. The nine-track internally developed library, when all of the denser functions where included, especially when compared with the usage weighting, beat the eight-track library by 15% assuming a uniform distribution and more than 18% assuming a weighted usage distribution, despite the architectural footprint being 12.5% taller.

Although this may not be repeatable for all technology nodes and for all vendors, it does illustrate that forcing a stdcell architecture based on matching a subset of the cells that need to be supported against the requirements metrics is a recipe for disaster.

6.2 PICKING AN ARCHITECTURE

Any stdcell library will be measured against three major axes:

- density,
- performance, and
- power.

How these three metrics are valued against each other by the design-center customer will largely determine how a stdcell library for that particular customer should be designed. In addition, although some design-center customers may specify optimizing all three axes together, this is somewhat akin to "concentrating across the board" and will yield only mediocre results in any of the axes. The reason for this is that the three are not orthogonal of each other and optimizing for one may entail compromising on one or both of the remaining two axes. Hence, if it is your great fortune to be asked to create a stdcell library from scratch, you should first go through a series of questions with the primary customer in order to determine how it views the prioritization of these three measures and how it will judge success in each.

But before there can be a discussion concerning such questions and what would be "good" architectures based on the answers from the design-center customer for these, it is probably a good idea to look at each axis in a slightly greater light.

First, for density, Table 6.1 gives a rough idea of how dense some of the best stdcell libraries designed and used by many of the leading library manufacturers (fabrication houses as well as microprocessor companies that have developed internal stdcell libraries over the years) for various technology nodes. The numbers in the table have been scrubbed so that no given company's data are compromised. Actually, this scrubbing is not necessary because the trend line is readily discernable. To a first-order approximation, density increases with the square of the technology-node shrink. There are limitations that slightly adjust this trend line such as some design rules not shrinking consistently with other design rules, the introduction of new design rules such as minimum area considerations or phase-shift considerations, or the loss of being able to route on active or polysilicon on deeper technology nodes. But for the most part, a decent approximation of what is a good density at a particular technology node is to scale up the best-in-class library at the next technology node.

TABLE 6.1   Edge of the Envelope Density as a Function of Technology Node. These are representative "raw" gate densities. Several key characteristics of the stdcell library, such as being designed for a performance goal as opposed to a density goal, can greatly affect these numbers. However, these numbers should be relatively close to the experience of the user community.

Technology Node	Known Clock Frequencies, General Purpose Bulk Process
.35 micron	
.25 micron	25 MHz
180 nm	
130 nm	333 MHz
90 nm	500 MHz
65 nm	
45 nm	
28 nm	1–3 GHz
20 nm	5–7 GHz

The curve in Table 6.1 and similar ones for the density of performance libraries and power libraries is a major contributor to the continued validity of Moore's law (the ability to place transistors on a piece of silicon doubles every 18 months) over the years.

Note also that the above curve is for "raw" gate density. This assumes a close packing of the stdcells on a design. APR tools cannot do this with any degree of success beyond a use of approximately 80% (and that is with an experienced routing engineer; less experienced engineering resources might only be able to get 65–70% utilization).

Each technology node brings a new set of stricter rules, so note that Table 6.1 is held to the "square of the shrink" trend not by just taking the same footprint and shrinking it repeatedly over the years. There is real creativity on the part of many people at many companies to overcome the increase in restrictions at every technology node and allow the trend in the table to go on "as is" as it continues to do so. However, at some point, the ability to maintain the trend will be unsustainable. At very deep submicron technology nodes, for instance, polysilicon is only patternable at a fixed pitch. Also, as alluded to, active and polysilicon routing is impossible at these deepest submicron technology nodes. However, this may not be as bad as it sounds. Just as the steep decrease in the cost of memory over the years has allowed the writing of large software packages, the fact that it is nearly possible to place 10 million gates per square millimeter may mean that the constant pressure for smaller and denser stdcells may also be relaxing.

Second, trend lines are a little harder to show in the performance axis over multiple technology nodes. The reason for this is not just that there are different processes such as bulk versus silicon on insulator (SOI) at any given technology node. The very term *performance* can be interpreted is many ways, and not just because most companies tend to protect this sort of information as proprietary. To some people, performance is the average delay through some typical cell; to others, it is the typical edge rate that a cell can produce for a given load; and to still others, it is the clocking frequency at which a system runs. All of these definitions are problematic. For the first, the definition of a typical cell is loosely defined; for the second, it is the definition of a typical load; and for the last, the clock frequency, as alluded to in the introduction to this book, largely depends on how many levels of inversion are in the longest cone of logic on the design. Certainly, none of these definitions has held for more than a technology generation or two over the years. Table 6.2 gives the rough clock frequency of typical nonmicroprocessor general-purpose

TABLE 6.2 Maximum Clock Frequency of Some General Process Bulk Designs as a Function of Technology Node. This is a much more difficult set of numbers to calculate because of the extreme divergence in design requirements. Typical ASIC design will have significantly longer cones of logic as opposed to high-speed multistaged processor design and therefore will have significantly slower performance capabilities at the same technology node.

Technology Node	"Raw" Gate Densities
.35 micron	25 Kgates/mm^2
.25 micron	50 Kgates/mm^2
180 nm	100 Kgates/mm^2
130 nm	210 Kgates/mm^2
90 nm	450 Kgates/mm^2
65 nm	850 Kgates/mm^2
45 nm	1800 Kgates/mm^2
28 nm	4000 Kgates/mm^2
20 nm	7800 Kgates/mm^2

process bulk designs that are known to have been designed at some of the listed technology nodes over time. The table is not fully populated because companies tend to treat such information as proprietary and do not want such numbers publicized. Once again, the data are somewhat scrubbed to protect the various companies' confidential information, but they are representative of what can be done.

Note that there appear to be step functions from sub-100 MHz to sub-GHz and to multi-GHz. This appears to be more of a function of what was in the design cycle at the time that the companies were active in design at these technology nodes as opposed to actual technical trends. Note, however, that deeper technology nodes do allow faster clock cycles.

Third, power is usually defined as power consumption, both static and dynamic. Dynamic power consumption actually tends to go down or to hold steady per gate as technology goes deeper. This is because faster technology nodes allow designers to run sharper edge rates, switching the succeeding stage of logic faster with every new technology node. By the way, note that dynamic power per transistor width probably does not follow this trend. Static power consumption, also known as *leakage*, is the exact opposite trend. Every technology node brings narrower channel lengths and thinner gate oxide, allowing as much as an order of magnitude increase in static power consumption for every technology shrink. This is why for technologies below 130 or even 180 nanometer, special cells are usually added to a library for special-purpose power-minimization techniques, multiple-Vt cells, clock-gating cells, dynamic frequency-voltage–adjusted cells, well isolation cells, and even state-retention FLIP-FLOPs. Chapter 2 lists some of these in more detail. A table for this section is worthless because the dynamic power consumption is relatively unchanged and static power consumption changes so dramatically over each technology node.

For architectural considerations, consider that all of the most used APR tools assume horizontal rows for placement; power and ground rails are connected by abutment. In addition, if it is possible to flip each row north–south, then N-wells can be shared between rows (as can P-wells). If wells can be shared, then it probably makes sense to have the power and ground rails at the very top and bottom, respectively, of the cells. In addition, deeper technology nodes have the first metal orthogonal to the polysilicon direction. In addition, routers like to assume alternating directions, horizontal and vertical between each metal routing layer, so if the architecture has internal routing beyond the first metal, make sure it is orthogonal to the predominant direction of the first metal.

Aside from the above, dense libraries tend to route signals internal to the cell with active, polysilicon, and metal 1, allowing "routing snakes" on each of these levels to complete these internal connections. By the way, as one moves to the deeper technology nodes, DRC rules, which are driven by phase-shift mask issues, tend to minimize the amount of active and polysilicon "snakes" but may compensate for that with creative additional layers such as local interconnect (which is a metal 0).

Performance libraries, tend *not* to allow routing on active and polysilicon because of the increased amount of capacitance and resistance on those layers as opposed to the various metal layers, and they do not allow routing snakes on any metal, especially on lower metals that may be thinner and hence have higher sheet resistances in terms of ohms per square. They tend to be taller, with wider transistors, but also have more internal routing tracks to allow unidirectional metal internal routing.

In addition to the aforementioned cells, power-consumption libraries may have strongly offset P-channel to N-channel ratios in an attempt to cause faster switching in one direction without introducing significant issues in the other direction. This is because of the reduced capacitance load on an output of a logic stage that would be caused by the reduced loading of the smaller P-channel or N-channel succeeding stage of logic (depending on which channel is reduced).

One related physical design kit (PDK) item should be discussed. Stdcells (and IOs, and to some extent larger physical blocks for memories and analog structures) are designed to "fit" nicely together. In the case of stdcells and IOs, this means abut together, left to right, and, specifically for stdcells, either top to top and bottom to bottom (for those libraries designed to flip and abut) or top to bottom (for those that do not flip). Such abutment will need to be tested to ensure that no DRC surprise is found through some inopportune placement once the library is released to the design center and is in use. One way to do this is to place each cell against each cell, including itself, in all possible arrangements and to run DRC on the result. Although cumbersome, it is possible to do this for all cells in a stdcell library (or all IOs in an IO library) side to side. However, for at least stdcells, doing so top to bottom or top to top and bottom to bottom, again depending on the architecture, for each possible and slightly left- or right-adjusted placement for each cell, although not completely impossible, is a much larger and more time-consuming activity that does become practically impossible.

So how does one get around such an effort? One way is to develop a *half-rule* DRC deck. Half-rule decks are DRC decks that determine how closely a polygon of a particular layer can be placed in respect to the cell boundary. Specifically, if there is a spacing rule for a particular layer, then a half-rule deck will define the minimum legal space of a polygon on that layer to the cell boundary as half the original minimum space rule for that layer. Similarly, if there is a minimum width rule for that layer, then the half-rule deck will define a minimum width with respect to the cell boundary as half of the original minimum width rule for that layer. The idea is that if not polygon is less than half the normal minimum space for that layer from the edge of the cell, then the full minimum space is maintained when any two cells are abutted. On north–south placement, rules are written for these half-rule decks such that the guarantees in the cell architecture that allows such north–south placement to occur DRC free. By the way, one additional benefit of such a deck is that for every new cell in a library, this DRC deck can be run once. Testing a new cell by abutting it to all the other cells in the extant library requires that a DRC deck be run several hundred times (once for each other cell in the library abutted to the new cell multiplied by the number of permutations of left or right placement allowed). As one goes further down through the technology nodes, the definition of such half-rule decks stretches. For instance, when a layer is required to be capable of *A/B coloring*, one solution is to ensure that the first polygon (for instance, the left-most polygon) on that particular layer within each cell begins with the same "color" and then restrict the polarity through each cell, thus maintaining such A/B coloring. A better and less restrictive manner is to ensure that either coloring of that particular first polygon produces a correct coloring polarity map in every cell, including abutment. In such a case, assurance of the left- and right-most polygons on that particular layer within each cell allows that particular coloring algorithm to become the realm of the half-rule deck. The reason to use the left- and right-most polygons within a cell of a particular layer that requires coloring as opposed to only using the left-most polygon is for the obvious reason that left–right flipping of cells during APR is necessary. If this is not a concern (for instance, when such flipping is restricted because of some standard alignment issue between cells), then the coloring algorithm is simplified by only requiring this from one side or the other (consistently through the library).

At some point, the external routing grid disconnected from the internal grid for the remaining features of the cells of the library. Specifically, down to about 90-nanometer technology, the router grid tended to remain consistent with the polysilicon placement grid. Some metal—for instance, metal 1 (the lowest "routing" metal in the backend stack)—could usually even be routed internal to the cell at tighter pitches than polysilicon. What this allowed was the more deterministic connection of the pins of a typical library cell. As long as the pins were placed on a specific grid within the cell, the chip-level routing could be better ensured, assuming that the chip-level routing grid could be aligned to the cell placements so that this chip-level grid matched the pin grid within each placed cell. However, at about 65 nanometers, the metal routing layers started to fall behind the curve that the internal features (specifically, polysilicon pitch) could achieve. That meant that the cells could shrink faster than the metal route. With the need to maintain certain shrink factor trends as technology nodes progressed, it was further assured that the cells would follow this shrink factor despite the lagging "route shrink" factor. As a result, the previous formal connection between specific pin locations on cells with the chip-level router grid was broken. At about 65 nanometers and below, new gridless router technology was developed. Here, although the majority of the route remained gridded, the actual connection to the cell was allowed to morph into a "unit tile" approach in which the final connection was accomplished with cell-specific knowledge within the router, the upper connections of which could be more easily connected onto the higher-level gridded route. What this means is that most routes can no longer ensure pin connection onto specific locations within a cell. Therefore, for deeper technology nodes, pins become stretched in one direction or the other (or both) to better ensure that any router unit tile really can find valid pin-connect points. Note, however, that it remains beneficial to ensure legal connection points (or "landing pads") on each possible gridded access location. Doing so allows the router to "operate in integer mode," following gridded route channels an integer number of intervals from each pin connection to each pin connection as opposed to having to use more time and resource-expensive pin-connection algorithms.

In conclusion, it used to be that stdcell layout at higher technology nodes was "creative." The DRC rules allowed for routing not just on metals, but also in polysilicon and active. As a result, dense layouts were only limited by how creative a layout engineer was in finding ways to snake signals around transistors. However, as technology nodes shrank, lithographic considerations began to affect and limit the ways that such creativity could be accomplished. Optical proximity correct (OPC) A/B coloring began to limit how first polysilicon and then active could be arranged, eliminating snake routing. Further still, lithographic considerations caused polysilicon to only be placed on exact pitches. At the deepest technology nodes, these restrictions had moved beyond such "fixed pitch" requirements to true template-based requirements. Layout is moving toward some discrete level of "correct by construction" templates, with standard transistor structures and highly structured interconnection, somewhat akin to tiny gate arrays replacing the traditional stdcell (but without the quick benefit of having banked blank arrays just before personalization). Hence, how applicable this section is depends on the technology node being addressed in the stdcell the library.

6.3 MEASURING DENSITY

The following discussion deals with "raw" gate density, which assumes a 100% use of cells in an APR region, something that is not achievable in practice. If I neglect to use the term *raw* when discussing density in this section, assume that it is implied.

With the discussion on NAND versus FLIP-FLOP measurements, not all cells in a stdcell library will have the same raw density. This is true for any given cell architecture. If a cell height is too small, then even with the size benefit for the simple cells such as the INV, BUF, NAND, and NOR, there will be a penalty in the size of the FLIP-FLOP cells and even the more complicated Boolean cells that is caused by the lack of routing channel resource within the cells. This will further cause these more complicated cells to explode in size (usually in width or possibly in double height layout development). Alternatively, if a cell height is too tall, the benefit of being able to efficiently route the complicated Boolean and FLIP-FLOP cells may be outweighed by the extreme inefficiency in the resultant nondense INV, BUF, NAND, and NOR cells. The bottom line here is that picking the optimum height for a stdcell library can be difficult. The reason why is that, contrary to a rather strongly held perception, the measure of a cell's raw density has almost nothing directly to do with the library raw density and even less to do with the density of a design done with that library, raw or otherwise. Any particular stdcell will represent just one of the 300 to 1,000 stdcells in the library; with the exception of some INV cells and some FLOP cells, it with be used far less than 1% of the time on any particular design (and the design will be further adjusted because of utilizations of significantly less than 100%). Therefore, the density of a particular cell will represent less than 1% of the average density of the library and usually much less than 1% of the density of a design. Only the density of a collection of representative cells bears any resemblance to library density or a design density.

Still, assuming that the library is complete, at some point either you will get curious about the density of the stdcell library that you have (either acquired or developed) or somebody else will get curious and ask you about it. Similarly, assuming that the library is not complete, somebody will ask you what the expected density of your library will be. In either case—reporting a posteriori results or estimating a priori results—there are a couple of valid methods of finding a correct raw gate density that is more than just a marketing number. The methods were alluded to earlier in this chapter.

The first of these is the *uniform distribution method*, and it can be very useful during experimentation on what the cell architecture for a library will turn out to be. Typically, during that experimentation phase, a small set of representational cells are created across several proposed architectures. What *representational* means and how the cells are picked for such experiments is up to the reader, but it can usually be defined "easily enough," although that fact does not necessarily correspond to "justifiably enough," and care should be taken when choosing these representative cells. Once the cells are physically completed for a given architecture, the individual densities can be calculated. A classic manner to do the individual density is as follows:

1. Count the number of transistors in a minimalist functional representation of the schematic (as opposed to the actual representation of the schematic, which may have additional inefficient extra transistors present (for any number of reasons). The minimal functional representation gives a standard to the proceedings that counting the actual transistors may not.
2. Divide the result by four to get a standard representational gate count (one gate equals four transistors, as in a typical implementation of a NAND2 or NOR2 gate).
3. If the cell has a higher drive strength than normal, multiple the entire gate count by the strength multiplier. This may not make immediate sense because one is not just multiplying the transistors in the last output stage of multistage functions such as

FLIP-FLOPS. If we count the entire number of transistors in a function when doing this strength multiplication, we are allowing for the higher strength cell being a substitute for multiple individual single-strength cells, which would be the situation if no higher strength cell existed. By the way, if the cell library has subinteger steps in cell strength such as half-strength INV or one and a half-strength INV, then it would probably be beneficial to "round up" these instances (half-strength cells should be multiplied by a factor of one because the cell count still exists), and the one-and-one-half–strength cell should be handled as a strength two cell because the one-and-one-half–strength cell would still need to be replaced by a single-strength cell and a half-strength cell if the one-and-one-half–strength cell did not exist.

4. Then divide the gate count by the area of the cell as expected to find a series of individual cell raw densities.

5. Finally, take a uniform average of these results; this becomes the estimated library raw density for that particular architecture experiment.

The above allows for no given cell to dominate the proceedings, so, assuming that the representative cells include both dense (FLIP-FLOPS) and nondense (NANDS and NORS) functions, then the *representative density number* will be somewhat close to the actual library raw density number should that particular architecture be chosen and the entire library produced in it. Note that the preceding also works for completed libraries where a library density measure is desired; just use the entire library's individual densities when doing the uniform averaging.

The preceding measure still does not bear any resemblance to how dense a library will be when used in an actual design. For that next step to be taken, some measure of how the cells will be chosen in actual use needs to be developed. If, by good fortune, the cell names of the various functions are maintained from previous technology nodes, and if there exist designs whose netlist are available for perusing that were done in those earlier technology nodes, then this is an entirely doable task. If this is not the case, then some amount of mapping of cell functions between previous technology-node libraries and the current library may be needed. That is to say, develop a map "previous function to current function," which may or may not be a one-to-one direct relationship, and substitute the resulting "current cells" for the "previous cells" in the following algorithm. This second alternative will lend a certain amount of uncertainty into the calculation. However, for the purpose of discussion, assume that either the naming is exact or the mapping is certain and unambiguous.

The above algorithm for uniform distribution needs just a few minor modifications to be applicable as a *weighted average distribution*.

- First, the strength multiplier might be adjusted downward (that is to say, because the actual transistor counts for each cell in the two technology-node libraries is known and because the new library is (assumed) in a deeper submicron technology node with smaller capacitive loading per node both in terms of input pin capacitance and in estimated and assumed APR routing, then it may be directly applicable to use actual transistor counts per cell for the larger drive strength cells as opposed to using the strength multiplier mentioned in the preceding algorithm.

- Second, instead of using a uniformed average of the cells of the library, develop some representative measure of the average use of the library cells in the previous technology-node designs. This is straightforward if the existing netlists are flat (nonhierarchical) such as in design exchange format or flat-Verilog format. If this is not the case and no translation tool is available, then a decent software developer should be able to build such and might be in demand for developing it. At any rate, once the flat netlist are available, a histogram of all the cells used is easily down with a few UNIX commands (sort the cell-type name field in whatever netlist format is used and then do a "UNIQ –C" to produce a list of how many times each cell type in the netlist). Compile this histogram for as many design netlists as are available (or appropriate) and then use that weighting as opposed to a uniform distribution to render the average.

• Finally, one additional step can be added here by determining the use for the previous designs, which is easily read from the make file (or possibly the resultant log files) that was archived when the design was previously completed, and by multiplying the resulting weighted average raw-gate density number downward by this utilization factor. Note that utilization rarely gets better and can easily get worse when crossing a technology-node shrink because of the increased need to take into account additional factors, such as intersignal cross talk at deeper submicron nodes, that force additional routing resource needs, which can force additional placement resource needs. Therefore, care should be given to choosing this utilization factor during the above calculations.

The above result, although still "not completely real," is a much more accurate representation of the real gate density that a particular library (and architecture) can produce. It reflects those types of designs that actually will be done by the users of that library. That is to say, if we assume that the same design team produced given designs in a previous technology node, then the scaled weighted average based on these previous generations of a given design will more likely be representative of those designs in a new technology node.

6.4 THE NEED AND THE WAY TO WORK WITH FABRICATION HOUSES

If you have ever taken a digital-signal sampling course, you are familiar with the term *Nyquist rate*, which sets the maximum frequency that can be detected by a sampling of data at half of the sampling rate. Something similar is applicable in terms of how tiny a pattern can be when printed on silicon. Images are patterned on silicon using mono-chromatic light. Although it is not directly equated with such a concept as a Nyquist rate, that monochromatic light frequency can be thought of as the sampling rate that deter-mines the maximum frequency or minimum size of geometry that can be patterned on silicon. As late as the 1990s, most ASIC development was at technology nodes at or greater than 350 nanometer, which was larger than the wavelength of light used in patterning it on silicon. As a result, with proper handling, as long as a pattern matched the minimum design rule checks, it was known that it would be drawn on the silicon correctly. As the dominant technology node moved deeper, down toward 180 nanometer, the wavelength of the monochromatic light was also dropped down to 193 nanometers; although larger than the patterns that were being drawn, this was sufficient to allow decent patterning. However, approximately at that particular technology node, special pattern additions were being added at tapeout or at the mask shop in an attempt to improve the ability to meet the patterning as desired by the designer. These special additions took the form of serifs at the corners of squares and rectangles, occasional crenellations along the sides of long rectangles, and OPCs interspersing rows of narrow structures or at the edges of it. Figure 6.1 illustrates these techniques. As mentioned, these were all designed not to be directly patterned on silicon but to help the patterning of what was originally drawn by the designer.

This would still be the case if the technology were such that the scaling of the wavelength of the monochromatic light, from technology node to technology node, could have continued downward at the rate that the size of the patterns that the light had to print had been decreasing. However, that has not been the case. The deepest technology nodes are now nearing 20 nanometers, whereas the wavelength of light is significantly larger. Lithographic experts have done some amazing things in order to make these 20-nanometer rectangles and squares appear but the situation has now progressed to the point even as if they had been written by light of a much smaller wavelength, but it is at the point where only certain patterns can be printed. At technology nodes at or less than 90 nanometers, polysilicon, among other critical layers, has to be processed with a phase-shifted mask, sometimes referred to as *A/B coloring*, which are masks where the optical thickness of quartz on one side of each particular pattern is different than that on the other side. The light on one side thus comes out a half-wavelength different, leading to destructive interference of higher-frequency side patterns caused by the narrow patterns. This is illustrated in Figure 6.2. Because of the phase-shifting mask operation, not all patterns of polysilicon are printable, especially those with a rectangular snake of polysilicon curving back on itself. Because of this phase-shifting mask operation, not all

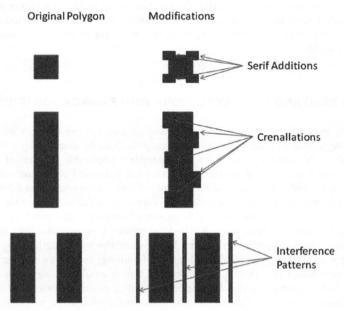

FIGURE 6.1 Examples of OPCs Used at Some Technology Nodes. These are usually attached during polygon "fracture" (mask prep), so the typical design engineer, although aware of the technique, usually does not see the extent of the effort of such additions.

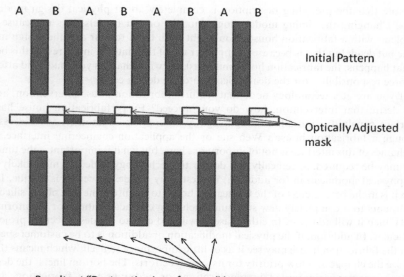

Resultant "Destructive Interference" between patterned rectangles

FIGURE 6.2 Phase-Shift Mask Cross Section. Also known as *A/B coloring*, this technique takes advantage of the potential destructive nature of interference. Still, such a technique is extremely limiting to the type of routing of polygons on layers on which it occurs.

patterns of polysilicon, especially those where a rectangular snake of polysilicon that curves back on itself is printable. At 65 nanometers, even this is not enough, and polysilicon has to be placed at a set stepping distance—and only at that distance—from each other. Therefore, a pattern is not guaranteed to appear on silicon even if it does pass DRCs.

Starting at around 90 nanometers, design for manufacturing (DFM) guideline decks were included in PDK releases. These DFM guidelines were usually one, two, and three manufacturing grid increases in the length of certain (and perhaps most) regular DRC rules, the concept being that adding one manufacturing grid extension beyond the minimum DRC rule would take care of 90% of any potential yield loss because of the DRC rule being insufficient. Adding two manufacturing grids extensions beyond the DRC rule would take care of 99% of that yield loss. Adding three would take care of 99.9% of the potential yield loss. In addition, with the addition of the guideline to use double VIA (that is to say, two adjacent VIAs between the same signal on two adjacent metal layers), this was the case. In addition, for a couple of technology nodes, passing these guidelines helped. At technology nodes below 65 nanometers, even this is insufficient. At technology nodes at the current deepest level—28 and 20 nanometers—the new norm is to use multiple-exposure techniques, with parts of the patterns on some levels exposed during a first exposure and other parts exposed during a second or even a third exposure. The bottom line thus becomes, if you are dealing with technology nodes at the deepest level, then it will behoove you to use the stdcell library physical views without modification. If you desire to add a special cell function or special IO for design-center use, then approach the fabrication house to do so, even if there is added cost. Conversely, if the technology node that the design center is asking about is higher than 90 nanometers, it is more likely that physical adjustments are safer to make on your own.

Note that the preceding description is completely about physical design modifications. Changing the timing models, logic models, or test models does not cause any problems with a fabrication house. You might push them too far and the design might come out dead, but that is because of improper use of the material in the rest of this book. If that happens, the fabrication house might ask for what changes you made and attempt to force responsibility for the dead silicon onto the design center.

If you are at a technology node that requires fabrication house intervention, and if you desire that intervention, how do you proceed? Every fabrication house has an interface capability, and all, although slightly different from another, are addressable through a fabrication house's Web site or the application-engineering interface. The mechanics of this interface is not of importance here, but what is important is the timeline that may be required. Specifically, the deeper the technology node, the more likely that the physical modification (or addition) requested by the design center is unique. If so, then it is in the best interest of the fabrication house to require some sort of test silicon. If it amounts to a completely new structure, including characterization (or monitoring at least), then it will add cost. In either case, it will add cost to the design-center project in particular. In addition, if the physical modification or addition is to be customer specific, then the fabrication house may see it as of little importance to itself, which means that it will see the change as a low priority for its support efforts. The bottom line: if the design-center customer is interested in modifications or additions to a fabrication stdcell offering, then it will behoove the customer to make it as broad-based as possible. This means that the premise of doing something to find excess margin in a third-party stdcell library by physical modifications diminishes with each technology-node shrink.

6.5 CONCEPTS FOR FURTHER STUDY

1. Dense stdcell library at 90 nanometer or above:
   - Above a certain technology node, interior-to-the-cell routing tricks can be made to ensure dense stdcells. These tricks include snake routing between transistor sources and drains in active, snake routing between transistor gates in polysilicon, and snake routing in general on low metals. The student or interested engineer might be inclined to investigate various cell height and router metal grid-direction architectures by picking a series of stdcells (some simple Booleans, some complex Booleans, some FLIP-FLOPs, and some higher-than-1X drive strength versions of these same stdcells. Once the list is determined, the student or interested engineer might attempt to lay them out in the various architectures to determine what density tricks in which architectures allow for meeting the density guidelines given in the chapter.

2. Dense stdcell library at 65 nanometer or below:
   - Below a certain technology node, lithographic considerations cause polysilicon to be laid out on a fixed grid or else not be printable. In addition, at even denser technology nodes, snake routing of active becomes impossible. The student or interested engineer could pick the same stdcells and drive strengths as those chosen for the previous concept for further study and attempt to lay them out by hand under these additional restrictions in the various previous architectures to determine what architectures continue to allow libraries to continue on the density curves given in the chapter.

3. Performance stdcell library:
   - Performance stdcell libraries require an altogether different arrangement of intertransistor connection. Specifically, snakes of any kind that add any additional parasitic capacitance to interconnect nets are considered poor. The student or interested engineer might be inclined to take the same stdcell list as for the previous two concepts for further study and double or even triple the transistor widths in the netlist of each cell. Then the student or interested engineer could double the height of the previously optimized density library for that technology node and attempt to lay out performance stdcells. Finally, the student or interested engineer could extract the parasitic of these to see if there are parasitic minimization techniques that lend themselves to further perform-ance improvement.

4. Power stdcell library:
   - Power-minimization stdcell libraries tend to have completely different approaches to laying out architecture as those taken for dense or performance stdcell libraries. One such approach is to ensure that routing pins for a stdcell are as high as possible in the backend stack so that routing can be accomplished one or even two metal layers higher up the backend stack (to minimize parasitic in the route). Again, the student or interested engineer could take the same stdcell as in the previous four concepts for further study and relay them in a given architecture previously chosen for one of those earlier concept studies. However, the layout should be such that all router pins are above metal 2, all routing in the cell is at or below metal 1, and the pins are at two times the normal router grid distance.

5. Lithography study:
   - If the student or interested engineer has access to a lithography tool, then it would behoove him or her to take a given stdcell library acquired by some means or another and run that library against the lithography tool. If the tool has the ability to simulate at other than a 193-nanometer wavelength, then it would further benefit the student or interested engineer to redo the study at multiple such wavelengths.

6. Engaged versus disengaged routing grids:
   - Before technology nodes of about 90 nanometers, APR router technology was such that stdcell architectures had to ensure signal pin access on exact gridded locations in the stdcell. Newer APR technologies no longer require such gridded access. The student or interested engineer might find it enlightening to take example stdcells from older stdcell libraries (perhaps the same list of cells as in the first concepts for further study in this section) and relay them out without the older router technology gridded pin access requirement.

7. Removing routing between shared nodes on high drive strength stdcells:
   - In many higher drive strength stdcells, there exist isolated nodes that normally stretch between two P-channel or two N-channel transistors that happen to be in that particular stdcells output stack. In higher drive strength versions of those same stdcells, the comparative connect, which is now between multiple parallel P channels and multiple parallel P channels (or multiple parallel N channels and multiple parallel N channels), can sometimes be better handled as being implied (or virtual), which means that these parallel nodes remain

physically isolated from each other as opposed to being connected by other wasteful and completely unnecessary material internal to the cell routing. The student or interested engineer could do a density study between cells that do not have such interior connections versus the same function cells where such connections remain implied (or virtual).

8. Local interconnect:
   - At deeper technology nodes, as the restrictions on active and polysilicon snake routing increased, the need to maintain density while minimizing the affect on APR router access led to the introduction of below metal 1 routing. The student or interested engineer might be inclined to take the same stdcells that were used in the earlier concepts for further study in this section and lay them out again with metal 0 (or local interconnect) that connects to active and to polysilicon by abutment and to metal 1 by means of CONTACTS. The goal would be to see what the effect on density for such technology might entail.

9. Statistical study of stdcells:
   - The student or interested engineer who has access to the netlist of several designs across several technology nodes could do statistical studies on which types and drive strengths were most often used on those designs' stdcell regions. This is most useful if the designs use libraries from the same vendor across those various technology nodes.

10. Automated stdcell layout tool:
    - The advanced student or interested engineer who has a background in software might find it beneficial to attempt to build a perfunctory automated transistor-placement tool. This is not the effort of a single weekend, month, or even semester. Successful completion, however, usually ensures employment at any number of semiconductor design companies on graduation.

CHAPTER 7

SPICE

7.1 LESSON FROM THE REAL WORLD: THE MANAGER'S PERSPECTIVE AND THE ENGINEER'S PERSPECTIVE

In the mid-1990s, I was asked to develop a 350-nanometer universal serial bus (USB) 1.0 input–output (IO) to support the addition of such a communication channel onto a series of parts. USB IO development is simple and straightforward, at least for the revision 1.0 specification, which is available from the USB consortium at usb.org. Two current-voltage (I-V) waveform envelopes define the loaded IO outputs. These were defined in such a manner as to allow any vendor's CMOS process to be capable of doing so with appropriately sized IO totem-pole P-channel and N-channel transistors. Even the rise-time and fall-time slope requirements introduced in revision 1.1 were not difficult add-itions to include. The only real issue was the need for a "hot plug" that allowed the IO to be powered for as long as a day with the chip itself remaining unpowered without damaging the device. We solved this through some isolation techniques on the IO power and IO substrate connections. However, when the register-transfer language (RTL) designers attempted actually to integrate the IO while designing the USB channel, they found that the synthesis through the IO and cone of logic to the first FLIP-FLOPs, internal, and the reverse of this path—that is, FLIP-FLOP, cone of logic, and IO—was longer than the specification could handle. Because the IO was new and had not seen silicon before, it became the suspect structure, and I was asked if I could improve the performance. Doing so without violating the two I-V envelopes, however, was demon-strably impossible. The eventual solution that was used was to develop a separate stdcell

Engineering the CMOS Library: Enhancing Digital Design Kits for Competitive Silicon,
First Edition. David Doman.
© 2012 John Wiley & Sons, Inc. Published 2012 by John Wiley & Sons, Inc.

library, including characterization, which was to be used only for the USB synthesis (this was before multiple Vt libraries existed, at least at that particular company). The new USB-only library was otherwise exactly the same as that used on the rest of the design, except that it was characterized at 2.5 standard deviations (2.5-sigma) worst-case process, 3-sigma worst-case temperature and voltage, instead of 3-sigma worst-case temperature, process, and voltage. Doing the actual characterization was easy at that particular company because the internal fabrication facility released the best-case and worst-case SPICE models with sigma-delineated adjustments from nominal SPICE, which is now a common but not universal practice by fabrication houses but was radically new then. I just had to define a variable that represented sigma as 2.5 instead of 3 in the SPICE characterization runs. The design passed timing with the new 2.5-sigma worst-case USB-only cells (in the USB circuit), and the remainder of the design being timing closed with the regular 3-sigma library. The chip came out of fabrication and reached certified immediately. The design was used in all of that company's USB 1.0 and USB 1.1 products.

It is important to emphasize what such cutting back in terms of sigma *tail coverage* really means. In some cases, it is a potentially beneficial activity; in others, especially in the memory section, it can be extremely detrimental because there often is a need to decrease the chances of extreme "tail events" in which a given feature is actually processed beyond the number of sigma's covered by the model (the feature represents processing in the tail). In those regions where such cutting back would be detrimental, the book will argue for the increase of sigma coverage. Table 7.1 gives the amount of product that can be expected in a Gaussian distribution out in the one-sided tail of the Gaussian curve (a similar amount can be expected in the tail on the other end of the Gaussian graph as well). Specifically, for a product that has a parameter that follows

TABLE 7.1   On-the-Tail Statistics for a Gaussian Distribution. Three sigma (standard deviations) is right at the knee of the curve in terms of parts that fall outside the distribution. Less sigma coverage yields parts-per-hundred (which first occurs at approximately 2.325 sigma) rates for parts falling in the tail. More sigma coverage yields parts per ten thousand (which first occurs at approximately 3.091 sigma). However, depending on usage, moving upward or downward in terms of sigma coverage may be beneficial. By the way, the chances of winning a "50 choose 5" style lottery (1 in 15,890,700) makes it around a 5.284-sigma tail event. It is fascinating to watch those who believe that a 3-sigma event is impossible for all practical purposes hurry to buy a chance at such a 5 + -sigma event.

Sigma	Percent, Mu to Sigma	Percent Single Sided Tail	Parts in Tail (1 per X)
0.5	19.1462461%	30.8537539%	3
1	34.1344746%	15.8655254%	6
1.5	43.3192799%	6.6807201%	15
2	47.7249868%	2.2750132%	44
2.5	49.3790335%	0.6209665%	161
3	49.8650102%	0.1349898%	741
3.5	49.9767371%	0.0232629%	4,299
4	49.9968329%	0.0031671%	31,574
4.5	49.9996602%	0.0003398%	294,319
5	49.9999713%	0.0000287%	3,488,556
5.5	49.9999981%	0.0000019%	52,660,502
6	49.9999999%	0.0000001%	1,013,594,236

such a Gaussian distribution and for a given number of standard deviations for that Gaussian distribution, the right-most column details the amount of material of that product in which the parameter would be in the single-sided tail (in terms of parts per X). Hence, for a 0.5-sigma measure, one part in three would be expected to be in the tail beyond the 0.5-sigma limit, and for a 6-sigma measure, one part in more than a billion would be in the tail beyond the 6-sigma limit. The mathematics of this is straightforward, and because I promised not to have such derivations in the book, it will not be given here. However, such derivations are readily available in numerous other books on statistics.

For a product, being in the tail beyond the chosen sigma limit just means that the product in question has a parameter outside that chosen limit. It does not necessarily mean that the product is broken. This is crucial to the previously mentioned USB illustration. A 2.5-sigma characterization says that for the few hundred transistors in the critical path USB channel cone of logic, approximately 1 in every 161 will be slower than expected (as opposed to the 1 in every 741 transistors that would happen if the cells of the channel had been characterized with a 3-sigma SPICE model). For the few hundred transistors in the path, two, three, or four will be expected to be slow. Again, let me emphasize, this is to say *slow*, not *broken*. In addition, a 3-sigma measure would mean that still perhaps one transistor in the path would be slower than such a characterization would indicate. Whether these two to four transistors are randomly placed, with an approximately uniform distribution throughout the communication channel, or are systematically located nearer to each other, the effect is still minimal and manageable. This is because there will be other transistors (in actuality, many more transistors because of the nature of the Gaussian statistical distribution) within the same long path cone of logic that are faster than indicated by the 2.5-sigma standard deviation liberty that was used in the USB-channel characterization effort. In aggregate, there is little penalty to this USB for the previously mentioned 2.5-sigma characterization.

Again, this will not be true for locations where a single transistor with a parameter being beyond expected measure means a more catastrophic failure. For instance, when it is one of six transistors in a bit cell, the parameter misalignment allows the bit cell in some manner or another to truly fail to read or store or write correct data within the allowed clock period. In such a case, if enough bit cells fail so that redundancy cannot be resolved, then a mere few hundred tail events can cause absolute device failure. As argued in Chapter 4 on memory, this is why the sigma coverage for memories needs to be in the 5- to 6-sigma range.

This experience registered with me and stayed in the back of my mind until about three years ago. I was again working with a company that needed a communication protocol that was nearly beyond the capabilities of that particular technology node. This time it was a double data rate, version 2 (DDR2) design. The company was using an outside vendor's library (including IOs). During development, communication-channel simulation experiments were run in order to see what sorts of memory-connection architectures were achievable. When those experiments were run, including the expected worst-case jitter for the system clock, the expected uncertainty of the memories, and the realization that the simulations could not ensure that the worst-case cycle had been in the simulation, it was determined that the communication eye diagram must be completely closed. After significant reworking of the system architecture and after involving the library IO vendor, nothing was found that could resolve the situation. I suggested the same technique that I knew had worked before. We then approached the IO and library vendor, but the proposal was rejected. The design center doing the design, however, asked if we could do the work-around on our own. If we were to break what is normally a golden rule—that is, "Do not change a vendor's SPICE models"—then we could. The IO vendor had given us nominal SPICE and worst-case SPICE (and best-case SPICE) models, but the best and worst cases were not sigma-delineated adjustments to nominal. We did know that the vendor considered these SPICE to be the 3-sigma worst case and best case. So I compared the normal parameters in the nominal SPICE model against the worst-case parameters in the worst-case SPICE model and built a third model in which all of the parameters were averaged between the two models (one-third nominal value, two-thirds worst-case value). I have since chided myself for doing this because many parameters really should be fixed by the physics and chemistry of the process, but we did not account for that in this effort. We characterized the IO and the stdcell libraries pertinent to the channel and used these to do timing closure. The chip came out and worked first in silicon. The customer built its system around this chip and is selling the product today. One aside: this further shows the loose margins that can be found in every step of the integrated circuit (IC) design cycle.

This second successful experience with adjusting SPICE taught another lesson. While researching the parameters that were changing between the nominal and worst-case SPICE, I found the several values that were supposed to be constant for a material that nonetheless had changed between the two models. Hence, it is always worthwhile to review the SPICE models for such changeable constant parameters.

I have since developed a personal tool that reads two SPICE models, together with their sigma values. Using this, I can produce any sigma SPICE for that particular vendor's process and technology node. I also use it to check the SPICE that vendors give me. Specifically, I question the vendor when the nominal SPICE model that the tool produces (by building it after inputs of the 3-sigma best-case and 3-sigma worst-case SPICEs) does not match the nominal SPICE that the vendor supplies.

It is also interesting to ask vendors why certain parameters in their SPICE that are supposed to be constant because of these describing exact chemical constants, such as the minimum and maximum mobility of electrons in a material doped by a specific chemical implant, are changing over the process. As a further extension on this particular topic, on a personal note, to build SPICE models that more closely match the vendor's supplied SPICE, my SPICE creation tool will adjust these "constant" parameters when I am interpolating between the supplied SPICE models but will hold these values when I am extrapolating beyond either supplied SPICE model.

As mentioned, and in conclusion, it is important to recall what doing such sigma adjustment means. Correctly determined sigma variances in a value determine the amount of processed material that will have measured values that fall between that value and the average value and the amount of processed material that will have measured values that fall on the tail. That knowledge, together with the cost of production, can allow the calculations that are needed to develop a profitable design. In addition, the 2.5-sigma range that can be used in many instances is nowhere near that bad in terms of allowing otherwise good material to be discarded. Finally, there is a good chance that the sigma measure of the standard deviation of a process is nothing but a guess by the developers of that process. Fabrication houses will tend to overestimate these standard deviations. Hence, the actual amount of material that will fall outside the 2.5-sigma range will be surprisingly small.

In this chapter, we will discuss what parameters can be changed and what parameters are truly determined by physics or chemistry and should not change.

7.2   WHY A TOOL MORE THAN 40 YEARS OLD IS STILL USEFUL

In the late 1960s and early 1970s, despite the existence of several company proprietary circuit simulators, a professor at the University of California at Berkley, Ron Rohrer, together with his graduate students, developed a nonlinear circuit simulator called *computer analysis of nonlinear circuits excluding radiation* (CANCER). CANCER allowed for several types of bipolar circuit simulations to occur, including AC analysis, DC analysis, and transient analysis. Luckily, from a naming point of view, by the time the simulator was released to the industry in the public domain for general use in 1972, it had been renamed *simulation program with IC emphasis* (SPICE). The program is written in the Fortran computer language. That means it was originally designed to operate on large, centralized mainframes. Although it is not applicable to CMOS stdcells, the first supported circuits in CANCER, all bipolar, where modeled by Ebers-Moll equations. Among other items, as SPICE quickly became the industry standard circuit-simulation tool, changes included the introduction of Gummel-Poon bipolar modeling equations. These model names are mentioned because they are probably just names to most CMOS library and CMOS design engineer readers, but Gummel-Poon models are significantly more complex than the earlier Ebers-Moll models. This illustrates the ability of SPICE to enhance in an almost modular fashion enhance its capabilities with ever more complex and accurate modeling additions, a process that continues to this day.

One key addition to this 1972 public domain release, however, that is applicable to the current topic is the support of metal-oxide semiconductor field-effect transistors (MOSFETs). Without this addition, modern IC design in general and modern stdcell library characterization in particular would be much more difficult.

SPICE is a powerful simulation tool because it is a specialized spreadsheet tool geared toward circuit-simulation capabilities. It uses this capability to do nodal analysis on the signals connecting the various linear and nonlinear elements that a design defines in the language used by the tool to describe these circuits. This is highly useful to the design engineer and is a key reason why SPICE is still being used four decades after its introduction. One further possible reason for this is that it can be seen as one of the earliest examples of open-source software. The source code is available, from Berkley, for anybody to download. Again, written in Fortran, it is easily readable for anybody with any nontrivial experience in a computer language. Downloading it and reading through the code can be an instructive way to see an example of good coding practice on something that continues to be widely used.

In 1975, one of the developers offered a much-improved version of the simulator, SPICE2, that allowed the inclusion of inductors, which were previously unsupported. SPICE2 has since become the basis of all modern circuit simulators. Since 1975, there have been some efforts to bring the tool into the current environment. These efforts include rewriting it in the C programming language to make it more consistent with the UNIX operating system on which many engineering work stations have run in the past, but these have been less than successful, partly because of the extent to which SPICE2 has entered the engineering design market. The code has not become obsolescent. Therefore, there is no reason to move on to later generations of the tool. In addition, since the mid-1970s, several companies have introduced commercial versions of the tool, all based on wrappers around the Berkeley SPICE2 version, these being integrated in the various IC companies' design environments.

As mentioned previously, one additional reason why SPICE has become the dominant circuit-simulation tool in the industry is that the tool's performance is based on ever-new development of models. Modern MOSFET support originally was based on equations named "SPICE level-1," "SPICE level-2," and "SPICE level-3," which were primitive but useful in the areas where they operated. However, these earliest metal-oxide semiconductor (MOS) models showed significant limitations, including discontinuities at the boundaries of the operations of their various modeling equations that caused, among other things, serious convergence issues during simulation start-ups. When these, plus the lack of consistency between simulation and actual operation at deeper technology nodes, became too serious, refined second-generation models named Berkeley short-channel IGFET model (BSIM and BSIM2) were developed, and these eventually led to BSIM3, which is the basis of all current MOS simulations inside SPICE. The various BSIM3 models can handle just about any simulation issue, including leakage, noise, and sub-threshold operation, and these severely reduce the previous convergence issues of earlier models (discontinuities remain, however, and these may become serious enough in the future to require a BSIM4 model development). Although that is yet to be seen, note that the first BSIM4 specifications where released by the BSIM Research Group in Berkley in late 2009. Also note that this first release of the BSIM4 specification appears to be problematic in that the parameters for the transistor models cannot be easily "fit" by data across large stretches of the design range. This will obviously need to be amended, or it will otherwise limit the acceptance of the new models within the design community.

BSIM3 models come in levels and versions. These have been chosen by the various suppliers of SPICE models, mainly SPICE tool vendors and fabrication-house technology vendors. Although the numbers may imply a level of maturity—for instance, a level-69 model being more advanced that a level-49 model—this is to at least some extent illusory, although the BSIM Research Group will disagree with this. Level numbers tend to just correspond to "the next higher available number" when that particular vendor chooses to publish its model. Vendors all have their own various slight modifications and, in many cases, more than slight increases in the number and function of parameters supported from the generic BSIM3 model. Some of these different parameters, which can run into the hundreds, can be adjusted by the user. Most of these different parameters cannot (or should not) be adjusted. Knowing which can and by how much can provide the library support engineer and the design-center user with significant ways to reduce the design margin left on the table by the fabrication house

Although it is not the purpose of this text to describe the functioning of a BSIM3 model or any SPICE model (there are several excellent texts available that do so), it is important to have a cursory familiarity of the model when dealing with the parameters that can and cannot (or should not) be adjustable. At any rate, the basic BSIM3 model containing increasingly complicated equations in each of its various levels describes the basic CMOS model shown in Figure 7.1. Furthermore, a list of pertinent parameters usually found in most of the BSIM3-level models are shown in Table 7.2.

The shaded items in Table 7.2 can be adjusted in order to interpolate a 2-sigma worst-case model from a 3-sigma worst-case model. Although they should not, the shaded items usually change between the various sigma-enumerated models that a fabrication house may release. As mentioned, care should be taken about scaling the shaded items backward from worst-case model to nominal model. In addition, as previously mentioned, the fabrication house should make inquiries when the worst-case model differs from the nominal model on those items that are not shaded.

The key parameters that can affect the performance improvement that can be expected in 3-sigma to 2-sigma worst-case change is because of oxide thickness (TOX) and those mentioned in the above figure, namely, gate source capacitance (CGSO), bulk

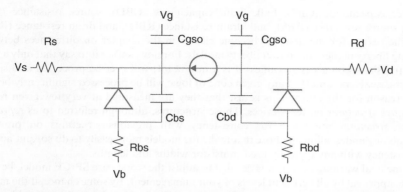

FIGURE 7.1   The Basic BSIM3 SPICE Model. BSIM3 currently is the most used CMOS SPICE model for deep submicron design. However, the BSIM4 model has been announced and is expected to overtake BSIM3 for accuracy in the deepest submicron technology nodes.

TABLE 7.2   **Various Pertinent BSIM3 Parameters. Expert knowledge of the BSIM3 model should be a requirement before any of these parameters are adjusted downward toward 2-sigma versions.**

Parameter	Description	Comment
AF	Flicker noise exponent	DO NOT TOUCH, SHOULD BE CONSTANT
CBD	Bulk-drain zero-bias p-n capacitance	Sometimes overly conservative since determined by test structures that represent only a subset of actual widths
CBS	Bulk-source zero-bias p-n capacitance	Sometimes overly conservative since determined by test structures that represent only a subset of actual widths
CGDO	Gate-drain overlap capacitance	Sometimes overly conservative since determined by test structures that represent only a subset of actual widths
CGSO	Gate-source overlap capacitance	Sometimes overly conservative since determined by test structures that represent only a subset of actual widths
DELTA	Width effect on threshold	Sometimes overly conservative since determined by test structures that represent only a subset of actual widths
ETA	Static feedback	DO NOT TOUCH, SHOULD BE CONSTANT
FC	Bulk p-n forward-bias capacitance coefficient	DO NOT TOUCH, SHOULD BE CONSTANT
KAPPA	Saturation field factor	DO NOT TOUCH, SHOULD BE CONSTANT
KF	Flicker noise coefficient	DO NOT TOUCH, SHOULD BE CONSTANT
L	Channel length	At older technology nodes factory Assumes that this will change, BUT at deeper sub-micron technologies this becomes fixed
MJ	Bulk p-n bottom grading coefficient	DO NOT TOUCH, SHOULD BE CONSTANT
MJSW	Bulk p-n sidewall grading coefficient	DO NOT TOUCH, SHOULD BE CONSTANT
NEFF	Channel charge coefficient	DO NOT TOUCH, SHOULD BE CONSTANT
RB	Bulk ohmic resistance	Sometimes overly conservative since determined by test structures that represent only a subset of actual widths
RD	Drain ohmic resistance	Sometimes overly conservative since determined by test structures that represent only a subset of actual widths
RG	Gate ohmic resistance	Sometimes overly conservative since determined by test structures that represent only a subset of actual widths
RS	Source ohmic resistance	Sometimes overly conservative since determined by test structures that represent only a subset of actual widths
THETA	Mobility modulation	DO NOT TOUCH, SHOULD BE CONSTANT
TOX	Oxide thickness	3 sigma worst-case number CAN be reduced, usually
TPG	Gate material type	DO NOT TOUCH, SHOULD BE CONSTANT
UCRIT	Mobility degradation critical field	DO NOT TOUCH, SHOULD BE CONSTANT
UEXP	Mobility degradation exponent	DO NOT TOUCH, SHOULD BE CONSTANT
VMAX	Maximum drift velocity	DO NOT TOUCH, SHOULD BE CONSTANT
W	Channel width	Factory assumes that this will change
WD	Lateral diffusion width	Sometimes overly conservative since determined by test structures that represent only a subset of actual widths
XJ	Metallurgical junction depth	DO NOT TOUCH, SHOULD BE CONSTANT
XQC	Fraction of channel charge due to drain	DO NOT TOUCH, SHOULD BE CONSTANT

source capacitance (CBS), bulk drain capacitance (CBD), source resistance (RS), bulk source resistance (RBS), bulk drain resistance (RBD), and drain resistance (RD).

The reason for the first will become clearer in the chapter on differences between timing liberty models and power liberty models. For now, suffice it to say that only a very few transistors on a design (unless it is a matrix lot) will ever be processed at 3-sigma worst-case. Hence, most of any given cone of logic will be processed significantly better. The reason for the other items is just that they are measured at very few given *representative structures* on test devices. Those structures, although referred to as *representative structures* are made for consistency with previous structures on previous technology nodes, allowing those that make the models more easily to do so, and not for consistency with any typically used transistor widths and lengths.

One final warning here. If it is decided to adjust the worst-case SPICE model, be sure to get approval by all pertinent levels of your management. By some chance, if the model adjustment was done incorrectly or was too radical, then dead silicon could easily happen. Going to management after the fact and confessing that the fabrication house is not at fault in such a case can be disastrous for a career.

7.3   ACCURACY, REALITY, AND WHY SPICE RESULTS MUST BE VIEWED WITH A WARY EYE

As discussed in the previous section, SPICE is a remarkable simulation tool. If its predecessor CANCER is included, it has been able to keep on the leading edge of simulation and accurate representation of what a circuit can actually do for more than 40 years. It has been so successful that it is ubiquitous within the design environment. There are several synthesis tools and several logical simulation tools on the market. There are several polygon place and route tools on the market. In the marketplace, these various tools can be used on even several operating systems. However, although there have been several attempts to replace SPICE as the circuit-simulation tool, none has achieved any level of widespread success at penetrating the monopoly that SPICE enjoys. As also pointed out in the previous section, the reason for this remarkable tenure has been that, as technology progresses, new transistor models have been repeated developed that can be directly used inside the SPICE framework. The SPICE BSIM Working Group mentioned earlier is continually working toward development of the next SPICE model that the fabrication houses can use to model the next technology node. There is nothing but success written all over this remarkable and ongoing record of accomplishment.

So, why does the title of this section suggest viewing SPICE results "with a wary eye"? Well, any time a tool is as flexible as SPICE, with all of the adjustments to the models that it can use and with all of the adjustments to the tool itself, there is a great chance that somebody will not set all of them appropriately. As a result, what can be produced in a simulation may bear little correspondence to reality just as easily as it may come arbitrarily close to representing reality. Although there is a serious effort to minimize this possibility during the design and development of some portions of the IC world such as analog design, this is not the case in other parts of the environment. This is the surprising case for digital design. The first case, analog design, is justifiably considered to be an intricate design activity. An analog design of a few hundred transistors, with an order of magnitude fewer resistors, inductors, and capacitors, is a large design, requiring a large design effort. The amount of design resource in terms of people and time that would be spent on such an effort would be significant. There is a repeated nature of SPICE simulation, that of:

1. setting up and running a simulation,
2. reviewing the results in order to see if they make sense, and
3. refining the conditions and variables of the simulation and repeating the sequence, usually over several cycles, before the desired accuracy is achieved.

This is still considered viable from an engineering and economic point of view. However, in the case of digital design, the design industry has bought into the promise of the engineering design automation (EDA) industry over the past several decades. Specifically, EDA tools have allowed the level of transistors that can be effectively managed by a single design engineer to be greatly and continuously increased during that same period. In this case, spending days to weeks on simulation of a few dozen to a few hundred transistors (depending on the tasks' level of complexity) is disastrous to a digital design cycle. Admittedly, the preceding analysis is for a digital IC device as opposed to a library development, but the mentality involved has translated itself to the digital library environment as well. When a new process, voltage, and temperature (PVT) corner characterization is requested by a design center, the next question is "When can I have it?" In addition, it is assumed that the delay in producing that characterization results from the limits of the computer resources required to produce the required simulations needed for the characterization as opposed to the engineering resources that could be needed to review and possibly rerun those simulations. The final analysis here is that the numbers in any given characterization file—whether timing arcs, power arcs, noise arcs, or capacitances—should be viewed cautiously by the user community.

So what kinds of issues can happen in an improperly tuned SPICE run? During a typical characterization effort, it could easily take more than 4 million or 5 million individual SPICE runs in order to make the measurements for perhaps a half dozen or more input edge-rate break points by a half dozen or more output load break points. This is repeated for each of the dozen or more unique conditions for each of the several inputs per each of the several hundred stdcells in a typical design. These are further joined by the longer time frame required for setup-and-hold calculations are accomplished by treating the memory element (e.g., flop or latch) as a black box during simulations. These are required for each sequential element in the typical library (Table 7.3).

Typical issues that can easily arise in such a case are:

- SPICE step-size mishandling;
- inappropriate SPICE, absolute error tolerance (ABSTOL), relative error tolerance (RELTOL), and charge tolerance (CRGTOL);
- simulation ringing;
- metastability; and
- missed edges.

SPICE step size usually manifest itself in terms of incorrect output edge-rate measures because of SPICE converging at time steps that straddle the more appropriate points. Doing so causes the interpolation of the value being measured at points too far away from actuality, this then allows incorrect (or inaccurate) measure. Figure 7.2 gives a quick demonstration of this issue. Although SPICE itself can override the step size

TABLE 7.3 Explosion of Required SPICE Runs per Characterization. In addition, the generation of SPICE runs for inter-break-point evaluation could easily push the total to far more than 10 million individual SPICE runs.

Input Edge-rate Break-points	7
Output Load Break-points	7
Average Number of Pins per cell	5
Input Conditions per Pin	16
Combinatoric Cells per Library	600
Combinatoric SPICE runs	2352000
Average Sequential Pins per cell	3
Setup Runs/per Pin	100
Hold Runs per Pin	100
Sequential Cells per Library	64
Sequential SPICE runs	1920000
Total SPICE runs per Characterization	4272000

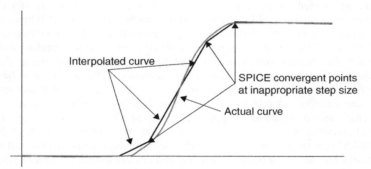

Interpolated curve

SPICE convergent points at inappropriate step size

Actual curve

FIGURE 7.2 Effects of Inappropriate SPICE TRAN Characterization Step Size. Such inexact interpolations can lead to false characterization within the resulting liberty file because the measurements will be either toward or from incorrect, poorly interpolated points.

requested in a SPICE transient (TRAN) card, it is not foolproof and can therefore be difficult to diagnose. However, once diagnosed, it is easily remedied by just forcing tighter steps in the SPICE TRAN card. Note that there then is a penalty for step size: the tighter the step size, the longer the run time. In general, it is probably a good idea to define step size as equal to the desired level of accuracy in the timing section of the characterization file. If the desire is for accuracy to 10 pico-seconds, then make the step size no greater than this.

The next issue that can arise is incorrect RELTOL, ABSTOL, and CRGTOL. The way that SPICE works in TRAN analysis is that the tool takes the current state of the circuit simulation at every step in the requested time span of the run, together with the changes to the inputs for the next step, and tries to converge on the next state of the circuit. It does this by repeated approximation using one of several sequential approximate techniques. In general, a state approximation is generated, and then the relative change and the absolute change in the voltage and current state of each node from the previous approximation of that same step in the run is determined. If the change is too large compared to the relative tolerance and absolute tolerance numbers given by RELTOL and ABSTOL, or if the change in total charge tolerance is too large as compared to CRGTOL, then the state of the circuit simulation is judged to be not yet converged and another successive approximate cycle is begun. This continues until such time that all nodes in the circuit do converge on voltage and current states that fulfill the RELTOL and ABSTOL limits. For many combinational circuits, this is not an issue, but for sequential circuits, convergence can become an important and time-consuming hindrance. The usual solution is to increase the RELTOL and ABSTOL numbers in the hopes of more quickly finding a convergence approximation. The issue with this is that the convergence approximation may be significantly off from the actual state that the circuit would be in under the same conditions.

Ringing will be discussed further in Chapters 9, 10, and 11, but suffice it here to say that it is demonstrable that any discontinuous change in a signal will cause a simulation of that change to oscillate over time. Although the oscillation should dampen out, it is still not real. There are real phenomena that produce the same thing in real circuits, but the two features are not correlated. Hence, any ringing in a SPICE simulation of a circuit that is not designed to cause such ringing is suspect. However, any measure of the various points of a characterization file will be around such signal discontinuities. Hence, every measure may have ringing-induced inaccuracies. The bottom line here is that this means that the characterization file is nothing but a simulation of reality. It is not reality. This becomes a liberating realization. This means that the values in the various characterization files can be adjusted as desired for other reasons without giving up on accuracy because the original characterization file is only a representation as well. The modifications that this allows will be discussed shortly.

Metastability is the next reason for inaccuracies in a characterization file. As mentioned in the section on sequential stdcells, metastability is the possibility of a storage element going into an indeterminate state. It is a real event and can produce circuits that remain in such states for extended periods. Parasitic resistances and capacitances within any circuit, however, will eventually cause the given metastable sequential cell to resolve. The bad news is that the time taken can be vastly longer than the normal operating clock frequency of the logic in which such a sequential cell in used. What is even worse is that metastability is almost impossible to simulate. The closest that can usually be reached is that the output of a sequential cell tends to grow as a metastable point is approached. However, the actual metastable point is significantly small. Nonetheless, metastability is real. The only real way to solve this is to keep the setup and hold region sufficiently far away from the region where metastability is suspected. Methods of doing this are extensively discussed in Chapters 9, 10, and 11.

One last reason that can cause inaccuracies in a characterization file are the actual missing or misinterpreted events within a SPICE result. Recall that a library characterization may involve 4 million or 5 million actual SPICE runs. Many times, two or three (or more) of these can be intelligently combined into multiple measurement runs. Doing such measurement run combination, however, can result in inappropriate measurements. Specifically, if a first or second measurement event in such a combined run does not occur as planned, for whatever reason, then subsequent measurements that occur in the same run should be questioned. A missed event, usually an edge on an output signal, could easily be interpreted as an edge event for an earlier missed event. As a result, several measures would be incorrect. These measures would be spread throughout any resulting characterization file. The one possible means to determine such failure, perhaps the only means, is to intensely inspect a resulting characterization file. Comparison of the results for a given pin on a given cell to similar pins in similar cells should be reviewed for consistency. Similar comparisons across the input edge rate or output load measures for a given pin can be reviewed for significant changes. Comparisons of similar pins within a cell should be reviewed. Finally, comparison across similar PVT characterization files also should be done.

Note that this process is not automatable. When a failure mechanism is identified, diagnostic tests can be automated and run on any resulting characterization file that is built from that point onward, but each new characterization file needs to be reviewed for previously unseen (or at least previously undetected but present) failure mechanism. This will be discussed further in Chapter 13.

Earlier in the section I mentioned that any characterization file, by the very fact that it is based on simulations that cannot be completely accurate, is nothing more than a representation of reality. This representation is designed to be useful to the chip-level synthesis and chip-level timing and power closure tools. Knowing this is the case, it should be easy to convince a design-center engineer that further modifications to a resulting characterization file, in order to allow for more efficient synthesis (as an example), will do nothing but allow for better or more efficient design-level synthesis. Chapters 9, 10, and 11 discuss ways to force monotonic behavior on a characterization file. There is a further discussion on ensuring that all measurement values are positive (or at least nonnegative), with the possible counter reasoning of the need for useful skew design techniques. The preceding argument suggests that these methods are valid.

7.4 SUFFICIENT PARASITICS

One of the first technology nodes I worked on was a 1.25-micron very-high-speed integrated circuit (VHSIC). The particular company that I was working for at the time, which was a VHSIC phase 1 contractor, defined a unit delay for a transistor as 3 nanoseconds. Hence, the devices that were designed for that particular contract to operate on a 25-MHz clock frequency had to have no more than eight transistor levels between each storage element in the design. When the parts came out of fabrication, they all worked.

It really does matter what level of technology a design center is using because that determination will tell what level of tool support the design requires. It certainly is nice to use the leading-edge EDA tools to do a design, but if those tools are meant to support 20-nanometer designs while the design center is making 180-nanometer designs, then it is equivalent to using a surgical scalpel to cut a loaf of bread: it is overkill. This is not just the case for EDA tools. It is also the case for the level used for the parasitic measure that is incorporated inside the characterization file. A secondary consideration when determining the level of parasitic capacitance and parasitic resistance that is used for characterization is how tight the estimated performance will be for a given design. If a design has been successfully completed at one technology node, then there is less (but not zero) need to increase the level of parasitic extraction at a deeper node. The assumption being that any given cell will be smaller and less capacitive in loading preceding stages of the various logic cones in a smaller technology node versus the older extent design. Hence, it should be easier to close timing on the same design at the same clock frequency. In general, this is true, but there are secondary effects that become dominant as technology nodes shrink. These newly dominant effects might mean that designs that are easily timing closed at one technology node might become more difficult at deeper technology nodes, even if just marginally so.

TABLE 7.4   Estimated Parasitic Level as a Function of Estimated Delay Through a Stdcell. As technology nodes decrease in size, the contribution of parasitic capacitance and resistance (and possibly inductance) increases.

Technology Node	Parasitic Needs
1.25 micron and above	unit delay
1 micron – .5 micron	no parasitic, just circuit delay (transistor dominant)
.35 micron – .25 micron	straight lumped C to ground
180 nanometer – 130 nanometer	lumped C between nodes and to ground
90 nanometer – 65 nanometer	RC between nodes
45 nanometer and below	RC between nodes plus stress effect of surrounding structures on transistors

So how do you decide the level of parasitic that a technology node requires? Part of the answer is historical knowledge—that is, what tools came online historically when that particular technology was being developed by the leading technology companies of the time? Part of the answer is the level of risk that the design center is willing to accept. Table 7.4 gives a rough but good estimate as to the level of parasitic required.

The reason for this increasing amount of required extraction is that as transistors decrease in size, the amount of contribution to the delay of a stdcell resulting from the parasitic effects of the surrounding structures (first, mainly the overflying metal layers but eventually the surrounding structures on the same routing layer) become more important. Eventually, these become the dominant factor.

As the required level of extraction required increases, the interaction of these parasitic elements also increases at the simulation level. This means that the amount of simulation that is required to characterize a library in terms of computer cycles also increases. Luckily, this increase in computer cycle requirement is usually held in check by the increase in performance of new processing engines.

7.5   CONCEPTS FOR FURTHER STUDY

1. SPICE parameter study:
   - The student or interested engineer might be inclined to take an existing SPICE model together with an extant netlist and see the effects on the simulations as each parameter within the SPICE model is independently adjusted up and down by some small percentage.
2. SPICE process model comparison:
   - The student or interested engineer might be interested in building a SPICE comparison tool that reads two SPICE models of the same process, together with their assumed standard deviations, and attempts to determine the variance of the various parameters between the models so that any standard-deviation SPICE model can be written as an output of that tool.
3. SPICE simulation study:
   - The student or interested engineer might be inclined to characterize a small stdcell library of a few cells, including at least one FLIP-FLOP, repeatedly as the tolerance and time-step variables that control the various SPICE runs are

adjusted in order to see the effects on the resulting characterization liberty files from making such tolerance adjustments.

4. Parasitic study:

- Modern parasitic extraction, at the extreme complexities of current deep submicron technology nodes, can produce a layout parasitic extraction (LPE) netlist with hundreds to thousands of small capacitors interspersed across and within hugely complex resistor networks, all constructed by the LPE run in order to more accurately model the "true" parasitic nature of the polygons that make up a given stdcell. This is sometimes the case even for the smallest stdcells. Such complex LPE netlists can easily hinder the performance of the EDA characterization tools (SPICE and SPICE wrappers) that are used. This hindrance can be enough to cause significant reduction in the throughput of the stdcells, which can further increase the time required to complete the generation of the liberty file being made. One solution is to attempt to reduce the parasitic file that is created by the LPE extraction. The student or interested engineer might be inclined to use existing tools to collapse these LPE networks to more manageable amounts (such as a maximum of three separate resistors in series on any signal and three separate capacitors between any two signals). This reduction should be accomplished in stages. Run SPICE characterization on the netlist and see if the loss of accuracy from the previous run is within a margin of tolerance. If it is, then a further decrease can be tested.

5. BSIM4 versus BSIM3 study:

- As mentioned in the chapter, the BSIM4 Working Group is actively working at the University of California, Berkley. The student or interested engineer might be inclined to acquire or develop comparable BSIM3 and BSIM4 SPICE models for an appropriately deep technology node and run characterization comparisons on a set of stdcells using the two models.

CHAPTER 8

TIMING VIEWS

8.1 LESSON FROM THE REAL WORLD: THE MANAGER'S PERSPECTIVE AND THE ENGINEER'S PERSPECTIVE

Early in my career, I was working for a defense contractor. We were doing a series of 25-MHz CMOS designs. Even as a junior-level engineer, I was put in charge of an entire memory map unit design. This was a large design by the standards of the time, perhaps 20,000 gates. I was proud. After I had the circuit designed and properly simulating and had completed initial placement and routing of the function, I commenced timing closure of the circuit, which was very rudimentary at the time. We could go in and substitute higher drive strength cells for lower drive strength cells, and we could move some placement and some routing, but modern-day tools such as those allowing useful skew had not yet been developed. The short paths were resolved quickly, a tribute, as I saw it, to my foresight during the circuit design. The long paths took a little longer, and even though there were not too many, I struggled. Finally, after several weeks, whereas some of the other designers had already released their chips to processing, I was down to one remaining long path. It was less than 40.1 nanoseconds long, including propagation, setup, and clock skew, and I could not figure out a way to get those last 100 pico-seconds out of the design. I worked at it day and night for two weeks and nothing I tried would reduce that last long path, Finally, one Friday morning, I went in to my boss's office and confessed and wanted to know if he had anybody more experienced to come and help me fix it. He asked me how far off I was. I told him that it was 100 pico-seconds. He looked at me for a second, then smiled, and said, "Your design is complete. Ship it."

Engineering the CMOS Library: Enhancing Digital Design Kits for Competitive Silicon,
First Edition. David Doman.
© 2012 John Wiley & Sons, Inc. Published 2012 by John Wiley & Sons, Inc.

8.2 PERFORMANCE LIMITS AND MEASUREMENT

Fabrication houses have a strong desire to margin the timing, power, noise performance of their library offerings. These fabrication houses usually have to market their manufacturing offerings with the caveat that if the customer-supplied design meets the customer's requirements while completely following the rules and guidelines set down by the fabrication house, then if that design ends up not working after it is processed, then the fabrication house will pay for the second pass mask fixes required in order to make that design work. Some, because of market and competition pressure, will even offer to pay some lost opportunity cost. So it is in their best interest to do what is necessary to make sure that the incoming design is as solid as possible. One way for these fabrication houses to do this is to margin the various views, including timing.

It usually comes as a surprise to a design team when it realizes that the best-case timing measures in the best-case liberty files are actually better than the SPICE runs indicate and that the worst-case timing measures in the worst-case liberty files are actually worse than the SPICE runs indicate. Surely, the reasoning goes, the fabrication house has gotten the liberty and SPICE versions mixed up in the current release. Further, this gives the design center the chance to get a round of free silicon. The design center will design a circuit, and if it does not work because of the mix-up in the liberty and SPICE versions, then the design center gets to see a round of debug silicon. In actuality, this is never the case. The SPICE numbers are always inside the margins expressed in the liberties. So if a design meets the customer long-path timing closure to the worst-case liberty and customer short-path timing closure to the best-case liberty, then the fabrication house is virtually assured that the silicon will meet the performance goals desired by the design customer. The fabrication house even gets two additional benefits with this process. First, the design center will have to add more and larger buffering for at least some of its long paths beyond what is really needed if a more accurate worst-case timing was available, and it will have to add delaying buffering to at least some of its short paths beyond what is really needed if a more accurate best-case timing were available. These extra cells will add to the size of the final design and may increase the actual size of the final chip, allowing the fabrication house to sell larger more expensive silicon to the design center. Second, the actual design will always be faster than the timing results indicate, thereby allowing the fabrication house to appear heroic to the design center.

Remember that the best-case and worst-case liberty files that are delivered by the fabrication house are not reality. They are envelopes surrounding reality. It is in the fabrication houses' best interest to have these margined envelopes. It is not the design centers' best interest to have them. Hence, recharacterizing them, even with the fabrication-house–supplied SPICE models, can create (or find) the margin needed in order to allow a design center to complete a better design than the competition. The following sections describe the what's and how's of doing this recharacterization.

8.3 DEFAULT VERSUS CONDITIONAL ARCS

A timing arc can be (1) written so that it is independent of the other signals in a function or (2) written so that it is dependent on the other signals in a function. In all instances, the second manner more closely matches the correct functional operation of any given cell. However, the first manner is what is needed for the synthesis engineers to work properly. The reason for this is that a synthesis engine does not have any concept of how the logic is

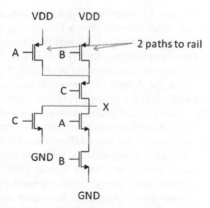

FIGURE 8.1 AOI21 Circuit Showing Different Resistance Between Power Rail and Output, Depending on the State of Inputs. This illustrates the need for full state-dependent characterization arcs ion every liberty file because some arcs will have more paths to rail than others will.

resolved after synthesis. It can only rely on a single representative rise propagation arc and a single representative fall propagation arc during the synthesis cycle.

To understand why the second of these is more closely matched to actual operation, just look at the circuitry of a typical Boolean cell, maybe a AOI21, as in Figure 8.1. There may be arguments as to whether the pair of parallel P-channel transistors should be located closer to the rail than the remaining P-channel transistor in the circuit or the opposite, but that is not the discussion here. (By the way, I prefer this connection if the name is to be AOI21, the opposite would be AOI12, but with the same functionality that the timing arc C falling edge to X rising edge is being measured.) As the signal C falling edge occurs, the N-channel transistor that it is gating turns inactive (digitally off) and the P-channel transistor that it is gating turns active (digitally on). This allows charge to transfer from the power rail through either or both of the P channels that are connected to signals A and B and through the P channel connected to signal C and finally out to the output signal X. This series of actions raises the level at X. However, the amount of resistance from rail to X is determined by signals A and B. If both are at ground (digital "low"), then there is twice the path between the power rail and the source side of the P-channel transistor gated by C than there would be if either A or B was at power (digital "high"). Hence, the timing for the C falling edge to X rising edge arc is going to be different if both A and B are at digital low than if either one is at digital high. In actuality, because of layout parasitic issues, there will be differences in the timing arc C to X that depend on which signal A or B is digital high. Similarly for C rising edge to X falling edge, dependence on the states of inputs A and B can influence the timing arc there as well.

However, during synthesis the tool does not know the state of the signals going into a function. So there is a need to determine *all* of the conditional arcs and choose the worst case among them for the default timing arc in the worst-case liberty files (which are used for long-path setup-time synthesis and analysis) and choose the best case among them for the default timing arc in the best-case liberty files (which are used for short-path hold-time analysis). Typically, a fabrication-house liberty file is not constructed in this manner. An inspection of the default timing arcs versus the conditional timing arcs will show that this is the case, with some conditional arcs on some pins in some cells actually being worse in worst-case liberty or better in best-case liberty than the default arc.

This can be attributed to the lack of understanding by the generators of the liberty files back at the fabrication house about how the liberty file is going to be used, so it is worth the design team's effort to inspect and adjust the liberty defaults as needed.

The way to determine if this is present in a liberty is to inspect and compare the default versus the conditional timing arcs for some pins on some cells. Could all of the preceding be avoided by removing either the default or the conditional timing arcs? The default timing arc is required, as mentioned, for proper synthesis. What about the conditionals? They just give the user the chance to see problematical construction of the liberty, correct? Besides, the large number of conditional timing arcs in a liberty can be cumbersome. Removing them could greatly reduce the size of these files. However, it is important to include all of the timing arcs as conditionals as well, because these can be used by postsynthesis dynamic timing tools (time-based logical simulation). Although static timing tools are definitely important for proper timing closure, dynamic timing tools can still be valuable for design teams doing cross-clock boundary and asynchronous-pulsing design techniques. Hence, the unfortunate answer is that the conditional timing arcs need to remain as well as the default timing arcs.

The lesson here is that it makes sense to write a script that inspects the liberty files and substitutes the worst-case conditional timing arc arrays for the default timing arc arrays for each pin of each cell in worst-case liberty files that you receive from a fabrication house. Similarly, the tool should be able to substitute best-case conditional timing arcs for default timing arcs in best-case liberty files that are received from a fabrication house.

8.4 BREAK-POINT OPTIMIZATION

Timing (and power and noise) measures in liberty files are usually represented as arrays of numbers keyed to certain input edge-rate and output load measures (known as *break points*). These arrays (actually, they are vectors of subvectors) represent the timing (or power or noise susceptibility) of a function in what is called a *nonlinear model* (NLM). The optimal choice of these break points that enumerate the NLM arrays is important to the optimal use of liberty files. In addition, correctly choosing these break points can help minimize the amount of time required to actually generate and validate recharacterization of stdcell libraries. Table 8.1 shows a typical NLM.

The way to read this NLM as a synthesis tool is as follows:

- Determine the input edge rate of the incoming signal and find the two input edge-rate break points along the side of the array that brackets it.

TABLE 8.1 Representation of a Liberty Nonlinear Model Array (Vector of Vectors) with Poorly Chosen Break Points. There exist large regions in Figure 8.1 that exhibit slight adjustments whereas there are concurrent narrower regions exhibiting significantly more-dramatic adjustments.

	1	2	3	4	5	6	7
1	0.15	0.16	0.2	0.26	0.3	0.4	0.5
2	0.16	0.19	0.2	0.31	0.33	0.41	0.6
3	0.2	0.2	0.23	0.34	0.4	0.5	0.67
4	0.23	0.26	0.28	0.39	0.44	0.56	0.7
5	0.6	0.67	0.71	0.75	0.89	0.98	1.12
6	0.67	0.69	0.73	0.81	0.95	1.09	1.23
7	0.7	0.73	0.8	0.89	1.05	1.19	1.41

- Determine the output load of the output signal and find the two output load break points along the side of the array that brackets it.
- Find the four measures in the array that represent where these four break points cross each other.
- Calculate the weighted average of these four measures, depending on how close the input edge rate and output load are to the four measures.

The result is a linear interpolation of how a signal would propagate through the cell with that edge rate and load. If the break points are too far apart, then the interpolation will not be close to the actual timing of the signal propagation through the cell. If the break points are too close together, then the size of the liberty will be larger than is required and take longer to generate originally and to compile load and use during design-center synthesis.

The preceding illustration is a slightly different look from an example of before monotonic and after monotonic later in the chapter. Here the difference is that some of the data illustrate some significantly nonlinear data steps. Tracing the numbers down through the table, the data appear to be fairly well behaved and nearly linear for all break points along the vertical axis down through break point four. In addition, again the data appear linear for vertical break points beyond five. However, there appears to be a step function of some kind between these two break points. As a result, if these two break points are the ones that bracket the range of either the input edge rate or output load, as discussed in the previous synthesis algorithm, then there is a significant possibility that the extrapolated delay (or power or noise susceptibility) will be significantly off from actuality, as can be seen in Figure 8.2.

It would be better to define fewer break points in the flatter ranges of the data and more break points in the nonlinear ranges. The correct way to do this is to evaluate the

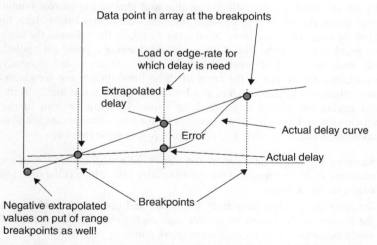

FIGURE 8.2 The Problem with Extrapolating Across Break-Point Step Functions. The allowed error rate of the liberty needs to be set beforehand and then the break points determined in order to ensure that the determined error rate is not exceeded during either interpolation or extrapolation processes.

delay, the power, or the noise susceptibility between the break points, using SPICE or some other simulator, versus the extrapolated value. When the error between the actual and the extrapolated is less than a certain amount, then the number of break points in that range is sufficient and the process can be repeated in the next region. The basic algorithm could be as follows.

- Determine the minimum and the maximum possible break points (this is a whole separate subject that I will come back to shortly).
- Evaluate the value halfway between the two break points and compare it against the extrapolated value. If it is sufficiently close, then recursively stop or else use the new halfway measure as another break point and then recursively repeat to the left (using the original minimum break point and the new middle break point) and then to the right (using the new middle break point and the original maximum break point).
- When all recursion stops, then the number of break points is minimal for the chosen error rate between the actual and the extrapolated values.

The preceding algorithm might choose just two break points or a much larger number. The prime determine of this number is how much error the design team is willing to accept in the characterization. The design team determines that by determining how much margin it is willing to put on a clock period. If it can accept 5% jitter from all sources, but the total measured jitter is 3%, then it can set the error limit in the algorithm to 2%. Many fabrication-house liberty files could be significantly larger than this. In addition, this extrapolation error rate is easily checked by actually doing the calculation between a few sets of break points for a few cells in the library. Finally, doing this for each pin for each cell can give some surprising differences. When you do this type of liberty file break-point optimization, do not be surprised if when you recharacterize a library (and in some cases for some pins and for some cells) that you get 2×2 arrays while for other conditions on other pins on other cells you need to produce 10×10 arrays.

How do you start the preceding algorithm and choose the correct minimum and maximum break points? This is a critical function. The liberty extrapolates for values outside of the range of break points in an array by taking the values in the array for the last two break points on the end of the array and drawing a linear extrapolation from them. If your two smallest break points for edge rate are at 0.5 nanosecond and 1.0 nanosecond, the values in the array at those break points are far enough apart, and the synthesis engine needs values at 0.1 nanosecond, then it is highly likely that the synthesis engine will give a negative delay value. This can be seen in the far-left extrapolation line in Figure 8.2. However, determining the values for the initial minimum and maximum is relatively easy. For edge rates, follow these two steps:

1. Simulate the largest inverter in the library with no load and with an infinite edge-rate input and measure the resulting edge rate, which is the initial minimum edge-rate break point.
2. Simulate the smallest three-stack P-channel and three-stack N-channel cell with the loading of the largest acceptable load on the output. Measure the edge rate, which becomes the initial maximum break point.

For output loading, the range definition is even easier. The minimum output load break point is the smallest input capacitance for any pin on any cell in the library, and the maximum output load break point is probably defined by electromigration rules [in the physical design kit (PDK)] for the expected frequencies in the library that the design team will attempt to use.

8.5 A WORD ON SETUP AND HOLD

As most library users realize, the concept that the storage element (FLIP-FLOP) grabs the data *at* the clock-edge is incorrect. Data must be stable for a certain length of time both before they are clocked into the storage element and after they are clocked into the storage element. This "before and after" stable time is usually known as *setup and hold* (*setup* being the required stable time before the clock edge, and *hold* being the required stable time after the clock edge).

However, a further complication needs to be addressed as technology nodes continue to shrink. Specifically, the interdependence of these two times has often been ignored in the past. That interaction should no longer be ignored.

Figure 8.3 shows how these two times used to be calculated. Specifically, an infinite hold time was assumed when the setup was derived (by walking a change on the data closer and closer to the clock edge until a failure criterion is achieved). In addition, an infinite setup time was assumed when hold was calculated (again, by walking a change on the data closer and closer to the clock edge until a failure criterion is achieved).

What this causes are insupportably aggressive setup and hold windows (setup plus hold). The reason for this being insupportably aggressive should be obvious. The measured independent setup assumes a stable data signal for a larger time after the clock edge than the measured independent hold time suggests. Hence, for designs done with such characterized stdcell libraries, it is possible to meet all the long-path and short-path timing requirements of the design and still find timing failures at test. In some

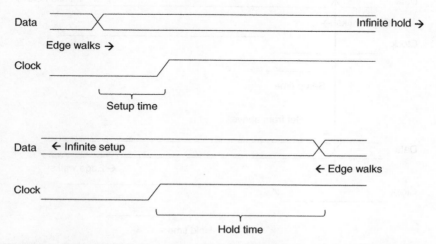

FIGURE 8.3 Independent Walking of Data Edges Toward the Clock Edge. A technique that most assuredly produces the smallest setup-plus-hold window, but it is so small as to be impractical in real circuits and is demonstrably broken, allowing minimum pulses on data. Tongue in cheek, the technique has also been characterized as "I can remove the data before I need to apply it."

technologies, this resulting minimum window can almost disappear, almost as if you can remove the data signal before you have to apply it.

There are other setup-plus-hold time-characterization methods. Two specific ones can be described as *dependent hold* and *dependent setup*. As the terms imply, in either case the measure of one of the times specifically depends on the a priori measure of the other. For dependent hold, first measure setup as previously described and then switch the data signal before the clock edge while walking the hold switch backward into the clock. Figure 8.4 illustrates the procedure.

The preceding method will cause longer hold times and maintain short setup times. Doing the exact opposite method, first walking the hold edge into the clock edge while assuming an infinite setup and then using that hold while walking the setup edge in will produce longer setup times and maintain short hold times. A comparison of the three methods is given in Figure 8.5.

The curved line illustrates all possible setup-plus-hold-time pairs. The dependent-hold and dependent setup methods given previously produce just the ends of this curve. Hence, there are many ways to generate legitimate setup-plus-hold pairs. However, notice that the insupportable minimum setup-plus-hold time can be significantly inside the line of viable minimum setup-plus-hold pairs.

Figure 8.6 shows another way to look at the same issue. Notice that the dependent-hold results have the same setup as the independent-setup-plus-hold method but with safer hold, and the dependent-setup results have the same hold as the independent-setup-plus-hold method but with safer setup.

Either dependent-setup or dependent-hold characterized libraries are safer than independent setup-and-hold libraries, but which method is best overall? Actually, both

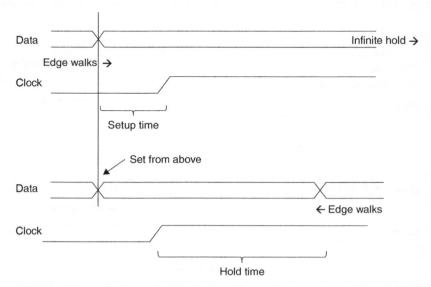

FIGURE 8.4   Independent Setup Edge Walking followed by Dependent Hold Edge Walking Toward the Clock Edge. This technique gives minimal setup and clean hold. An alternative approach would be independent hold with dependent setup, producing absolute minimal hold times but larger setup times (useful in scan-path situations).

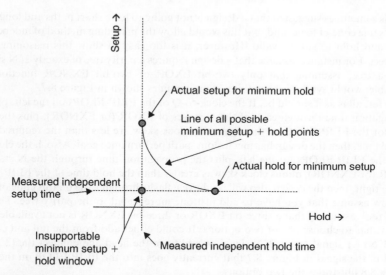

FIGURE 8.5 The Setup and Hold Plane, All possible setup and hold pairs of points occur on this plane, but the only valid pairs occur either along the curved line or to the upper right of it, whereas the insupportable minimum setup plus hold occurs to the lower left of the curved line.

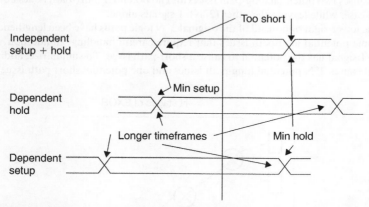

FIGURE 8.6 Comparison of Three Setup-and-Hold Calculation Methods. For valid minimum pulses on data inputs, minimum setup causes extended hold whereas minimum hold causes extended setup.

should be used together in the same library. First, data inputs (and control inputs) into FLIP-FLOPs and other storage elements tend to need to have absolute minimum setup times. This allows them to be used most aggressively against design long paths. Scan inputs, on the other hand, tend to come directly from adjacent FLIP-FLOPs in a layout and are subject to hold-time violations. They need absolute minimum hold times. Hence, a good library will have the storage element setup-plus-hold tables built as a merger of the two dependent methods.

It is sometimes suggested that a design is not going to have short paths and long paths in the same cone of logic, and that this would allow the preceding method of independent setup and hold to still be valid. However, it is too easy to show this reasoning to be incorrect. For instance, assume that a design requires a parity tree of exactly 2^N signals. Such a tree, assuming that only two-bit EXOR or two-bit EXNOR functions are available, would synthesis something along the lines shown in Figure 8.7.

So far, all is as it should be. If the clock-to-Q of the FLIP-FLOPs on the left, plus the propagation time through each of the N stages of EXOR (or EXNOR), plus the setup time for the FLIP-FLOP on the right, plus clock skew are less than the required clock period time, then the design has met its long-path performance goal. Also, if the clock-to-Q of the FLIP-FLOPs on the left, plus the propagation time through the N stages of EXOR (or EXNOR), minus clock skew is greater than the hold time of the FLIP-FLOP on the right, then the design also will not have a flash-through issue on this cone of logic.

Now assume that you have to add just one more signal to the parity tree. The synthesis tool, assuming that a three-bit EXOR or three-bit EXNOR is not available in the library, has to choose one of two options. It could either add from the tail and connect the $(2^N)+1$ signal to the 2^N signal or add from the head and connect the $(2^N)+1$ signal to the signal in Figure 8.7 that currently goes into the FLIP-FLOP on the right. Figure 8.8 illustrates the two options.

In the upper-left option, $(2^N)-1$ logic paths from FLIP-FLOP to FLIP-FLOP remain unchanged. If they were going to meet timing requirements in the original 2^N signal case, then they would still meet time requirements. Just one of the original 2^N signals, plus the new $(2^N)+1$ signal, could end up being long-path problems. To summarize, adding from the tail causes two potential long-path issues and no potential short-path issues beyond the original case, while leaving the other $(2^N)-1$ signals alone.

In the lower-right option, all of the original 2^N logic paths have been lengthened. They all become potential long-path issues that require careful handling. In addition, the new $(2^N)+1$ signal has the potential to cause a short-path issue. To summarize, adding from the head causes 2^N potential long-path issues and one potential short-path issue beyond

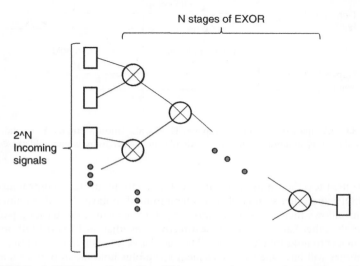

FIGURE 8.7   A 2^N Binary Parity Tree. This structure is common in many state machines.

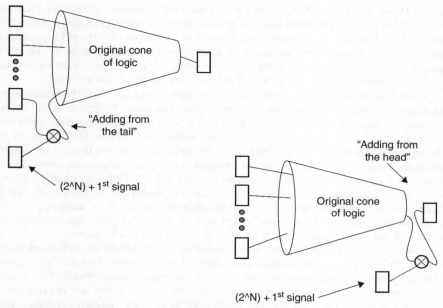

FIGURE 8.8 Adding a $(2^N + 1)$st bit to the Tail (Left) or the Head (Right). With such examples (and there are many more), it is easy to show that valid short-path and valid long-path cones of logic can exist to the same terminating FLIP-FLOP.

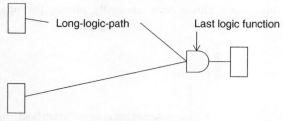

FIGURE 8.9 Why FLIP-FLOPs Can and Do See Long-Path Setup Followed by Short-Path Hold Issues. This shows why the dependent-hold methodology is needed. This is the generic version of the long- or short-path cone of logic illustration. If the last logic gate, where the long- and short-path cones of logic merge is unate, then the potential minimum pulse is only either minimum high or minimum low (which remains a problem anyway). But if that last gate is non-unate (for instance, a MUX or EXOR or EXNOR or ADDER), then the minimum pulse can be in either direction, depending only on the data inputs to the two cone-of-logic paths.

the original case while touching all cones of logic. In general, adding from the tail appears to be the better of the two options. Most synthesis tools on the market today add from the head.

This description of parity-tree synthesis is pertinent to the discussion of dependent versus independent setup-and-hold characterization because it shows how easy it is to have long-path and short-path timing in the very same cone of logic. The preceding parity-tree example, while legitimate, is just one example. Any logic that can be synthesized as in the following Figure 8.9 is potentially a problem. It can happen if the last logic function is an ANDING or a NORING function (causing potential minimum positive pulses) or if it is an ORING or a NANDING function (causing potential minimum negative pulses). If the last logic function is non-unate, which means that the output of the logic functions can go in either direction on the change of an input, dependent on the status of the other inputs, then the minimum pulse could be either positive, negative or both (in the same cone at different times). EXOR, EXNOR and MUX are examples of non-unate cells.

Let us look at the original "curve" illustration again. Figure 8.10 reproduces this curve. However, two pieces of information have been added. These are a 45-degree tangent line to the curve of legitimate setup-plus-hold pairs and a point marking an absolute minimum setup-plus-hold pair, which can be defined as where the 45-degree line is tangent to the setup-plus-hold curve. It can be noted that the dependent setup and the dependent hold methods of storage element characterization might not give an absolute minimum window. Rather, by finding several alternate setup-plus-hold pairs and then plotting them, one could draw the previously mentioned tangent along the plotted line. By the way, the line is illustrated as a 45-degree line but such angle could be different if the weighting toward better setup or better hold was desired. By picking the point where the tangent touches the curve, the optimal absolutely minimum setup-plus-hold pair can be found.

A scarce resource issue defeats this optimal setup-plus-hold characterization method. Both setup characterization and hold characterization are black-box operations because they are constraint measurements as opposed to length measurements. For length measurements, a characterization exercise can be accomplished by appropriately arranging a circuit (perhaps but not necessarily in SPICE), toggling an input, and measuring the length of time until the output switches. For a constraint operation, this has to be repeated dozens or even hundreds of times (depending on the accuracy desired). One starts by switching an input significantly far away from the clock edge that the output is expected to not fail. Then one repeats it closer to the clock edge and determines if the output is still correct. This is done repeatedly, moving the input switching ever

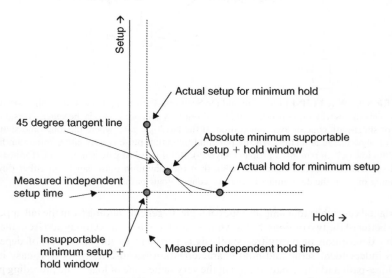

FIGURE 8.10 The Setup-and-Hold Plane Revisited. This deterministic but extremely time-consuming method is used to find the "true" absolute minimum setup-plus-hold pair.

closer to the clock edge until the output does reach some failure criterion. There are recursive search methods that can track this critical failure constraint faster, but the need for repeated simulation and testing in order to determine constraint measures remains. As a result, creating one set of the setup-and-hold constraints for the storage elements in a given library usually takes one half or more of the time required to accomplish all of the other length measures on all of the other cells in a library. To determine the absolute minimum setup-plus-hold window constraints would require the repetition of this constraint method dozens to hundreds of times beyond this current effort. Furthermore, in the end, the vast majority of the collected measurements would be thrown away as unneeded. Hence, although the absolute minimum setup-plus-hold measurements would be useful, accomplishing those results are impractical.

For those still interested, however, in generating smaller setup-plus-hold windows, it could be pointed out that the minimum setup-plus-hold curve given in Figure 8.10 does appear to be concave. Perhaps it is not necessary to get to the absolute minimum setup plus hold. Perhaps it is just sufficient to get to a better setup plus hold than either the dependent-setup or dependent-hold method can give. It might be possible to first use the dependent-setup or dependent-hold method, then back the independent side off a few time units before measuring the opposite setup or hold constraint, to get a smaller opposite constraint at a minimal cost. This is, however, a mirage: no physics forces the minimum setup-plus-hold curve to be concave. Yes, it usually is concave when it can be determined (and to the accuracy that it can be determined), but it does not appear to have to be concave. Hence, backing off the dependent-setup and dependent-hold measure constraints might cause larger setup-plus-hold windows. Figure 8.11 illustrates why this

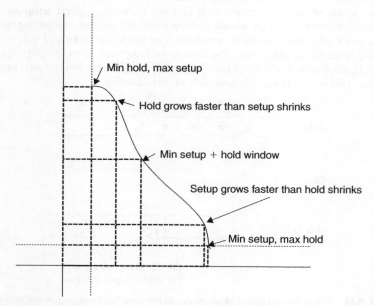

FIGURE 8.11   Back Off from the Characterization Corners Is Not a Guarantee of Smaller Setup and Hold Windows. There is no proof, just engineering conjecture, that the minimum setup-plus-hold curve is truly concave at all points along it. Until there is, back-off techniques are not assured to produce smaller setup-plus-hold windows.

122 TIMING VIEWS

could be true. If the setup-plus-hold curve is not concave, then backing off either end of the curve could be detrimental. Without stronger evidence, the concave setup-plus-hold curve observation together with the supposed benefit is suspect.

8.6 FAILURE MECHANISMS AND ROLL-OFF

In the previous section, we described the point that determined when a constraint had been reached as a *failure mechanism*. That was on purpose. One does not necessarily run the constraint all the way to the point when the storage element (or FLIP-FLOP) fails correctly to grab the data. This is for several reasons, two of which are obvious.

First, the closer that you get to the clock edge, the more likely the circuit will go metastable. Recall that metastability is very difficult to see in a simulation (SPICE or otherwise) but can be shown in a failure lab with just a small amount of test equipment and time. Because, as I mentioned in the metastability section, *no* circuit can be made immune to metastability but only more resistant to metastability than the next circuit, the real way to get away from the danger zone is to prevent the designer from approaching it. That is done through making sure the setup-plus-hold window is strictly outside of the metastability point (or zone) of operation.

The second reason is a little more subtle. However, that subtleness gives the means of developing a useful failure mechanism. In the past, at nodes above 0.5 micron, people would walk data setup and hold edges into the clock until the output showed that the storage element did not grab the data correctly. Doing so would not degrade the cell's propagation time significantly (or even noticeably). In addition, even metastable zones remained inside of these extremely tight (for the technology level) setup and holds. However, as technology nodes started to drop below 0.5 micron, the propagation time after the clock edge began to show noticeable lengthening as setup and hold edges got closer and closer to the clock edge. The same trend happened on the output edge rate as Figure 8.12 illustrates. The cell will take longer and longer to resolve the state in which it should be. This is, in a sense, the definition of metastability.

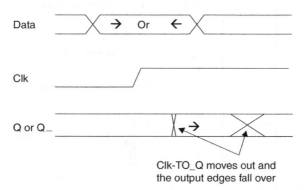

Clk-TO_Q moves out and
the output edges fall over

FIGURE 8.12 Roll-Off as a Failure Mechanism. As the most common failure mechanism in use during characterization (that of walking synchronous input edges into the clock until the output edges decay from "best results" by a certain percentage), roll-off replaced run to hard failure around the 0.5-micron technology node. The preceding technique also helps to keep setup and hold windows away from potential metastability regions.

This performance degradation allows a way to measure setup and hold constraints. Specifically, determine the output delay from the clock edge (positive and negative) and the output edge rates (positive and negative) for large (infinite) setup and hold. Then walk the setup and hold into the clock (see the previous section on dependent hold for an example) and measure the same delay from clock edge to output and the same output edge rates; when any of them go beyond a previously determined degradation percentage, also known as a *roll-off percentage*, call that a failure mechanism. This allows for a safe setup-and-hold constraint as opposed to pushing these constraints all the way to cell failure. The cell will be less likely to fail when it is used on a long path or a short path because it is *not* near the absolute limit of functionality.

This method of defining a failure mechanism also illustrates a way of adding or subtracting risk to a project without requiring fabrication-house approval. Specifically, fabrication houses have a stake in not producing failed silicon, so the timing files they deliver tend to have wider-than-needed setup-plus-hold windows. They may have defined their failure limit as a 5% degradation (or *roll-off*). That is certainly safe, but there can be additional margin in the FLIP-FLOPs. Recharacterizing at a 10% roll-off could allow for a tighter setup-plus-hold window, which would cause easier timing closure of both short paths and long paths, giving a smaller and less-power-hungry chip. Conversely, perhaps your user environment is extremely sensitive to metastability concerns. By recharacterizing for a smaller roll-off, maybe 2.5% instead of 5%, you are moving your allowed setup-and-hold edges farther from the critical metastability regions, making designs done with your modified library less susceptible to metastability. It might make sense to add certain special FLIP-FLOPs (characterized at 2.5% for strict use in communication-channel blocks of designs) and other special FLIP-FLOPs (characterized with 10% roll-off for high-speed blocks of the design) while leaving the regular FLIP-FLOPs for other parts of the designs. In addition, this definition, already fluid, can be further loosened by separating the roll-off percentages for setup and hold. Perhaps a 10% propagation roll-off for setup calculations and a 5% propagation roll-off for hold calculations could give faster long-path timing closure without much decrease in hold risk. By the way, the reason that this is doable without feeding back to the fabrication house is that the view that is being changed is a timing view only. No physical structure is changing. Whatever polygons the fabrication house has allowed, specifically the FLIP-FLOPs in question, are not altered in any way. You are changing just a timing view of them.

So how much roll-off is too much? Each FLIP-FLOP that is characterized in this manner now has a longer (or shorter) than simulated clock-to-output propagation time (and edge rates). Therefore, a good method of determining a correct roll-off would be to compare the absolute increase in propagation versus the absolute decrease in setup (or hold). If the decrease in setup is longer than the increase in propagation, it makes sense to accept this new timing. If the increase in propagation time is greater than the decrease in setup time, then this is perhaps too much penalty to pay and the roll-off percentage for that specific cell could be unacceptable.

Finally, by recharacterizing the FLIP-FLOPs like this with various roll-off, you really are causing the propagation and output edge rates to change. Hence, you need to be careful how these propagation and edge-rate numbers are recorded in the timing view. These need to be measured assuming the setup and hold times that caused the specific roll-off. That is usually not the case for these in a timing view. The reason is that propagation measure methods do not assume that there is a critical need for such measure. Hence, during roll-off, adjusted setup-and-hold simulations measure the resulting propagation as well and use this in the propagation fields.

8.7 SUPPORTING EFFICIENT SYNTHESIS

8.7.1 SPICE, Monotonic Arrays, and Favorite Stdcells

When the usage of stdcells is examined, there often appears to be a strong reliance on just a small number of cells in the library. Typically, this percentage of predominantly used cells is usually around 20% of the total number of cells in the library. When the list of highly used cells is compared from design to design, there is usually a high correlation between them. This is usually the argument given to suggest that a pared-down set of functions is better than a larger set because the extra functions are rarely if ever used.

One aside: this discussion brings up a good idea for a library vendor, which is to build histograms of the various cells from a stdcell library that are used across several designs. Once accomplished, examining the cells on the list that are used more than 1% of the time is a good way to determine which cells should be optimized (those on the 1% list) and which are good enough as is (the remaining 99%). The optimization effort, be it area, power, performance, design for manufacturing (DFM) (yield), or all of these features, will focus on the cells that take the majority of area on and contribute the most to power and performance and manufacturability of designs coming from a design center and thus have the largest impact on future efforts by that design center.

Although there is a preference in any synthesis tool to certain functions, it is to the function and not the cell. And although there is a preference in any synthesis tool to some drive strengths, it is to the minimum required while allowing correct performance targets to be met and not to the strength itself.

Why are certain functions preferred by synthesis? Some functions are obvious. Designs will always need FLIP-FLOPs and latches (which one is used depends on the register-transfer language, or RTL, coding). Some functions appear to be obvious but are in reality a little less so. In hand-generated schematics, inverters are often used, contributing several percent of the cells in a design. It is normal for humans to think in terms of "I need to invert that signal" and place an invert appropriately. For machine synthesis, however, the Boolean equivalents can often be propagated back to the FLIP-FLOPs at the beginning of the particular cone of logic, making "I need to invert that signal" into "invert the cone of logic" and further making the placement of an inverter less likely. However, inverters remain highly used cells in any synthesized netlist because they are used for drive strength staging. The machine adds the inverter, the Boolean converts the cone behind it and gets a higher drive strength for the next stage of that particular cone of logic for minimal extra area or delay or power. Some functions are inherent in the algorithms used by the synthesis tools. Enabled AND, which is an AND function with one or more inverted inputs along with the regular inputs, and enabled OR, which is an OR function with one or more inverted inputs along with the regular inputs, are good examples of this. The leading synthesis tools on the market actively look for such constructs in the RTL from which they work.

Why are certain drive strengths predominant in the cells used in a design? The answer is obvious: it is the function of the synthesis tool to produce efficient designs, that efficiency being area or performance or power or a combination of all of these. Using the smallest drive strength for a cell that allows the cone of logic to meet performance goals will accomplish such efficient design.

However, there is another reason that some cells are chosen over others. The abundance in a design of some functions and strengths—or, more specifically, cells—is because of how the cells are defined in the timing (or power or noise) models combined

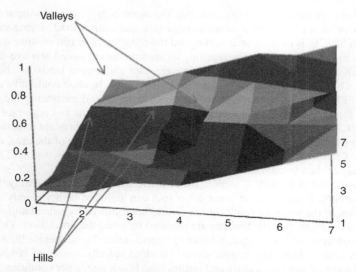

FIGURE 8.13   Graphic Representation of a Nonmonotonic Liberty Array (Vector of Vectors). It is too easy for logic synthesis reverse-hill-climbing algorithms to get lost in local minimum valleys with such offensive liberty arrays.

with an unfortunate manner in which the synthesis tools work to optimize the synthetic netlist. Specifically, characterization can improperly generate hills and valleys in the characterization of cells, whereas the synthesis tools tend to use "reverse hill climbing" algorithms during optimization. *Reverse hill climbing* is a descriptive term for how the algorithm works. Once a proposed cone of logic for a function exists, the synthesis tools try to "walk down the valleys." The synthesis tools do this by looking to see if there are possible substitutions for any of the proposed cells, any reductions in the output load of any of the proposed cells, or any increase in the input edge rate to any of the proposed cells that would shorten the total propagation time of that particular cone of logic. Hills and valleys in the timing descriptions of the cells in a library, as shown in Figure 8.13, can produce local minima that can hinder better optimization by the synthesis tool. By the way, Figure 8.13 is an actual 7×7 rising-edge propagation array for a cell (with all information on function, edge rates, and output loading removed because it is not pertinent to the discussion). Even worse, there is a tendency for some synthesis tools to allow an increase in output loading or a decrease in input edge rate, either of which would normally cause an increase in delay of a cone of logic if that will push the cone into a local minima. Doing so can cause the synthesis tool to attempt to force the use of certain cell that have such local minima. Nothing can be done at the user level to change this reverse-hill-climbing algorithm beyond weighting some variables more than others during the setup of the synthesis run. This results in a blind attempt to avoid improper reverse-hill-climbing efforts. However, there are techniques that can catch the small hills and valleys in the characterization data and remove them in order to prevent the issue in the first place.

First, an explanation is in order about how characterization of hundreds of stdcells with scores of conditions and more than dozens of characterization break points is accomplished. Inverters and buffers tend to be easy to characterize. The process, voltage, and temperature corner is chosen and then the input edge rates and output loads are chosen. For each combination of input edge rate and output load, a characterization (maybe in SPICE) is then set up and run, and the resulting propagation-time and output edge-rate time are measured. The resulting measurements are stored in a two-dimension table (with coordinates based on the input edge rates and output loads) as visualized in the Figure 8.13. This is then redone for the opposite edge. In all, it could take as little as running SPICE (or other characterization tool) 18 times and measuring the results in order to generate a set of 3×3 tables that give the propagation and edge-rate times for such a cell. If 5×5 tables are desired instead, 50 runs and measures are required. Larger arrays give larger numbers, increasing with the square of the size of the area, of runs as well. (See Chapter 9's section on break-point optimization for a discussion of how large of an array of data is sufficient.)

As mentioned in the SPICE section, the characterization tool is not reality. Whatever version of SPICE you have or some other tool can simulate reality closely, but close observation of the first pass results of the simulation followed by refined adjustments to the various controls that the tool has are required to refine the simulation. This cycle of simulate, observe, and refine can and should happen several times before the simulation resembles reality. However, across several hundred stdcells, scores of conditions per stdcell, dozens of input edge rates and output load break points per condition, the sheer number of required simulations becomes too cumbersome to adequately examine and refine. As a result, even with the best guess initial setting of the SPICE (or other circuit-simulation tool) controls, some improper simulations will occur. The gross errors can be detected easily. For instance, a missed output edge in a simulation will result in infinite time for a point in a timing arc array. However, some errors are hidden. Figure 8.14 illustrates a well-known phenomenon of SPICE. The simulator will allow some oscillation to occur near discontinuities in the output of a design. This ringing is caused by the simulator first overestimating and then underestimating the extent of the discontinuity. It can be controlled and somewhat minimized on a case-by-case basis by adjusting

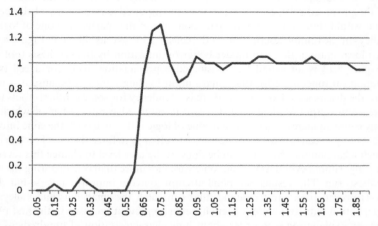

FIGURE 8.14 Simulation-Based Ringing Not Correlated to Reality. Although signal ringing is an actual phenomenon in circuits, ringing of signals in many simulations may or may not be correctly correlated representations of it, so they should *not* be viewed as such.

time-step controls in the SPICE run. Although a similar phenomenon actually does occur in many circuits, the two are not correlated. Hence, the ringing in the simulation is not real. In the illustration, the rising edge appears to go well above 1.2 V, whereas in actuality the edge ended nearer to 1 V (some actual ringing can still occur). The 0.2-V overshoot was caused by the simulation environment not matching actuality. Because of the simulation-caused ringing, the actual edge might have been slightly sharper or slightly less sharp than reality. So if a measurement was made that was based on this edge (for instance to the midpoint of the rise), then that measurement has a slight uncertainty embedded. These embedded errors are not systemic because the uncertainty in one simulator run in one condition might not be the same for a different measure in a different run and condition. In addition, these can be difficult to catch because of the vast number of characterization runs that need to be made.

Thus, it is important to realize that the resulting timing files that are built from these embedded uncertainty measurements are not reality. They are a simulation of reality—hopefully close to reality, but still not reality. This means that changing the resulting files is not denying reality, it is just making the model of it easier to use as opposed to "wrong." That is what I am proposing to do as a means of preventing the reverse-hill-climbing algorithms in the synthesis tools from being stuck in favorite cell local minima.

Certainly, if using a timing closure tool that is "based on reality" is desired, then keep an untouched copy of the characterization file. It should be used rather than the following massaged characterization files for *timing closure* purposes. However, using the following adjusted files for *synthesis* can force the synthesis tool to better optimize the synthetic netlist. In addition, the preceding warning that the characterization file really is not reality must be remembered even here.

The way to remove hills and valleys from an array is to adjust it so that it becomes monotonic. For every location in a propagation or edge-rate array, this can be accomplished by making any given point in the array greater than or equal to any point in the array "above and to the left." More specifically, any timing or edge-rate measure for a given break point must be greater than or equal to any measure for a break point of a smaller output load or sharper input edge rate. Table 8.2 gives the before-and-after illustration of this for the timing array given in Figure 8.13. The graphical representation of the same adjusted array is given in Figure 8.15. Note that the valleys have been filled in.

The preceding algorithm will need to be slightly adjusted for constraint arrays because these arrays are about the inherent race between a clock input and a synchronous input. Hence, there are times when the constraint array should not be made monotonic from the upper left corner. It is better to observe how the majority of the constraint array appears to want to be made monotonic and then to do so from the appropriate corner, which could be any of them. A description of the algorithm would be along the following lines:

For each corner of the Constraint array, if that corner is the minimum of the four corners and neither adjacent corner is equal to it, then force the array monotonic from that corner. Otherwise, if an adjacent corner is equal to it then if the other adjacent corner is a maximum of the array, then force the array monotonic from the equal adjacent corner, else force the array monotonic from the current corner.

The preceding algorithm says that to find the corner of the constraint array that is the minimum of the four, force the array monotonic from that minimum corner unless one adjacent corner equals it and the other is the maximum of the array. On the other hand, force the array monotonic from the adjacent minimum corner. Following this algorithm

TABLE 8.2 **Nonmonotonic Liberty Array Converted to Monotonic Liberty Array. Notice that the changes are extremely localized and rather small in character. Such is the common character of monotonic changes inside properly characterized liberty.**

Before Being Forced Monotonic

	1	2	3	4	5	6	7
1	0.1	0.1	0.2	0.2	0.3	0.4	0.5
2	0.1	0.2	0.2	0.2	0.2	0.4	0.6
3	0.2	0.2	0.1	0.3	0.3	0.5	0.6
4	0.3	0.3	0.3	0.4	0.4	0.5	0.7
5	0.4	0.4	0.3	0.4	0.4	0.5	0.6
6	0.4	0.5	0.5	0.5	0.6	0.7	0.8
7	0.5	0.5	0.6	0.7	0.8	0.8	0.9

After Being Forced Monotonic

	1	2	3	4	5	6	7
1	0.1	0.1	0.2	0.2	0.3	0.4	0.5
2	0.1	0.2	0.2	0.2	0.2	0.4	0.6
3	0.2	0.2	0.2	0.3	0.3	0.5	0.6
4	0.3	0.3	0.3	0.4	0.4	0.5	0.7
5	0.4	0.4	0.4	0.4	0.4	0.5	0.6
6	0.4	0.5	0.5	0.5	0.6	0.7	0.8
7	0.5	0.5	0.6	0.7	0.8	0.8	0.9

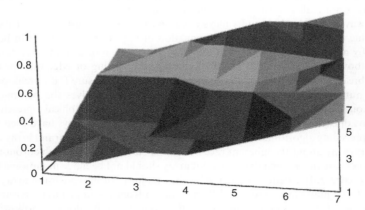

FIGURE 8.15 Graphic Representation of a Resulting Monotonic Liberty Array (Vector of Vectors). Comparing this figure to Figure 8.13, the changes are minimal and hardly noticeable, but now the 7 allow proper reverse hill climbing during logic synthesis.

on all sequential cells, constraint arrays will cause a monotonic array in all of those sequential cell cases, and those arrays will always cause the cone of logic produced during synthesis not to become stuck into using only specific FLIP-FLOPs. When combined with monotonic propagation and edge-rate arrays, this will allow the synthesis tools to choose from a larger proportion of the stdcell library offering, despite the reverse-hill-climbing nature of the synthesis optimization algorithm used. The reason for this is that all local minima valleys in the data have been removed.

There is a difference between setup stdcell library characterizations and hold stdcell library characterizations in terms of how they are used. Specifically, setup characterizations are used to check for long-path failures, and hold characterizations are used to check on short-path failures. The previously explained valley-removal algorithm adjust the data in the characterization arrays by making each data point in the array monotonically larger than any point representing a lower output load or sharper input edge rate. This means that all data are equal to or larger than the characterization tool's measured data, with all valleys in the array being removed. This means that the timing of the cone of logic will always be an overestimate (or equal estimate) and never an underestimate. This is specifically what a setup stdcell library characterization would require for most data, the exception being the clock buffer cells described earlier. For these, the data in the array should always be equal to or less than the characterization tool measured data. This ensures that no long path is overlooked during synthesis and evaluation. As said, all data paths are equal to or longer than actual, and all clock paths should be equal to or less than actual. Hence, the clock buffers in a setup characterization should be forced to be monotonic downward (decreasing) from the lower right corner of the array—that is, all data are less than or equal to measured data for larger output loads and less sharp input edge rates, with all hills in the array being removed (as opposed to the valleys). The algorithm for doing so is a straightforward adaptation of the valley-fill algorithm previously mentioned. The specific algorithm is left for the reader. A hold stdcell library characterization would require the exact opposite. All normal stdcells should be forced monotonic downward and from the lower right, whereas the clock buffers are forced monotonic upward and from the upper left. This forces all data paths shorter than or equal to actual and all clock paths longer than or equal to actual, which is precisely what the hold-time calculations require.

8.7.2 SPICE, Positive Arrays, and Useful Skew

Useful skew is a term that describes the process of shifting the clocks going to some FLIP-FLOPs in such a way as to "borrow" time from shorter cones of logic. This is done in order to lengthen the allowable time for a longer cone of logic and to complete propagation. Figure 8.16 illustrates a rudimentary example. In the example, a long delay cone of logic is bracketed by two shorter delay cones of logic, one (or more) before it and one (or more) after it. By shifting the clock edge that is attached to the initiating FLIP-FLOP(s) of the long path earlier and by shifting the clock edge that is attached to the terminating FLIP-FLOP of the long path later, the allowed time for the long delay cone of logic can

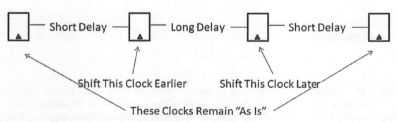

FIGURE 8.16 A Simple Application of Useful Skew. Most modern physical synthesis tools allow such techniques, assuming the design center permits it. Hence, stdcell libraries have to support this. Doing so through poorly characterized cells, however, is dangerous.

be lengthened, possibly allowing timing closure on what is otherwise a problematic issue. This is a design technique being used by more and more design centers, and engineering design automation (EDA) tools that facilitate this are now in the second generation of development.

One way in which the EDA tools facilitate this design technique is by looking for library elements with negative propagation times or negative setup times and substituting them into the cone of logic. The second of these two categories of cells is useful and actually can have negative setup times when characterized correctly. Recall that setup is a constraint measure. If the path that the clock input takes to the latching mechanism in the FLIP-FLOP is longer than the path that the data input takes, then the data can have actual negative setup. However, the first of these two categories is a remnant of improper characterization of an improperly sized function (Figure 8.17).

The issue involved is that the sizing of the P and N channels is such that the switching threshold in the cell is significantly off from the assumed switching threshold. Thus, in either the rising edge or the falling edge (but not both), the cell will start to switch before the tool being used to measure the timing arc expects it to switch. It can actually go beyond the next switching threshold value before the instigating edge appears to do so. This results in a negative delay on one edge of the rise or fall timing arc. Most timing tools used today do not catch this as an error and merrily go on and report it as accurate in the liberty file. However, no effect can ever occur before the cause of that particular effect. These negative times are illusionary. They need to be removed from the liberty file or they can cause to rather bad effects. First, during synthesis, the tool can and does occasionally attempt to solve a long-path issue by adding such a negative delay cell into an already too long path. The tool believes that it is helping the situation, but it is actually

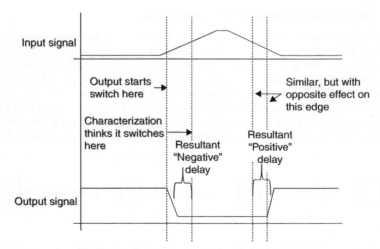

FIGURE 8.17 Negative Delay Through an Inverter Showing Effects of a Poorly Chosen Switch Threshold. Because the switch threshold of this inverter is so far away from the assumed switch point (possibly assumed to be at 50%–50% of rail), the resulting liberty will show negative delay times. This is one reason to push thresholds during characterization away from 50%–50% to either 40%–60% or even 30%–70%. It is also an indication of a poorly designed inverter, which should be reviewed for better effort.

making the situation worse. Second, during timing closure the design engineer that is responsible for closure might not catch the issue of such cells being present in a cone of logic during either static or dynamic timing analysis, which will eventually cause failed silicon. The easy way to remove such times is to "zero them out." Does this mean that the resulting liberty file is no longer accurate? After all, the "zeroed out" value is arbitrary. Well, it is no less arbitrary than the inaccurate negative delay, and it is much less harmful, so the answer I would suggest is that the resulting liberty is more suited for both synthesis and timing closure efforts.

Not all switching thresholds on every timing arc of every cell at every pressure, volume, and temperature (PVT) would be at the 50% mark. Hence, if you decide to recharacterize a library, it is not generally considered valid practice to assume that all switching thresholds occur at 50%. Liberty files, and the characterization EDA tools that can be used to generate them, allow for characterization from points other than 50%. Consider assuming that things start to switch at 40% and 60%, or at 30% and 70%. Of course, the further out these values are from 50%, the more timing margin is "left on the table," but the safer the assumption becomes that the cell switches on or after the point that is assumed.

8.8 SUPPORTING EFFICIENT TIMING CLOSURE

Over the last 30 years, EDA tools have matured in several amazing ways. One physical side example is that router technology, along with the inclusion of items such as cross-talk avoidance algorithms and efficient layer-resource-allocation algorithms, now allows for gridless routing, whereas even 10 to 15 years ago, all pins had to be defined and accessible on specific router grid locations. Of course, moving to a gridless methodology is expensive in terms of computer cycle time, so maintaining gridded pin locations is still preferable. Another EDA tool advancement through the years is that synthesis tools have become much more flexible in how they use the data supplied to them (for instance in a liberty file). Specific items, such as alignment of break points and array boundaries, are not as necessary as when the first-generation synthesis tools were coming on the market in the 1980s. In many ways, this is a welcome development. This is especially true with regard to break points because they should be matched to the actual regions of high nonlinearity of the arrays of the cells in which they exist (as defined in the section on break points). However, just as with gridded versus gridless physical routing, there is a penalty in terms of computer cycle time that is to be paid for many issues in synthesis and timing closure. In addition, just as with the physical side keeping pins on grid, there are some easily attainable minimizations of this computer-cycle time cost that can be captured with certain easy-to-comply-with precautions.

First, in the section dealing with break points, the actual edges of the arrays (minimum break point and maximum break point) were defined as a function of each individual cell (and even each individual timing arc). From a description of and perspective of the efficiency of the individual cell and timing arc, this is altogether accurate. From the perspective of the efficiency of the synthesis process, each cell and timing arc having array extents that are not consistent (although overlapping) becomes problematical. It means that as the tool attempts to try to close timing through optimization of the cells in a cone of logic, there are going to be input edge rates and output loads where proper *interpolation* can occur using some cells but improper *extrapolation* is required for other cells (Figure 8.18).

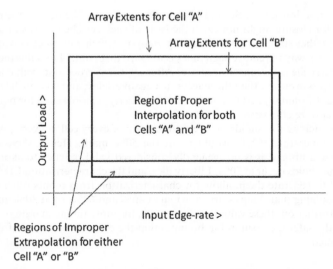

Array Extents for Cell "A"

Array Extents for Cell "B"

Output Load >

Region of Proper
Interpolation for both
Cells "A" and "B"

Input Edge-rate >

Regions of Improper
Extrapolation for either
Cell "A" or "B"

FIGURE 8.18 Incomplete Overlap of Array Extents Leading to Regions of Proper Interpolation and Improper Extrapolation During Timing Closure. This is caused by poorly chosen (inconsistent across library) minimum and maximum break points.

If the event illustrated in Figure 8.18 occurs during timing closure, then there is a chance that significant delay will result from the thrashing that could occur as cells that are not properly characterized for a region of the input edge rate and output load plane are substituted for other cells that are proper characterized. When that occurs, an improper extrapolation occurs and the resulting timing is wildly altered for the cone of logic. Hence, at least at first thought, it might benefit the characterization effort to adjust the break-point algorithm in such a way as to allow the minimum and maximum break-point array extents to be a superset of all the minimum and maximum break-point array extents. However, doing so really causes issues with the number of break points needed within each individual cell and timing arc and ensuring that any extreme nonlinearity in those individual arcs are properly covered.

So how can we optimize both the individual cell and timing arc and the library as a whole? The solution is not as bad as might be first thought. First, one axis of the array should already be fairly, if not exactly, consistent. The algorithm on defining the break points gave the input edge-rate minimum and maximum as being defined by the minimum possible edge rate for a given library in a given technology node and maximum allowable edge rate for a design. This does not change per cell or per timing arc (although it might and probably will change per PVT). This is a discussion only on the minimum and maximum extents of the array; the number and location of the actual break points within the array are still adjusted as defined in the optimized break-point algorithm previously given in this chapter. However, because the input edge-rate minimum and maximum extents are set, only the output load axis should be different for different cells within a library. Because it would seem rather suboptimal for the output load ranges to be defined consistently across the range from low drive strength versions of functions to high drive strength functions, there must be a different way to force the synthesis tools away from improper extrapolation while maintaining efficient substitution during cone of logic optimization. The answer is to adjust the output load minimum and maximum array extents for all the cells within a given drive strength to be the superset of the minimum and maximum extents for each of the individual cells in the drive strength. Proper tokens within the liberty file are then added to set the minimum and maximum load to which that particular set of drive strength cells can be applied.

A second set of liberty file adjustments should also be made that can greatly enhance the efficiency of cone of logic optimization. Specifically, this falls into two categories, neither of which is easy to accomplish. The cell timings need to be inspected for consistency. First, the easier of the two is to inspect the timings of the various drive strength versions of each given function. Figure 8.19 offers a one-dimensional representation of the timing.

In Figure 8.19, each succeeding drive strength has a region of the output load axis in which it is the minimum output delay (or edge rate), with the exception of drive strength 4×. Notice that the 4× drive strength cell in the figure is never less than both the strength 3× drive strength cell and the strength 5× drive strength cell. Hence, there is no region of the output load axis in which it will give the best solution. This is a poorly defined cell and should be either redone or removed from the library. Leaving it in the library "as is" will lead the synthesis tool to occasionally attempt to use it but ultimately reject it in any given cone of logic. As an aside, there are valid reasons to have such inefficient cells in a library—for instance, they may be the only acceptable connection point (having been characterized to be such) for some particular interface. In such an instance, they should be relegated to a special cell sublibrary and removed from the general-purpose synthesis library.

Although the preceding did not seem that difficult, please recall that the illustration was a one-dimensional output delay based on output load. The drive strength analysis

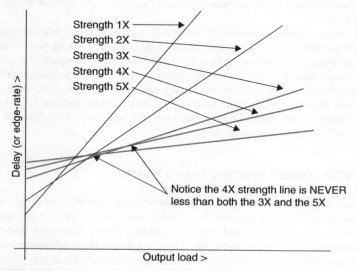

FIGURE 8.19 One-Dimensional Representation Output Delay or Edge Rate as a Function of Output Load for All Drive Strengths of a Function. Each drive strength of a function should have loading regions in which it is the best driver. If such regions do not exist, then the need for the cell in the library should be questioned.

should be done over the two-dimensional output load and input edge-rate surface. There are visualization tools available in the EDA market that can help this effort, but the actual inspection will tend to be manual.

The second of the two difficult library file adjustments is to repeat the preceding effort but for interfunction cases. In general, buffers should be faster than staging two successive inverters—pr why have a buffer at all because it could be faster to use two inverters? For similar reasons, AND functions and OR functions should generally be faster than NAND functions followed by inverters and NOR functions followed by inverters, but slower than plain NAND functions and NOR functions. AND functions should not be faster than NOR functions with inverted inputs. OR functions should not be faster than NAND functions with inverted inputs. Noninverted output Booleans should be slower than inverted output Booleans but faster than inverted output Booleans followed by inverters. This will be explored further in Chapter 14.

Along with these (and other) timing comparisons, similar analysis can be done in terms of area. Buffers should take up less (or the same amount of) space as two inverters. NAND functions and NOR functions should be smaller than AND functions and OR functions. Along bit-width considerations, two-bit NAND functions and NOR functions should be smaller than three-bit NAND functions and NOR functions. Along driver strength comparisons, unless there are some special reasons, 2× drive strength cells should be no more than twice the size of 1× drive strength cells, and 3× driver strength cells should be no more than three times the size of 1× drive strength cells.

In addition, similar comparisons can be made across the cells of the library in terms of power consumption. Functions with more transistors and more stages of inversion and thus more transistors potentially switching at any given input change should consume more power. Higher drive strength cells should consume less switching (dynamic) power than lower drive strength cells at equal output loadings. Cells with more or wider transistors should consume more leakage power than cells with fewer and narrower transistors.

All of these comparisons are meant to allow the synthesis tools' inherent assumptions on the comparative cost of functions in terms of delay or area or power to be more accurate, allowing for synthesis that is more efficient. These comparisons, along either the delay or the area or power-consumption axis, are all easily described, but defining all such comparisons, across an extensive library can become nearly intractable. As mentioned, some library analysis tools exist in the EDA market that can help this type of analysis. Many of these tools include extensive (but still user-extensible) list of which functional comparisons should be done.

8.9 DESIGN CORNER SPECIFIC TIMING VIEWS

When a fabrication house delivers a stdcell library, it usually delivers three PVT liberty files. These three PVT will typically be a worst-case low-voltage and hot-temperature liberty file; a nominal-voltage-and-room-temperature liberty file, and a best-case high-voltage-and-cold-temperature liberty file. These are fine if the design center's definitions of worst-case temperature and voltage and best-case temperature and voltage correspond to the fabrication house's definitions. But what if this is not the case? Aside from asking for extra liberty file development at the fabrications house, which could cost several hundred thousands of dollars, there seems little that is possible. Even more so, what if the temperatures and voltages match but the design center cannot hit some performance target using the worst-case process liberty files as defined by the fabrication house? Even at the normal cost of liberty file development just mentioned, the typical fabrication house might be disagreeable with developing a "not as worst-case process" or "not as best-case process" liberty file. In addition, there are other aspects of the characterization process such as the thresholds and edge rates requested by the design center that might or might not be supported by the fabrication house. What if the design center just does not agree with how the fabrication house has modeled a FLIP-FLOP? For instance, perhaps the design center depends on the outcome of the race condition when asynchronous set and asynchronous reset are removed from a DFFSRP (asynchronous set, asynchronous reset, positive clock-edge scan D FLIP-FLOP) and the model in the stdcell library release is undefined there (by the way, ignore design centers that do rely on such race conditions).

The correct answer to any of the above conditions or any of the others mentioned in this chapter, assuming the cost of asking the fabrication house to redo its effort is too high to bear, would be a "roll your own" characterization effort. However, how is this done efficiently?

The first item that needs to be considered is the parasitic extraction. Just running the layout parasitic extraction (LPE) deck out of the PDK directory is actually insufficient here. This LPE deck is for the design teams to extract post–automated place and route (post-APR) parasitic numbers from a routed design, so it does not assume any polygons that could be but are not actually present in the route. Library recharacterization cannot assume this. The physical views of the various cells in the stdcell library are stand-alone. When one is called up inside of a graphical display standard (GDS) viewer, the wells, active, polysilicons, depositions, contacts, metals, and VIAs that might be part of any given cell are present, but there is no assumption of overlaying route that might be found above the cell in an actual design. Also, there may be missing overlaying metal tiling (or underlying active tiling, although because of the nature of how much active needs to be in a stdcell, this is less likely). However, the characterization of the cell should represent how that cell is used actually in a design. In a design, a stdcell usually has some amount of routing or tiling over it. Depending on how the LPE deck does the extraction, a representation of this overlaying routing or tiling, on whatever layers that they can occur, needs to be added to the cell before the parasitic can be properly extracted. For best-case extractions used for best case recharacterizations, it might be sufficient to assume no overlaying of any route or tiling on any layer, and the PDK LPE deck can be used as is. For nominal-case and worst-case extractions used in nominal and worst-case recharacterizations, perhaps the following automated algorithm can be used to extract this representation of overlaying routing or tiling without causing the significant amounts of work of having to actually place parasitic metal over each cell before LPE extraction.

As in Figure 8.20, for all layers that potentially will have routing or tiling on them up to one metal layer above the highest metal layer used in the particular stdcell:

- oversize the polygons on that layer by the minimum space between polygons of that layer,
- invert (negate) this result,
- AND it with CELL BOUNDARY, and
- attach GND to the resulting construct.

Use these constructed layers along with the original cell during nominal-case and worst-case LPE extraction. In effect, what is being done is the addition (in the extraction only)

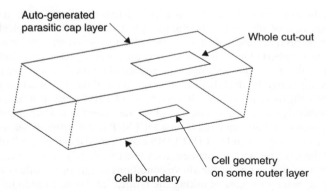

FIGURE 8.20 Graphic Representation of LPE Capture of Worst-Case Route or Tiling Parasitic Algorithm. The resulting parallel plate created parasitic ensures a worst-case extraction that will never be superseded in practice for a given LPE deck.

of a capacitive metal wrapper on the closest layers to the internal routing of the cell. After LPE extraction, the result will be a most aggressively worst-case parasitic extraction that will be used for nominal and worst-case recharacterization.

For those interested in either building new or validating extant LPE decks, this method should be used for extracting all layers but with the further assumption of requiring a three-dimensional extraction tool, several of which are available. In that case, the mode would be to define a nearest neighbor layer above, below, and to the side for each layer of the back-end stack. Note also which actual layers are the limiting features. For more complex processes, for instance, the limiting feature for contact-to-contact spacing might be contact-to-unconnected-polysilicon spacing (common in self-aligning contact processes) as opposed to contact-to-contact spacing.

However, beyond just these differences in the addition of a parallel plate extraction, it is important to realize that the actual extraction deck may be contradictory to the characterization corner desired. The reason for this is that best-case parasitic extraction decks usually assume some amount of excess thickness on the oxides and dielectrics of the back end (which gives the smallest parasitic capacitance), with the routing layers being either thinnest (giving best sidewall capacitance but highest resistance) or thickest (giving the lowest resistance but highest sidewall capacitance). However, the best-case model assumes among other things that the oxides are thinnest. The worst-case parasitic deck is likewise counter to the assumption in the worst-case model (in the opposite sense of the best-case contradiction). One may leave the decision as to which metal thickness gives the best- and worst-case resistances to the PDK build team. However, it remains that the library support team, if it is to do a recharacterization, has to determine if it makes sense to use a best-case extraction as the parasitic for a best-case characterization and a worst-case extraction as the parasitic for a worst-case characterization.

Next, beyond the generation of the parasitic enhanced SPICE netlist, there is a need to build an accurate driver environment. The reason is that characterizing a cell assuming that the input signal is a straight edge is going to produce different results than characterizing it with a signal that more closely represents the type of signal that the cell will most likely see in actual use. Figure 8.21 shows the difference between these two signals.

Generating such a signal is not a difficult problem. SPICE allows for voltage-controlled voltage sources (VCVSs), also known as E elements (the format of the E card

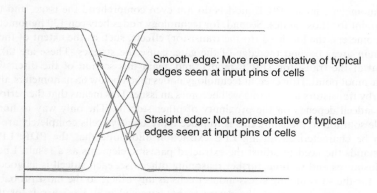

Smooth edge: More representative of typical edges seen at input pins of cells

Straight edge: Not representative of typical edges seen at input pins of cells

FIGURE 8.21 The Difference in What a Stdcell Input Sees in Situ Versus a First-Pass Characterization Edge Rate. Cells will react differently in the chip than they will appear to react in such characterization techniques. Hence, more realistic characterization techniques need to be used.

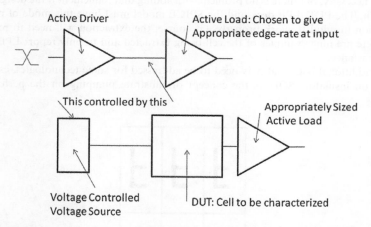

Active Driver

Active Load: Chosen to give Appropriate edge-rate at input

This controlled by this

Appropriately Sized Active Load

Voltage Controlled Voltage Source

DUT: Cell to be characterized

FIGURE 8.22 The Voltage Controlled Voltage Source (VCVS) from SPICE That Can Resolve the Improper Edge-Rate Issue. As the "E" card in SPICE, the VCVS allows more natural input curves without allowing device under test (DUT) loading to backward affect driver performance.

can be found in any SPICE manual). That is all that is needed. Set up an active driver and active load and tune it for the various input edge-rate break points desired (as defined previously in this chapter). Then use the resulting voltage curve of that pair as the controlling signal for the VCVS, with the unity gain that is used to drive the input of the device being characterized—that is, the device under test. Figure 8.22 shows this configuration.

The concept of stressed induced changes to the transistors of the stdcells is another issue that may or may not need to be addressed, depending on the technology node being recharacterized. Stress is captured in the recharacterization process in one of two main areas, depending on the technology node in question. First, for technology nodes above

130 nanometers, most SPICE models do not even comprehend the issue, and this is a moot point for those nodes. Second, for technology nodes between 130 nanometers and 65 nanometers, the localized (to the transistor) effects, such as the extent of the source and drain areas beyond the edge of the gate, cause the stresses. These are taken into account at the transistor SPICE model, and again this section of the discussion is a relative moot point. However, at technology nodes much below 65 nanometers, the effect of nearby transistors in the same well becomes an issue. This means that the performance of the stdcell depends on the proximity of other stdcells. The only way to find a reasonable solution to modeling this is to place representative cells completely around the cell to be characterized during LPE extraction and hope that the PDK LPE deck understands the need to adjust the extracted parasitic elements as a result. Luckily, in most instances and without further reasoning otherwise, each stdcell is as good a candidate for the surrounding stress-inducer cells as any other, so the most probable optimized extraction is to place the device to be extracted in a 3×3 array for the LPE extraction (see Figure 8.23 for an example). Note that the north–south arrangement is flipped in this example because it is common practice to build stdcells so that they can share common wells in flip and abutted rows. As far as the east–west abutted cells, the example has them not left- or right-flipped. For the extraction of stressed transistors, this is not a necessity, but there is no problem with adding that refinement if the design center wishes it. The PDK LPE deck and the SPICE model must have this mode of parasitic extraction in mind to work properly. Of course, the extraction will need to recognize that there are nine examples of the cell being extracted and to only report LPE on the center instance.

An additional issue that may need to be addressed for some technologies, especially silicon on insulator (SOI), is the concept of substrate pumping on the performance

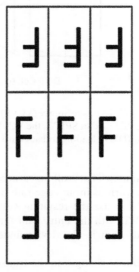

FIGURE 8.23 Three-by-Three Extraction Placements for Proper Deep Submicron Recharacterization. Local neighborhood stress is extractable in most deep submicron LPE decks and does cause changes to the SPICE model and hence to characterization.

capability of the various stdcells. Specifically, bulk transistors actually exist in wells that are connected to some rail or another, power or ground. Sometimes for power-consumption reasons, as mentioned in Chapter 3, this rail is at some interrail voltage. However, SOI transistors do not exist in these rail-connected wells. The surrounding material around the transistor effectively floats, so it can be charged by the action of the stdcell. The best thing to do in this case is to charge and discharge this material by toggling each input signal before the edge rate that is meant for actual timing or edge rate or constraint measure. SPICE models for SOI understand this charging and discharging and will produce different results on measures of first input edge to first output edge as opposed to measures of second input edge versus second output edge. Figure 8.24 shows typical waveforms for bulk measure versus the same for SOI measure combinational logic. The sequence is pulse the input and then change the input and measure the timing arc. The pulse on the output caused by the pulse on the input may or may not resolve completely, depending on the length of time that the input pulse remains. Waveforms for synchronous cells are only slightly more complicated. A description of the sequence would be to hold the data at one rail, pulse the clock, and pulse the data; then change the data edge for the setup constraint measure, walking this edge closer and closer to the clock edge; and then change the clock and measure the constraint. Hold time is similar, but the last two stages are reverse. Finally, there are such signals as asynchronous set and rest on sequential cells. For these, the sequence gets only slightly more complicated: while the data are held to the opposite rail, the asynchronous signal will force pulse the asynchronous signal; then pulse the data (these first two steps can be done in either order); then pulse the clock; then pulse the asynchronous signal and measure the timing arc.

This could be just as effective for bulk measures. Specifically, many of the more complicated Boolean circuits have several intertransistor nodes where charge can be stored. When only a small number of inputs are changed for a measure—and for only a single time during the actual timing measure—those intertransistor nodes may or may not contribute to the timing measure.

Finally, one more item should be addressed when the decision to recharacterize has been made. When a liberty file is referred to as a specific PVT characterization file, then a

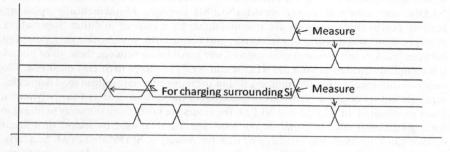

FIGURE 8.24   Proper Bulk Characterization Waveforms Versus Proper SOI Characterization Waveforms. Substrate charging of SOI is required because it does not have well ties. The best means to accomplish this is through the circuit itself. Figure 8.23 illustrates typical waveforms used in characterization. The top two show an incoming waveform and the resulting output waveform, together with indications as to where the measurement would be taken. This is typical of bulk characterization. The next two waveforms show similar characterization for SOI characterization. Notice the toggling of the signal before the actual measurement. This allows for the previously mentioned substrate charging. No measurement is taken on these edges.

specific voltage is assumed and even specified inside the file (along with the specific process and temperature). However, the actual environment that the stdcell will experience when placed inside an APR area in a design will not be at the actual voltage. Hence, it makes sense to account for this in the characterization (using slightly lower voltage than worst-case voltage for worst-case liberty, and slightly higher than best-case voltage for best-case liberty). This is not the case for the process axis because the process people already guard band the SPICE model, and it is not the case (or at least not much of a case) for the temperature axis because cross-die temperature is unpredictable for any given cell during the characterization process. However, if desired by the design center, it might make sense to guard band the temperature slightly as well. How much of a voltage (and possibly temperature) margin should be added is a decision of the design center.

8.10   NONLINEAR TIMING VIEWS ARE SO "OLD HAT"…

Unless the size of the nonlinear arrays is "nearly infinite," there is an inherent error in the delay and slew calculations based on that particular set of nonlinear arrays. The basic premise of this book is that these errors in the various views are useful once found because they can be removed to increase the performance, area, or yield of the resulting design. We know that these errors exist. They exist in the liberty timing views (and similarly in the liberty power and noise views discussed later), so it should not be prohibitive to remove them. It is so easily the case, when combined with the obvious nature of the error source, that the various EDA vendors attempt to reduce or remove the errors on their own without much input from the library intellectual property (IP) engineer.

Synopsys has generated several formats in an attempt to reduce or remove this error. In an attempt to reduce the initial format of a single set delay through a stdcell, the first of these formats was the nonlinear delay model (NLDM) format, replacing this initial set number with a calculated number based on input edge rates and output loads (as described earlier in this chapter). The next version of this error-reduction reformatting was the introduction of scalable (or standard) polynomial delay (SPDM) and scalable (or standard) polynomial power (SPPM) models. These remain heroic extrapolations of the NLDM and nonlinear power model (NLPM) concept. Mathematically speaking, because polynomial equations are manipulatable by means of addition, subtraction, multiplication, and division of each other, and because the straight-line extrapolations between break points can be considered as *linear polynomials* between those given points, it is straightforward to expand the NLDM and NLPM concept to SPDM or SPPM. The only real difference is that the numbers within the input edge-rate output load delineated array are not set delays or slews but coefficients for the various orders of the polynomial being represented. In NLDM and NLPM, the values are taken from the array in question for a given input edge rate and output load and are then used by means of a linear extrapolation to determine the delay or slew (or power) in SPDM or SPPM. Similarly, the values are taken for the various coefficients of a given polynomial representing the delay or slew (or power) based on the input edge rate and output load; those coefficients are then used within an easy calculation to determine that delay or slew (or power).

There are at least three concerns, however, with this approach. First and simplest is that SPDM or SPPM arrays are not as easily or as directly read as NLDM and NLPM arrays are. One can open an NLDM- or NLPM-based liberty and by inspection determine if a given delay is reasonable. In other words, for a given edge rate and a given load, a given delay or slew is directly read from the array. For SPDM and SPPM, all that can be directly read is the coefficients of the polynomial; the delay or slew is then determined by a further secondary calculation. Hence, error checking of a SPDM and SPPM liberty is a more complicated process.

Second, calculated polynomials are inherently unstable beyond the given range of the input values used in the original construction calculations. To expand on this, assume that one takes a set of SPICE measurements of a delay over a given range of edge rates (or loads). Assume that there are N such measures over that range. One can exactly fit a polynomial of order N – 1 onto those N measurements. However, that order N – 1 polynomial, while exact over the specific measured points, quickly grows away from any curve (other than itself) for edge rates (or loads) beyond the extreme ends of the original range.

Third, over the range measured, the polynomial tends to oscillate up and down around a stable trend line. In other words, calculating a polynomial of order N – 1 based on N measures results in a an equation that approximates a given curve that exists in a given SPICE model but does not exactly match that curve. Although I am sure there are N – 1 order effects in deep submicron transistors, I am also sure that there are no SPICE models that accurately represent those higher-order effects. Figure 8.25 illustrates both of these last two issues. A means of getting around both of these issues is to cut the range into multiple subranges and then use spline approximations of the polynomials. In other words, instead of a single polynomial over N measures, have L ranges representing M measures (where $L \times M = N$), each range represented by a more complex SPLINE polynomial, the various SPLINE approximations set so that they are contiguous (or nearly so) at the boundaries between the L various regions. Multiple design centers have developed various patented methods to efficiently determine these various regions and these various SPLINE coefficients, so the user should be wary before choosing to develop such methods.

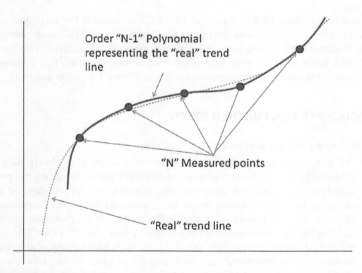

Order "N-1" Polynomial representing the "real" trend line

"N" Measured points

"Real" trend line

FIGURE 8.25 Polynomials Do Not Model Real Trend Lines. Standard polynomial models can easily introduce inaccuracies. When combined with the less than directly readable format of SPDM and SPPM models (as compared to NLDM and NLPM), the user can easily lose undetectable errors.

If SPDM or SPPM is not the answer, is there a better model? At least according to two different EDA companies, the answer is "yes." Unfortunately, the formats for this alternative model as proposed by these two companies are different. For the sake of the level of extraction in this book, however, these differences are minimal.

These two formats are the composite current model (CCS), the next model proposed by Synopsys; and effective current source model (ECSM), which was proposed by Cadence as an answer to Synopsys. ECSM and CCS are dramatically improved methods of calculating delay or slew (or power) through a given timing arc. Both ECSM and CCS take a much-refined approach to loading conditions of a real circuit (such as the Miller capacitance of a load as opposed to a simple one-number representation of the output load, such as multiple voltage drain drain (VDD) regions across a cone of logic and long interconnects between a source and a sink, among other conditions. Surprisingly, however, ECSM and CCS are, in the IP view build process, extremely close to NLDM and NLPM (and SPDM and SPPM). The only real difference is that ECSM and CCS measure (and the resulting arrays represent) the output current of a stdcell as a function of time, as opposed to the output voltage of that stdcell as a function of time. This output current as a function of time is then stored in the array. When combined with one more measure, input reference time (which represents when an input crosses a threshold), these measures can be used more accurately to represent timing (or power) through a stdcell timing arc. Therefore, the bottom line here is that liberty generation for ECSM or CCS is of minimally more difficulty than for the original NLDM or NLPM model. ECSM or CCS almost begs the question of the need for error measurement in that the current curve represented in the array is an N point measure still represents the continuous curve of current as a function of time. The more points in the array, the more accuracy of the representation extrapolated numbers. Hence, do not be surprised, that future ECSM and CCS development might lead to polynomial-based ECSM and CCS, but this is just a supposition of the author at this time.

One real benefit that ECSM has over true CCS is support for statistical or Monte Carlo timing closure. ECSM has the concept of individual VDD-based, process-based, and temperature-based timing arcs. As a result, per cell adjustments can be made in various EDA tools to allow true Statistical base timing closure. The addition of the VDD, process and temperature constructs into the liberty format is minimal.

8.11 CONCEPTS FOR FURTHER STUDY

1. Improper best-case and worst-case synthesis:
   - As mentioned in the chapter, the best- and worst-case liberty files that are released by the third-party IP vendor are envelopes surrounding the proper and more accurate best- and worst-case characterizations. The student or interested engineer should be challenged to build accurate best- and worst-case liberty files for a small subset of library cells for which a vendor-supplied pair of enveloping best-case and worst-case liberty files already exist. The student or interested engineer can then use the four such files to test synthesize several netlists to absolute maximum performance in order to see how the vendor's enveloping margin additions hinder maximum performance of a given library in a given technology (and simultaneously cause added buffer insertion on the best-case hold side of the timing closure experiment).

2. Default arc substitution script:

   • The student or interested engineer might be challenged to build a script that reads a liberty file and determines per pin for each particular cell which of the state-dependent timing arcs are the true longest (for worst-case liberty) and true shortest (for best-case liberty). This script should then substitute the correct state-dependent arc array for the default arc array for that pin in that cell.

3. Scan-substitution script:

   • In many instances, design-center users of a library incorrectly attempt to close timing before scan insertion. As a result, when they substitute the scan flops for the regular nonscan versions during scan insertion, they find that both the area of the design and the long paths of the design are severely impacted. One way to alleviate this is to build a scan-substitution script that reads a liberty file, determines which scan flop and nonscan flop correspond to each other, and then substitutes the area and regular signal timing power and noise arrays of the scan flop versions into the corresponding nonscan flop arrays. In this way, if that resulting liberty is used during synthesis, the resulting netlist is optimized beforehand to accept the regular nonscan signals in such a manner as to more easily still meet those timing constraints for timing closure after "real" scan insertion. The student or interested engineer might find it useful to build such a scan-substitution script.

4. Break-point optimization algorithm:

   • As mentioned in the chapter, many third-party-vendor liberty files contain arrays with edge rate and output load break points that are not optimal for the desired maximum error possible within the array. The student or interested engineer might find it useful to build a script that repeatedly runs characterizations within SPICE and then compare the results of those runs with the extrapolated values of an ever-larger liberty array, the break points chosen as given by the algorithm in the chapter text. The script would run until the error between the SPICE runs and the extrapolation is within a desired limit.

5. Synthesis outside the array:

   • Sometimes the minimum and maximum limits for a given array for a given pin for a given cell in a given liberty file are too large or too small for the application required. Either the application uses edge rates that are much smaller or larger than the liberty generator had assumed or the output load is much smaller or larger than the liberty generator had assumed. In those cases, as mentioned in the chapter, the synthesis engines tend to use linear extrapolation from the closest edges of the array. This can cause significant error. The student or interested engineer might find it instructive to attempt to synthesize (the original RTL into a gate-based netlist) and then to adjust that netlist until it is "timing closed" to a required frequency using an array that is artificially limited so as not to reach the edge rate or load desired and then to run SPICE simulations on the resulting netlist in order to determine how error prone such an incorrectly limited liberty timing arc can be.

6. Independent setup-and-hold SPICE experiment:

   • All currently available characterization tools allow for independent setup-and-hold calculations. The student and the interested engineer are highly encouraged to use a liberty file that has been characterized with independent setup-and-hold constraints to determine what those minimum setup-and-hold numbers actually are. Then the student or interested engineer should build a SPICE netlist includes the sequential element in question. Data should be applied to change the data input of the sequential cell at the minimum setup constraint before the clock edge and again at the minimum hold constraint after the clock edge, as given by the independent liberty, in the same SPICE run. The goal is to see how the output of the sequential actually resolves.

7. Parity-tree synthesis experiment:

   - The student or interested engineer is encouraged to build a 2N input complete parity tree and synthesize it with a liberty file that does not have a three-input EXOR and EXNOR and then to repeat the effort with a (2N) + 1 input parity tree to determine if the example given within the chapter is correct.

8. "Nearly absolute" minimum setup-and-hold window algorithm:

   - The student or interested engineer might find it educational to attempt to build an algorithm that chooses a "nearly" absolute minimum setup and hold window as opposed to the minimum-setup and dependent-hold or minimum-hold and dependent-setup windows described within the chapter. The assumption on concavity of the minimum setup-plus-hold window should be tested as part of the algorithm. Employment by characterization-tool EDA companies is almost ensured if successful.

9. Roll-off versus metastability experiment:

   - At deeper submicron technology nodes, roll-off values of 10% become dangerously close to causing metastability with many sequential cells. The student or interested engineer who has access to a metastability analysis EDA tool (that is better at detecting metastability than SPICE) might be inclined to test how much roll-off during constraint characterization is too much.

10. Monotonic and positive adjustment script:

   - The student or interested engineer might be interested in developing a script that correctly adjusts liberty timing arrays to be monotonic and pure positive as per the algorithms given in the chapter.

11. Useless stdcell script:

   - The student or interested engineer could develop a script that compares stdcells of the same function but of varying drive strengths in order to determine which cells, if any, do not have ranges of output loads at which they show minimal performance. These drive strengths of those cells are generally worthless and can be safely removed from the liberty offering. The script may be extended to compare such other function combinations as NAND-INV versus AND and NOR-INV versus OR and any others that the student or interested engineer finds important.

12. Parasitic top-plate addition Boolean adjustment to LPE extraction deck:

   - The student or interested engineer should determine the Boolean constructs within a LPE extraction deck that allow for the automatic inclusion of a series of per layer top plates in order to best extract a true worst-case LPE deck.

13. Bulk-to-SOI characterization vector writer script:

   - The student or interested engineer might be inclined to build a bulk-to-SOI SPICE vector converter script that takes valid bulk characterization vectors and automatically converts them such that extra rise and fall edges are included on the SOI vectors so that these can appropriately charge to SOI substrate.

CHAPTER 9

POWER VIEWS

9.1 LESSON FROM THE REAL WORLD: THE MANAGER'S PERSPECTIVE AND THE ENGINEER'S PERSPECTIVE

All circuits will burn power in order to accomplish the functions for which they are designed. This is true for digital circuitry as well as for other circuits. Digital circuits will burn power in order to carry out the logical functions they are designed to accomplish. Although there are exceptions to every rule, larger circuits generally burn more power than do smaller circuits. In addition, faster circuits will burn more power than slower circuits. And, generally speaking, circuits that drive heavy loads will burn more power than circuits that drive lighter loads. This all seems to be inherently obvious, but experience can suggest differently.

Unfortunately, because chip-design engineers sometimes misinterpret the power number with in a liberty file, the amount of power that a given circuit will actually consume is seriously overstated. This is unfortunate because in order to care for such a supposed large amount of power consumption, chip designers add significant amounts of power-rail routing. They thereby significantly reduce the percentage of the core area budgeted for stdcell routing, thus increasing the size of the core, which pushes the stdcells farther apart than required, and causing the need for larger stdcells that in turn consume even more power. Further, the chip designers supply that power-rail routing with increased numbers of power and ground inputs–outputs (IOs), thereby increasing the size of the device. In addition, because this misinterpretation of the power numbers is always multiplicative, device packaging capable of handling such assumed large amounts of

Engineering the CMOS Library: Enhancing Digital Design Kits for Competitive Silicon, First Edition. David Doman.
© 2012 John Wiley & Sons, Inc. Published 2012 by John Wiley & Sons, Inc.

power burn and heat dissipation must be used. Sometimes, all of these concerns are real and the need to account for them is required. Sometimes, however, they can be avoided by correctly understanding how to use a stdcell library.

I was once asked to investigate why some designs developed by a particular design center appeared to burn so much power, even though they were relatively small designs of a few hundred to a few thousand FLIP-FLOPs and done at a 250-nanometer technology. I found that the typical clock tree used in these designs was anywhere from 12 to 20 inversions deep. Even assuming that each stage of the clock trees in those designs involved a minimal bifurcation branching structure, these clock trees meant that there could be as many as 4,000 to 1 million leaf nodes on those trees. Thus, these clock trees could supply anywhere from 1 to more than 250 leaf nodes per FLIP-FLOP.

It turned out that there were several reasons, in the eyes of the design engineers who worked on these designs, for such deep clock-tree design. These included attempts at extremely tight clock balancing, especially with internal third-party intellectual property (IP) (which was not procured through my organization) that exhibited significant variability in clock insertion delays. However, a significant contributing issue was that the design engineers were strongly afraid of "crowbar" current, which occurs when the clock buffers in the FLIP-FLOPs are momentarily in some unknown intermediate state. Hence, they had a design goal of making the clock edges at the plethora of FLIP-FLOPs and registers as sharp as possible, in a belief that doing so would limit the system's turn-on power drain. I had to point out that, assuming that the unknown inputs to the various FLIP-FLOPs were digital and stable, the only circuitry that would be changing on the initial clock edges were the double inverters on the clock input to the FLIP-FLOPs. These are generally small circuits. Hence, to minimize the power consumed in a few hundred to a few thousand flip-slop clock inverters, the designs had added several hundred to several thousand large inverters on the same clock. In the end, most of these designs were redesigned to remove these extended clock trees among other goals.

A second instance of wasteful power management resulting from misinterpretations of power numbers in a clock tree was the time I was requested to supply 50X symmetric inverters for a design. I try not to supply such large drive strength cells. The need for proper floor planning of such cells, together with the power rails required to supply them with adequate power, is generally prohibitive. I fought the customer on these for awhile but eventually gave up and supplied them, together with as much "Use at your own risk" documentation as possible. I could not see how anybody could add such circuitry in any manner that would make sense. A week after I supplied these 50X inverters, I received word that they did not resolve the timing issues that the particular design had requested and was using, and the design center was going to request 100X and 150X inverters as well. I asked for details and was told that the design center had added several hundred of the 50X inverters to the design in question. It had improved the clock-tree insertion by just a few pico-seconds. It turned out that instead of using the 50X inverters only a few times near the base of a clock tree, the design center was using them at each leaf node of the clock tree, each driving the clock inputs to a few FLIP-FLOPs. As noted, in the end the design was redone in order to build a significantly less power hungry but clock-skew and clock-insertion delay minimizing clock tree.

The bottom line is that designs need to be educated in what circuits burn power (and when they burn power) and what can be done to minimize this.

9.2 TIMING ARCS VERSUS POWER ARCS

In Chapter 8, it was pointed out that it was important to include all state-dependent timing arcs inside a liberty file. The given reason for this was that, even though during synthesis there was no knowledge of the states of other signals within a cone of logic, use of a liberty file for prime-time simulation would be more accurate with the use of state-dependent timing numbers as opposed to the use of default timing numbers. It was further pointed out that whichever state-dependent timing arc was to become the default timing arc was probably different for each process, voltage, and temperature (PVT) corner. This was so because the various PVT corners would have some timing arcs that had longer times than the same timing arcs in other PVTs because of being more or less susceptible to that particular process, voltage, or temperature corner. In addition, however, it was also because of the implied use of that corner as a long-path setup timing closure corner or as a short-path hold timing closure corner. That is to say, the default timing arc for a data signal inside a nonclock cell will be different in a liberty used in long-path timing-closure as opposed to a data signal inside a nonclock cell in a liberty that is to be used in short-path timing closure. A good rule of thumb is as follows.

- Long-path PVT corner liberties:
  - Data inputs into "nonclock cells" should have the longest timing arc as the default timing arc.
  - Clock inputs into clock-tree cells should have the shortest timing arc as the default timing arc.
- Short-path PVT corner liberties:
  - Data inputs into nonclock cells should have the shortest timing arc as the default timing arc.
  - Clock inputs into clock-tree cells should have the longest timing arc as the default timing arc.

All of that remains true for the power-arc section of the liberty file. In the first case, that of state dependency, although prime time-based power simulations of modern chips tend to remain statistical at best (because of the impossibility of knowing all possible uses of the circuit in the real-life environment in which it will operate), there is a legitimate amount of such power simulation—for instance, during the powering up or down of various internal-to-the-device regions of the chip that are designed as such in order to reduce power consumption. Such instances will require state-accurate power arrays within the liberty file for that particular PVT corner. In the second case, default power arcs are as useful during synthesis as default timing arcs. As previously mentioned, the synthesis engines will attempt to reduce long-path timing (and increase short-path timing) by using these default timing arcs within each cell when so directed. This is true for power synthesis as well. These synthesis engines will also attempt to reduce power by substituting cells (or cell combinations) that have lower default power for cells (or cell combinations) that have higher default power numbers (assuming that these engines are instructed to do so). This is especially the case for logic cones that are not constituent long-path limiting or short-path limiting cones.

However, there is no correlation between what constitutes a worst-case power arc and a longest-path (or shortest-path) timing arc. Recall Figure 8.1, in which an AOI21 is shown. Assume for the time being that we are only interested in the N-channel pull-down side of the circuit. The longest path on the pull-down side for that structure is probably through the two stacked N-channel devices (as opposed to the single N-channel parallel to them). The default timing arc would thus probably be the bottom N-channel transistor in the stacked pair turning active (a logical rising edge on its input B) while the N channel above it is already active (the input A) and the parallel N channel inactive (the input A). However, this may or may not be the most power-consuming transition. It is more likely to be the discharging of the node inside of the P-channel stack by a rising edge on the C input with the A and B inputs remain inactive. Such a transition would cause a the discharge from the internal node of the P-channel stack (possibly a significant amount because of the capacitive load on that node caused by the three P channels attached to it as well as the output node to be discharged. Although the above analysis is tractable, it becomes less so when the Boolean becomes slightly more complicated. As a result, best guesses as to which transitions will produce the default worst-case power arcs become less important. Luckily, the solution is extant in the characterization database, requiring no further analysis than to compare each state-specific power arc in order to determine which is the most power hungry and thus use that as the default.

Although state-specific timing and power arcs are similar in that all of them need to be present inside of a liberty (to ensure accuracy during prime-time simulations), the default arcs differ significantly in both the definition of default and the application of those default arcs.

Nobody should conceive of putting a best-case power arc as default in any liberty. One always wants to minimize power and should do so by assuming maximal power usage, no matter if the rest of the liberty is conceived for use as a long-path setup or a short-path hold liberty. The default power arc that should be used in both long-path setup and short-path hold liberty files should always be the most power-intensive arc for that particular PVT.

9.3   STATIC POWER

Decades ago, it was frequently observed that CMOS technology had a significant advantage over other types of technology used for integrated circuits because digital CMOS circuits only burned power when they were actually switching. As technology nodes shrank, the number of digital structures increased, and the resistance of the signals nets running between these digital structures increased, the length of time that a driving signal stayed within the switching threshold range of the P-channel and N-channel combination proportionally increased. Thus was born the need for dynamic power-measurement tools. But also, as technology nodes shrank, an additional current consumption component emerged that led to the contradiction of the original premise of CMOS stated previously that it does not use power when not switching. Static (or leakage) power, sometimes known as *wasted power*, for the deeper technology nodes exists.

Leakage power, which is also known as *static power*, or the power that is dissipated even when a circuit is not involved in a logical operation, can grow by up to an order of magnitude (or more) for each technology node step below 180 nanometers. The reasons for this is because such items as gate oxides decrease and channel length decreases, both of which allow for the charge to enter the substrate or to cross the gate area as illustrated in Figure 9.1. The bottom line is that the exponential growth in such static leakage power

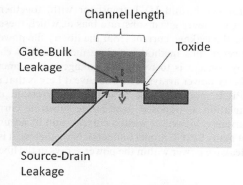

Channel length

Gate-Bulk
Leakage

Toxide

Source-Drain
Leakage

FIGURE 9.1 Transistor Cross-Section Major Contributor Factors to Static Leakage. As technology nodes shrink, decreases in gate oxide and channel length allow for an ever-increasing leakage current, which debunks the old adage about CMOS not burning power when it is not switching.

means that severe low-power design techniques, as mentioned in the boutique library section in Chapter 3, become necessary.

What does this mean in terms of the representation of this static leakage power consumption in actual cell views? There are no separate static-power and dynamic-power arrays in the current liberty Backus-Naur format (BNF); the current format only allows for one set of power numbers for any given input-to-output power arc (arrayed by input edge rates and output loads). So how should this static leakage power be handled? This depends on the technology node that is being represented and on the desired design techniques employed by the design center that is being supported.

If the design center that being supported is not going to incorporate any power-reduction design techniques, then the inclusion of static power measure within the liberty is wasted effort. The only static current information that will be gleaned from the liberty could be more easily calculated. A better solution is to document the fact that some amount of current leakage is present. This leakage is measured as a function of total channel width. This knowledge can then be supplied to the design center with a typical, best-case, and worst-case measure of that static leakage as a function of voltage and temperature. Further, another tool can be added to measure total channel width used in a design [possibly as a token within a liberty file or with a library exchange format (LEF) reading tool that understands the entire amount of transistor width]. As such, for these design centers, it is common to substrate out any assumed static leakage current from the liberty power arrays. Although no design center that is developing products at or below the 130-nanometer technology level should be in this grouping, designs of 180 nanometer and higher (and even some 150-nanometer designs) can still find this approach useful.

On the other hand, if a design center that is being supported is going to develop low-power design alternatives to at least some of the circuitry that is going on a device, then static as well as dynamic power measurement is required. A design engineer will definitely require knowledge of how much static power will be reduced, at the cost of performance, by, for instance, changing a Low Voltage Threshold (LVT) cell for a High Voltage Threshold (HVT) cell or adjusting the substrate voltage. Surprisingly, many design centers can sometimes suggest that handling static leakage as above still makes sense, even in the technology node realm where such active low-power design techniques are required. That is to say, just give them a tool that will calculate the total amount of LVT transistor width, SVT transistor width, and HVT transistor with, together with the estimates amount of leakage at whatever given substrate bias at which these cells are operating. One can then remove the leakage current from the liberty file power arrays is adequate for their needs. However, this method is failing more so with each technology node shrink. A more likely concept is to retain the leakage measurement during characterization within the liberty power arrays. The one caveat here is that such measurement is based purely on SPICE models. These SPICE models remain more geared toward accurate digital simulation as opposed to accurate substrate currents (after all, such substrate leakage current is viewed as wasted current, and its accurate modeling is thus not a high priority of many SPICE model builders). Nonetheless, it is more common than ever to retain static leakage power within the power arrays in the liberties for these lower-technology nodes.

9.4 REAL VERSUS MEASURED DYNAMIC POWER

Figure 9.2 illustrates the current paths that may be found in a simple but typical stdcell. These, plus one additional current source to be mentioned shortly, will be encountered during power characterization. How one handles these that can be problematical during the building of the power arrays.

For liberty power-array characterization purposes, one can easily run a simulation of the previously described circuit and carefully measure the various current components shown. So what is the major concern? The issue is not the actual measurement of the various constituent parts of the current used in the circuit, but how they are then manipulated in order to build the correct power array in the resulting liberty files for the various PVT corners. Consider the following characterization circuit in Figure 9.3. (Note that the actual input should be generated consistent with that in Figure 8.22, but for simplification purposes the VCVS source is removed in this illustration, so also note that the output load may be more complicated, such as an resistance capacitance (RC) network for characterization in some high-speed applications.)

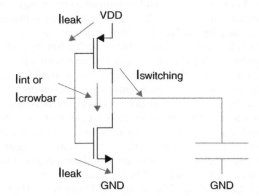

FIGURE 9.2 Some of the Current Paths Encountered During Characterization. Careful handling of the various components illustrated here is required in order *not* to end up either overestimating (by double counting some of the power consumed) or underestimating (by dismissing apparently unimportant currents) the power consumed within a stdcell.

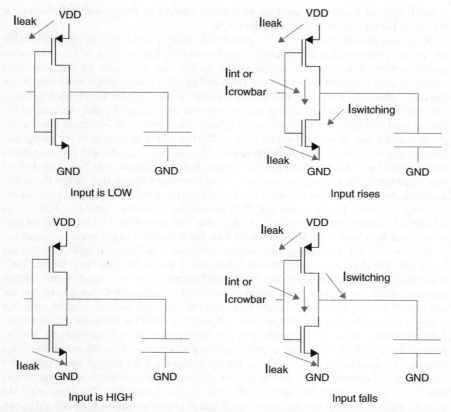

FIGURE 9.3   Current Measurements as a Function of the Input Signal. In every case, note that the measured current is actually a combination of different types of current (I): dynamic I-internal, static I-leakage, and dynamic I-switching. As a result, what is used to compute the actual power consumed by the circuit's response to the input stimulus needs to be based on a calculation that attempts to isolate these components properly.

When the input signal is low, a leakage current flows from the power rail to the input signal. Unless there are exceptional circumstances, this current is rather small, even for the deepest technology nodes (because of the assumed small size of the P-channel transistor), but it is not zero. When the input signal is high, a leakage current flows from the input signal to the ground rail. Again, and assuming a relatively small N channel, the amount of this current is also small, but not zero, and it is probably different from the P-channel leakage current just mentioned. When the input transitions from low to high, there is a switching current, primarily determined by the width of the N channel (and current-carrying capability of that transistor width) and the size of the output load. In addition, however, an internal crowbar current flows short circuit between the rails during the time that input voltage allows both the N and P channels to be active. When the input transitions from high to low, there is a switching current (as before) that is primarily determined by the width of the P channel and the size of the output load, as well as the previously described internal crowbar current. In addition, unless the output load is ideal, there is probably a continuous leakage current to the ground rail through it whenever the charge on it is greater than zero.

What will actually be used for the power calculations that are to be included within the power arrays within the liberty? Straight substitution of what is measured with each transition will yield rising-edge power arrays that are based on I-leakage plus I-internal plus I-switching current; similarly, falling-edge power arrays will be based on I-leakage plus I-internal plus I-switching current. However, this means that some of the current, specifically I-switching, will be counted twice (once for the charging of the output load and again for the discharging of the output load) and I-intrinsic (once when the P-channel and N-channel totem pole is short-circuited during both the rising and falling input transitions. Although the second of these two is correct and accurate (there actually will be a short-circuit crowbar current flowing in each transition), the first of these is counting the same electrons twice. What is needed to accurately reflect the power array numbers, is somehow to isolate the I-switching from the I-internal current (and the I-leakage current). Further, one can then divide the switching current in half before adding the I-internal crowbar current back in (the addition of the I-leakage current may or may not be added back, in accordance with the discussion in the previous section).

So how can one accomplish this current component isolation? If SPICE netlist are being used for characterization, then this becomes a matter of identifying the pertinent P-channel and N-channel transistors in the final output totem-pole stage of the cell that are affected during a given input signal transition. For an inverter, this is easy, but for any stdcell that is designed logically to merge two or more signals, usually only two possibly folded sets of transistors are affected, with the others in the same output totem-pole remaining stable. If successful, one can measure the current through the source, gate, drain, and bulk (assuming a bulk transistor as opposed to a silicon-on-insulator transistor) connections, along with the charge buildup on the output load; categorize these currents correctly; and do the proper mathematical operations on them before converting to power for the appropriate arrays in the liberty files. Just measuring the current through the rail connections of each cell at characterization time is insufficient.

At the beginning of this section, I promised one additional current path. Figure 9.2 illustrated a simple inverter. More complicated Boolean functions not only have multiple transistors in the final output totem pole as just mentioned but also have nodes that are internal to the stdcell that depend on the input signals being either isolated or not isolated from the rest of the output-stage totem pole, depending on the values of the various inputs. A classic example is the AOI22 shown in Figure 9.4. It has two nodes that may or may not be isolated from the output totem pole, depending on the current state of the inputs to the cell and that further may or may not have charge on them that needs to be dispersed during possible output transitions. This potential charge will need to be included in some manner within the liberty power arrays. However, depending on the ordering of the input transitions, any particular intertransistor connection (which amounts to a charge storage node) may be previously fully charged when a transistor that is between it and the power rail is turned active. This cannot divert current away from the totem-pole output, or it may be fully discharged when a transistor between it and the ground rail is turned active, thus not delaying the totem-pole discharge. But such activity means that the accurate current measures for such cells is unlikely without proper scheduling of which vectors will be run in which order during power characterization. An alternative to this vector-scheduling issue will be discussed in a later section.

One additional current path needs to be discussed. Multistage stdcells, even as simple as a buffer cell, have complete P-channel or N-channel totem-pole stacks that are internal

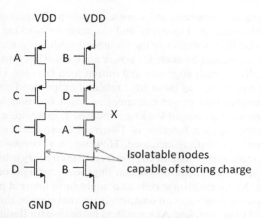

FIGURE 9.4 The AOI22. Many Booleans such as the AOI22 has nodes that are internal to the stdcell (highlighted) that may or may not be isolated from the output totem pole and may or may not have stored charge that requires dispersion during proper library power characterization.

to the cell. The good news is that the I-switching current for these internal to the cell nodes will show the current flow correctly, so the previous discussion in this section on output totem-pole switching is not required on those internal totem poles. The basic reason for this is that the totem pole that is internal to the cell is driving an active P-channel or N-channel load as opposed to a characterization load. Further, it is connected through some combination of P-channel or N-channel transistors to one rail or the other as opposed to being held by some artificial output load. When the signals driving those transistors change, the totem pole internal to the cell will accurately reflect real life in how it operates. This is not the case for the output stage of the cell, which drives an output storage load. Thus, as mentioned, the I-switching current on these totem poles that are internal to the cell does not require modification.

The preceding discussion does not apply to stdcells with transmission-gate outputs or pure transmission-gate internal nodes. These cells hold a state at the output or the internal node in question (note that this is not the case for FLIP-FLOPs because the internal transmission-gate nodes are fully defined so that when one input to the internal in question is held inactive, the opposite transmission-gate input to that same node is active). This is another good reason not to have such devices in a library offering.

9.5 SHOULD POWER BE BUILT AS A MONOTONIC ARRAY?

Chapter 8 on library timing discussed the need to build monotonic timing arrays. The basic argument was that the liberty-timing model was nothing but a model of the actual timing of the cell and, as such, any slight adjustment to the timing numbers in the arrays to facilitate the reverse-hill-climbing cell-synthesis algorithm was well worth the effort. In addition, any nonmonotonic behavior that occurs within an array of measured values is most likely a remnant of the measurement process. Besides, for those design centers that are not willing to use the always more conservative liberty in the assumption that there was no measurement error, there is always the choice of using those original measured liberty files. The massaged liberty is always more conservative. Best-case data-path signal numbers are always faster than or equal to measured numbers, and worst-case data-path signal numbers are always slower than or equal to measured numbers, and vice versa for clocking signal numbers.

With one exception that is similar to the timing-constraint argument, this is the exact argument that can and should be made for power numbers. That is to say, power should change monotonically as input edge rate and output load changes. However, the direction of that monotonic change as input edge rate and output load changes might be an issue. It seems probable that power consumption, as a function of I-switching, will monotonically increase as an output loading is increased. In addition, it seems probable that power consumption, as a function of I-internal (I-crowbar) will increase monotonically as an input edge rate is increased. However, it is conceivable that a library provider can build some sort of load or edge-rate sensing circuitry into the more complex stdcells (and IOs) such that added polysilicon fingers are brought onto line as load or edge rate increases. Many multistage cells also might have internal nodes that perform better because of slower edge rates on competing external signals, thus possibly causing the monotonic trend to be opposite. As a result, as in the case for timing constraint arrays within sequential elements, the monotonic trend is conceivably not always increasing from the upper left corner. With that in mind, it would benefit the library support engine that forces monotonic trends onto a power array to use a variation of the constraint monotonic algorithm. The description of that algorithm is repeated here:

> For each corner of the constraint array, if that corner is the minimum of the four corners and neither adjacent corner is equal to it, then force the array monotonic from that corner. Otherwise, if an adjacent corner is equal to it, then if the other adjacent corner is a maximum of the array, force the array monotonic from the equal adjacent corner or else force the array monotonic from the current corner.

The basic reason why this algorithm is inadequate for power is that for most deep submicron technology nodes, power array trends are remarkably flat over the typical input edge rates and output loads at which the stdcells are characterized. It is common to have the minimum and maximum power numbers in an array be within a few percentage points of each other. As a result, any slight measurement error can easily cause what should otherwise be a minimum value to be measured as a maximum value or vice versa. Hence, one must be careful that the local minimum that is chosen from which to force monotonic trends actually is the most likely minimum corner. With that in mind, it is beneficial to not just rely on the values of the four corners of the array but to average the values nearest to the four corners of the array and use these average values as the indicator as to which direction to force monotonic trends.

With the above in mind, the algorithm becomes:

> For each corner of the power array, if the average of the four values that are closest to that corner is the minimum of the similar averages of the four nearest values to each of the four corners and neither adjacent corner average is equal to it, then force the array monotonic from that corner. Otherwise, if an adjacent corner average is equal to it and if the other adjacent corner average is a maximum of the array, then force the array monotonic from the equal adjacent corner or else force the array monotonic from the current corner.

The preceding will allow the reverse-hill-climbing algorithm used by the synthesis engine to better and more efficiently minimize the power used within a synthesized design or at minimum better allow for the understanding of the trade-offs between a further increase in performance versus some resulting increase in power that might be had during some timing optimization.

As to the other data-manipulation techniques that are prescribed in Chapter 8, such as adjusting all timing arrays and timing-constraint arrays to be purely positive, they are much more improbable to occur within the values of power arrays. No stdcell is going to be designed to use negative power over some portion of its input edge rate and output load range. Hence, if a cell tends to show a negative value anywhere within the power array, it will tend to show it everywhere within the power array, most probably indicating that the nodes or device across which the SPICE current measure command was set in the characterization SPICE netlist are connected opposite to the current flow. If something like this occurs, it is likely fixable by negating all the numbers in the array (thereby, converting the negatives to positives). At any rate, it should never be necessary to convert some subset of negative measure power numbers to positive (or zero) power numbers.

9.6 BEST-CASE AND WORST-CASE POWER VIEWS VERSUS BEST-CASE AND WORST-CASE TIMING VIEWS

There is a basic difference in how timing measures and power measures should be handled for a design. What is surprising is that the reason for this difference is that process errors, which are the underlying cause for changes in both timing and power and are based in random instances, manifest that affectation in fundamentally divergent means. Hence, how those process errors need to be interpreted also diverges.

Figure 9.5 shows a representation of how these instances might be randomly found across a design. The blue regions are large arrays of digital or analog transistors or

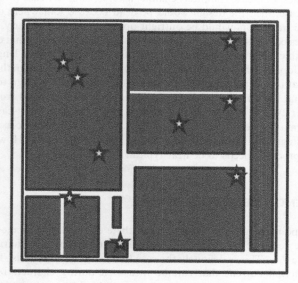

FIGURE 9.5 Possible "Fault" Distribution Across a Design. Because any cone of logic could have an extremely fast or extremely slow gate, all cones should be run against best-case and worst-case performance liberties. But because the small number of extremely fast or extremely slow transistors is swamped against the huge number of near-nominal case transistors, running anything close to best-case or worst-case power liberty results in extremely unrealistic power-usage numbers.

maybe RAMs, the vast majority of which are processed to typical process standards for that fabrication house. The yellow stars are random small process errors (or best-case or worst-case processed transistors). The modified processing could be the result of random faults in the crystal or endemic issues, perhaps lithographic, that occur whenever one set of polygons happens to be adjacent to others on the design. Nonetheless, these random instances might be over dozens or even hundreds of transistors, but they are relatively small localized regions. They can occur anywhere in a design. Because any particular cone of logic might contain some best-case processed transistors whereas the clock tree might contain some worst-case process transistors, then every cone needs to be examined for short-path hold-time violations. Similarly for worst-case processed transistors in any particular cone of logic with best-case transistors in the clock tree, every cone of logic needs to check for long-path setup-time violations. This is why best-case and worst-case timing views are needed by design centers and why these best-case and worst-case timing files must contain real best-case and worst-case timing measures, however those limits are defined.

However, that analysis does not hold for power views. Because the instances of process errors are random and their numbers exceedingly small compared to the entire design, if actual best-case and worst-case power measures for each transistor are used then the designs that use these files will see unrealistically large and unrealistically small power numbers. Instead, the change in the power consumption caused by the small number of process error transistors in a design, assuming that the design does function, will be swamped by the typical corner power-consumption numbers for the vast majority of the transistors on a design. Specifically, using best-case power measures to calculate the power of a design assumes that every transistor in the design has best-case (thinnest) oxide thicknesses, best-case (thinnest) channel lengths, and best-case (lowest) dopant. If this were the case, then the design would be dead because somewhere one or more of those best-case transistors would be outright opens or shorts. Similarly, for worst-case power measures being used across a complete design, which assumes that every transistor is worst-case (thickest) oxide thickness, worst-case (thickest) channel length, and worst-case (highest) dopant, which would mean that somewhere on the design there is assuredly a transistor that is nonfunctional which, would cause the entire design to be dead.

I suggest that the power views being included with best-case and worst-case timing views be either typical process or just slightly best-case and worst-case process (for instance, 1-sigma) power views. Do these power views still need to be best-case and worst-case voltage and temperature? Of course, the answer is yes. However, the third axis—that of process—needs to be near typical.

9.7   EFFICIENTLY MEASURING POWER

It was earlier pointed out in this chapter that there exist charge-storage–capable nodes internal to the cell that are potentially isolatable and potentially chargeable in many stdcells (Figure 9.4). This is definitely the case for classic Boolean cells. This means that the total I-internal current that can be measured during characterization on such cells can vary, depending on the stored charge on these internal nodes. As mentioned in that discussion, this is because sometimes the discharged node will divert some current that would otherwise be used in the resolution of the resulting logical state in order to charge up. If the isolated node already had a charge, then it would not have needed to diverge some of the current. Similarly, sometimes isolated but charged nodes, when the inputs change, now disperse the charge adding to the measured I-internal current that would not be present otherwise.

How much current is seen during characterization depends on the ordering of the changes of the input signals in the test vector applied during that characterization. Hence, what is reported in the typical liberty file power might be inaccurate. What is needed is to build the power test vector for each state in a manner that ensures that nodes are charged or discharged in the most pessimistic manner for worst-case liberty power arrays. Similarly, test vectors should be built for each state in a manner to ensure that nodes are charged or discharged in the most optimistic manner for the best-case liberty power array, consistent, of course, with the manner in which the cells will tend to be used. Typical case liberty tends to be characterized such that the nodes are charged or discharged as in the more pessimistic worst-case manner. The usually attributed reason is the natural tendency of engineers to expect the worst.

How this is accomplished whenever it is recognized as an issue that requires addressing is usually by brute force. Take the AOI22 illustrated in Figure 9.4. To measure the largest worst-case (maximal) current discharge through a rising edge on either signal A or signal B, it is first important to allow the node between the N-channel transistors gated by signal C and signal D to be charged to whatever value it can hold. Strictly after that node is charged, then simultaneously switch signal C and either signal A or B, whichever is the signal whose power measure is to be made. In this way, the added charge of the node between the two N channels controlled by signals C and D will travel up to the output node before being channeled down through the stacked N-channel transistors controlled by signals A and B. On the other hand, to get the best-case (minimal) current discharge through the same input controlling signal, it is typical to leave signal C static, thereby not allowing that internal node from contributing charge to the draining current path.

In general, the following guidelines can be applied to such power characterization efforts.

- For inverters (INVs) and buffers (BUFs), there simply are no isolatable nodes, hence the vectors are sufficient.
- For simple NANDs, NORs, ANDs, and ORs, the test vectors "by simple good fortune" usually are sufficient to ensure that all isolated nodes are correctly charged and discharged for proper power characterization.
- With classical Booleans, for worst-case (and typical-case) power characterization:
  - Care must be taken to charge isolatable nodes in the N-channel transistor stack during power-charging vectors by precharging the nodes appropriately.
  - Care must be taken to charge isolated parallel P-channel nodes during the power-charging vectors by predischarging the nodes appropriately and then simultaneously discharging them during the measurement test vector.
  - Care must be taken to discharge isolatable nodes in the N-channel transistor stack during power-discharging vectors by predischarging the nodes appropriately (this is the case illustrated in the text).
  - Care must be taken to discharge isolated parallel P-channel nodes during the power discharge vectors by precharging the nodes appropriately and the simultaneously charging them during the measurement test vector;
- With classical Booleans, for best-case power characterization:
  - Isolate any isolatable nodes.

- For sequential cells power characterization and worst-case (and typical case) power characterization:
  - The issue, although not completely intractable, is usually conceptually difficult enough to prevent adequate analysis (thus causing the analysis usually to be "swaged" by scaling the measured generic numbers (generated by generic test vectors) by a small factor.
  - For best-case characterization, again measure with generic vectors and use as is.

How does one supply such power-measurement vectors? Sufficient staring at the various schematics of the various cells will allow for the manual building of the appropriate vectors. This is one reason why library vendor offerings tend not to be too dramatically different from technology node to technology node. Any new schematic change to a stdcell requires reanalysis of the vectors. If, on the other hand, the schematic structure remains consistent between technology nodes, with only the transistor widths and lengths changing, then the power-measurement test vector that was previously used at the earlier technology node will remain valid at the next technology node.

It is possible to build all-encompassing multiple-signal test vectors for performance characterization, sequentially setting the other nodes appropriately before applying an edge to the input signal in question. However, the fact that there is such extensive setup toggling of the other signals (in order to appropriately charge or discharge isolatable nodes) during power characterization usually means that these vectors cannot be stacked. Hence, each input to output power-arc characterization tends to be run in single SPICE runs (albeit with potentially multiple-input edge rate and output load subcircuits). The astute library design engineer says an ever-increasing list of such test vectors in their back pocket for future use.

9.8 CONCEPTS FOR FURTHER STUDY

1. AOI21 timing versus power characterization:
   - As mentioned in the chapter, various stdcells, by their very nature, have various different paths between the power and ground rails connected to them and the functional output of the circuit. As a result, some transitions are inherently faster or slower than others because those transitions permit some but not other of these various paths to conduct or not. This is also true for the power characterization of these cells. However, in general, the paths that are fastest or slowest from a performance point of view are not necessarily the ones that are most or least power intensive. The student or interested engineer might be inclined to study this by taking a broad spectrum of Boolean stdcells—for instance, starting with the AOI21—and running performance and power characterization SPICE runs on them and analyzing the results.
2. Static power as a function of substrate voltage:
   - The performance of stdcells and the power they consume depend on the substrate voltage of the bulk material. Assuming that four terminal transistor models are available, the student or interested engineer might run a series of performance and power characterizations in which the substrate voltage of the N well and substrate are varied and then analyze the results.

3. Real versus measured dynamic power script:
   - There is an inherent doubling of the power consumed at the output of the characterized stdcell because of the nature of the output load. The student or interested engineer might build a script that decomposes the current measurement of the stdcell into its constituent parts and corrects the issue caused by that output load and then reconstructs the accurate timing. Care should be taken not to divide the loading component on internal nets for cells that are multistaged.

4. Isolated node charging and discharging experiment:
   - The student or interested engineer might find it fascinating to build SPICE power-characterization decks that specifically charge and discharge otherwise isolated nets that are internal to the stdcell. The student or interested engineer should then activate these isolated nodes in ways that when they later become nonisolated, either contributes to the power or performance of the stdcell or conflict with the power or performance of the stdcell. The student or interested engineer can then analyze the results. As mentioned in the text, such vectors should be held in a list for future use without having to regenerate them each time.

5. Three-sigma worst-case power liberty:
   - No chip that has a majority of the transistors on it centered at 3-sigma best-case performance (short channel length, thin oxide, low dopant) can be expected to work. The reason given was that statistical variance of these transistor characteristics across the device would all but ensure that some transistors would fail, thus producing dead silicon. The implication was that 3-sigma worst-case power (best-case performance) numbers should not be used in liberty files. The student or interested engineer should attempt to affirm this by taking a 3-sigma worst-case power liberty characterization file and running it on a design and comparing the resulting power estimate from the liberty analysis versus the bench-testable power consumed by the actual device.

CHAPTER 10

NOISE VIEWS

10.1 LESSON FROM THE REAL WORLD: THE MANAGER'S PERSPECTIVE AND THE ENGINEER'S PERSPECTIVE

Pulses die out. In the 1980s and 1990s, that was the mantra of system-on-a-chip (SOC) designs, at least when it came to electromigration (EM) calculations. The capacitive load on a signal was thought to have a sufficiently dampening effect on glitches that combinations of short-path and long-path minterms within a complex cone of logic would not cause considerable toggling of the various signals within itself that might otherwise force EM considerations on the signal. Figure 10.1 shows a worst-case parity-tree example. If all of the inputs to such a tree change at the same time, each signal (with a different delay to the output if the loading was not present) could cause the last stage to toggle up or down before finally settling back to the original state of the output. However, because there was such a load on the various signals, any glitch on an output of a XOR in the tree would tend not to reach a voltage level sufficient to change the next stage within the tree. Back in the 1980s and 1990s, before signal-integrity failures started to occur with much frequency, this was usually not questioned (or checked). Although the preceding analysis was for EM, back in the 1980s and 1990s it also held for noise. Any aggressor net-induced pulse on a victim net would tend to die out as well. For years, typical SOC design engineers did not bother with calculating noise immunity (sometimes known as *signal integrity*) on their circuits. This is true at least to the point of not exploring much beyond possibly looking at the rails of the IOs. Hence, noise additions

Engineering the CMOS Library: Enhancing Digital Design Kits for Competitive Silicon, First Edition. David Doman.
© 2012 John Wiley & Sons, Inc. Published 2012 by John Wiley & Sons, Inc.

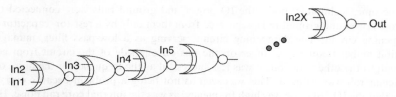

FIGURE 10.1   A Worst-Case Parity Tree. If each input changed at roughly the same time, the output would not change (parity is invariant if an equal number of inputs toggle state). However, without sufficient capacitive loading, such a mass change would cause each stage to glitch, which would cause each succeeding stage to glitch, cascading to a significantly large number of glitches on the final stage before settling back to the original state. This was felt to be unlikely at older technology nodes. Similar but causing a real and measurable effect is the capacitive coupling of aggressor nets toward victim nets.

inside of the liberty file to accomplish internal-signal aggressor to internal-signal victim analysis was not done.

If each input changed at roughly the same time, the output would not change (parity is invariant if an equal number of inputs toggle state). However, without sufficient capacitive loading, such a mass change would cause each stage to glitch, which would cause each succeeding stage to glitch, cascading to a significantly large number of glitches on the final stage before it settles back to the original state. This was felt to be unlikely at older technology nodes. Similar but causing a real and measurable effect is the capacitive coupling of aggressor nets toward victim nets.

One of the first sets of designs on which I worked was a 1.25-micron multiple-chip effort for a government-sponsored program. The program involved the development of seven designs, ranging from a processor and math co-processor to a memory map per crossbar switch. On the math co-processor design (as mentioned, one of the seven individual devices in the entire program) was a 64-bit output bus. It also had a power-on reset (POR) that was used purely to synchronize some controlling flops inside of the Joint Test Action Group (JTAG) test access port (TAP) controller that was used on the device. JTAG had yet to be accepted as an industry standard, so the TAP controller that exists in the standard today is not the one that existed in the company version of the standard a generation ago. At any rate, the JTAG TAP controller that was used was not guaranteed to come up out of start-up in a known state, nor could it lock over some known length of time. As a result, there was an option to have a reset connected to the periphery of the chip that could be used to force a reset on the TAP controller in order to ensure that it settled into a valid initial condition. Although the circuit designers had no concern over such an addition, even suggesting that having such a reset simplified the synthesis effort, the system's engineers abhorred it. They took great pride in not having a need for asynchronous resets. As a result, in many instances, the system engineer would prohibit the addition of the JTAG reset pin on his or her designs. To simplify the design of the JTAG TAP controller, however, many circuit design engineers took to the policy of adding a POR surreptitiously to the design of the TAP, said POR just controlling the pertinent registers in the TAP controller. That was the case for the engineer who built the math co-processor JTAG TAP controller. Unknown to the rest of the design team, a POR was present. The unfortunate coincidence, however, was that the JTAG pins (and, as a result, the controller circuit) was adjacent to the 64-bit output bus. Further, because of a concern over rail-to-rail electrostatic discharge (ESD) failure at such an advanced, for the time, technology node, the IO power and ground rails were connected to the internal power and ground rails, separated from them only by a resistor–capacitor (RC) dampening circuit. The dampening circuit, serving as a low-pass filter, managed to prohibit higher-frequency changes on the bus on one side of the circuit from getting through to the other side, but it was less successful as the frequency of the noise on the particular rail was reduced. This was seen as not being a problem because everybody knew that the IO rail noise was high frequency, as was the internal core rail noise. Hence, the RC dampening circuit was known to be adequate. One more item, because the 64-bit bus had to route to multiple locations on the system, the IO driver strength that was chosen was rather large, even for the time, at 48 milliamps. Knowing this, the layout engineer added extra power and ground IO interspersed within the 64-bit bus, ensuring that there would be no EM concerns and that the rail would not droop downward more than 1 V (it was a 5-V design) even if all the IOs simultaneously went high. Similarly, the ground rail was ensured not to bounce upward by more than 1 V even if all the IOs simultaneously went low. Although this was judged to be bad, it was accepted. Knowing this was a problem, the test engineer built an initial vector in which all 64 of the IOs were simultaneously "bank" switched from low to high and from high to low early in the system-level test, immediately after system reset. The final piece of the puzzle was when the processed design was put on a test head, with the power supplies not close to the device but at the other end of the tester. As can be imagined, a perfect storm on issues had been found.

The voltage (IR) drop across the tester arm was sufficient so that when the test vector that caused the simultaneous switch of all of the IOs just after reset.

- It was adequate to pull the rails far enough that the previously unknown JTAG POR would reset itself.
- This caused it to send out a system-reset notification (a nonstandard event even for the company nonstandard TAP controller).
- This caused a full system reset.
- As a result, the tester had to wait until that reset finished and reapply the bank-switching test vector.
- This caused the loop to repeat continuously.

Although the real cause of the issue was the inclusion of the POR without system engineering knowledge, something that never occurred again on that program, the happy outcome was the idea that there really are noise events within a design that need to be modeled.

The device, when tested with tighter power-rail restrictions, did not show the same level of susceptibility to noise. The tester assembly was redesigned in order to allow for less IR drop on the power supplies to the test head. What should have happened was the redesign of the chip to be less susceptible to power-rail droop or ground bounce. The chip was released as is.

10.2 NOISE ARCS VERSUS TIMING AND POWER ARCS

For timing, there will be some input-to-output arcs through a stdcell that are faster than others. For power, there will be some input-to-output arcs through a stdcell that are more power hungry than others. In addition, the default worst- or best-case performance path may or may not coincide to the default worst- or best-case power path. Similarly, then, there will be some inputs to a stdcell that are more susceptible to noise pulses and some that are less susceptible.

One can easily see this for sequential stdcells. Noise on a sequential data input, although being required to settle before the clock window sufficiently in order for the data signal to be registered correctly during that clock window, will not otherwise affect the output of the sequential stdcell. However, similar noise on the clocking signal into that same sequential cell, if it were of a sufficient level to cause an actual clocking of data, might easily cause incorrect results to appear on the output of that cell. However, it is also the case for combinational cells (Figure 10.2).

In the previously illustrated AOI22 cell, an aggressor signal coupled to the signal attached to the C input will produce a large but diminished glitch on the output X. An aggressor signal coupled to the signal attached to the B input will produce a smaller and also diminished glitch on the output X. An aggressor signal coupled to the signals attached to A and D will produce a glitch on the output X somewhere between these two levels. These glitches on X can be viewed as increases in the delays (or at least uncertainties in the delays) of the timing arcs from the given input to the output that represent how long such an input glitch takes to settle on the output. However, in addition, the level of the glitch on the output needs to be represented. Figure 10.3 shows the required relationships.

More so, if it is in the same direction as an underlying input change, this delay adjustment can actually affect how the change from input state to output state occurs, either speeding it up (in an additive manner) or slowing it down (in a subtractive manner) (as in Figure 10.4).

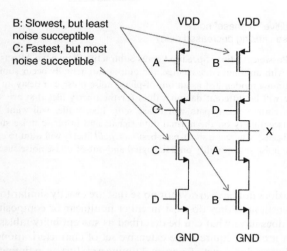

FIGURE 10.2   Inputs closer to the rail are inherently slower to respond. For this AOI22, because the B input controls a P channel and an N channel that are far from the output, it will be slower in response than the C input, which controls a P channel that is adjacent to the output and an N channel that is adjacent to the output. However, the slower response of input B, in an inertial sense, means that random aggressor noise hitting it will victimize it less than a similar amount of aggressor noise hitting input C.

Input "A" glitch height and duration

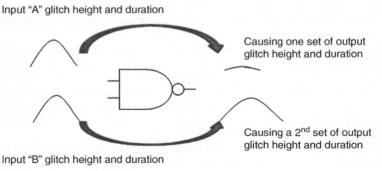

Causing one set of output
glitch height and duration

Causing a 2nd set of output
glitch height and duration

Input "B" glitch height and duration

FIGURE 10.3 Input-to-Output Noise Relationships. Each input-to-output noise-immunity arc will need to comprehend two items: input delay uncertainty to resulting output delay uncertainty and input glitch height to output glitch height.

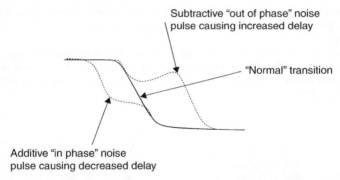

Subtractive "out of phase" noise
pulse causing increased delay

"Normal" transition

Additive "in phase" noise
pulse causing decreased delay

FIGURE 10.4 Possible Additive Speedup Versus Subtractive Slowdown Affects Noise. If the noise pulse is in phase with an input state change, the output can actually occur sooner than otherwise indicated in the timing model, just as an out-of-phase noise pulse can delay an output transition. This means there will be different default values for the liberty files that are slated for different functional design corners. Long-path worst-case *setup* liberty files will want to emphasize out-of-phase noise susceptibility on the normal data signals and in-phase noise susceptibility on the clock signals (and cells), whereas a short-path best-case *hold* liberty will want to emphasize the exact opposite in-phase noise susceptibility on data signal and out-of-phase noise susceptibility on clock signals (and cells).

A liberty file does not have arrays for noise that are exactly similar to those for timing and power (even though they do exist in either nonlinear or composite current model format), but it does have what can be described as susceptibility tables. Specifically, an input-to-output arc will require some extensive set of characterizations that effectively judge the speedup or slowdown of a normal timing arc. Just as in timing and power arcs, each state-dependent set of inputs can affect how large or small this delay adjustment is. Further, as is shown in Figure 10.2, above, these noise susceptibility (or signal integrity) arcs are different from each other, depending on the state of the other inputs to the cell. Hence, the default for each cell will generally be different from the default arc for either performance or power.

10.3 THE EASY PART

Let's look for a second at the output response curve of a typical output of a stdcell to a noise pulse on the input of the same stdcell as a function of the intensity and duration of that pulse (see Figure 10.5). The illustration shows an intuitively obvious set off features of the response of stdcells to changes on the inputs to said stdcells.

- If a pulse does not hit threshold, then it does not affect the output of the stdcell.
- If a pulse, even if it is of significant intensity, is too short in duration, then it does not cause a change in the output of the cell
- Further, the longer the duration of the pulse, the smaller the intensity of the pulse that is needed to produce some amount of output glitch.

A signal-integrity, noise-immunity measure will be a combination of determinations of all three of these features.

Noise pulses, over much of the range of possible input voltages and durations, can actually attenuate. That is to say, not all noise pulses make it through even one stdcell before they get absorbed; stdcells really are inertial for small signal fluctuations.

To determine the first of these three illustrated regions (that region being the bottom of the graph), all that is required is to run a single direct current (DC) sweep SPICE characterization for each input-to-output arc under each meaningful combination of logical values of the remaining inputs of the cell. When this is accomplished, find the threshold voltage for that particular input in that particular condition for which the cell output starts to change. That value determines the amount of noise that can be tolerated on that input without causing any signal noise pulse to propagate any further. This is a good first-pass signal-integrity measure and one that is used commonly in many library

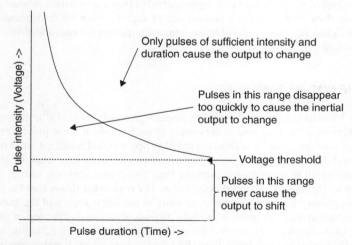

FIGURE 10.5 Output Response as a Function of Input Signal Pulses. Because a stdcell is actually an amplifier circuit, an output of a stdcell exhibits two key features applicable to noise susceptibility characterization. First, the output does not do anything until some threshold is reached. Second, the circuit is inertial and does not respond as a queue. If a signal disappears before the output is affected, then if the output ever does respond, that response is highly attenuated.

offerings' liberty models. More so, such an offering for signal-integrity noise suscepti-bility can be sufficient for many design applications in many technology nodes. Only designs that actually push a technology node in terms of performance are really beyond such a level of signal-integrity measure; for these, a noise-induced delay (out-of-phase noise pulse) can cause an intermittent long-path failure and have tight, dense designs with many opportunities for aggressor net coupling.

A good second phase of the above signal integrity, noise-immunity measure within a liberty is to determine the curve in Figure 10.5 that determines the division between the two remaining regions. This division is between the signal levels above the voltage threshold that do cause some noise pulse to propagate to an output versus those signals above the voltage threshold that do not cause such noise pulse propagation. This requires a large (but bounded) set of SPICE alternating current (AC) runs, the input vectors for those SPICE runs being each combination, obviously subject to the granularity possible, of input intensity and duration in order to determine which do and do not produce any such propagated output noise pulse (of any intensity and duration). Although there is a large set of such runs that might be required to determine this, it is a resource-versus-requirement-bound issue. One determines the granularity of the voltage intensity that is desired as compared to the amount that can be supported (characterized in the time frame available) and, similarly, the granularity of the duration that is desired as com-pared to the amount that can be supported (again, characterized in the time frame available). The multiplication of these two sets of granularity determines the upper bound of the number of AC SPICE runs that are required. This can be significantly higher than the actual number of SPICE runs required, because the actual runs only have to fall along the curve illustrated. Once a SPICE run of a certain input intensity shows that a noise pulse does propagate, no further run is needed of a larger intensity and a same duration, a larger duration and the same intensity, or a larger intensity and the larger duration. Once these various SPICE decks are determined and run, they can be divided by classic methods into two regions divided by a curve with a defined equation. This can be then used to finish a second tier of slightly more valid measure than that previously defined of easily defined noise-immunity regions for each input-to-output arc within a stdcell.

10.4 THE NOT-SO-EASY PART

Figure 10.5 illustrated the small signal-response regions of a typical stdcell circuit. It did not illustrate what the response at the output of such an input noise pulse would be. As one crosses over the curved line from lower left to upper right (increased input noise pulse intensity and duration), there is not a sudden rail-to-rail swing sized pulse on the output. Further, for small input voltages, even for long pulse durations, the output might not reach even to the level of the input noise pulse. The reason for this is that for those low-input noise pulses, the circuit is operating more in the linear region of the transistors as opposed to the saturation region. As a result, the stdcell circuit, in that region really is not operating as an amplifier. Hence, the curve in the illustration is, in actuality, not a hard and fast division of regions but a fuzzy line that indicates that a pulse does propagate from an input to an output when crossed but is typically attenuated in intensity or duration or both. Hence, Figure 10.5 should actually be redrawn as Figure 10.6. Admittedly, as one gets further into the region, this does not hold. Eventually a high-intensity long-duration pulse on an input will cause a high-intensity long-duration pulse

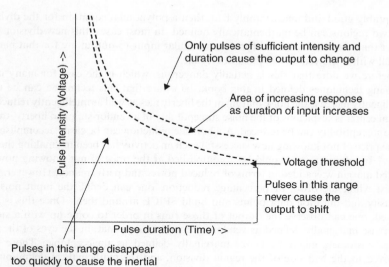

Pulse intensity (Voltage) ->

Only pulses of sufficient intensity and duration cause the output to change

Area of increasing response as duration of input increases

Voltage threshold

Pulses in this range never cause the output to shift

Pulse duration (Time) ->

Pulses in this range disappear too quickly to cause the inertial output to change

FIGURE 10.6   The Curve Is Fuzzy. This is a redrawn version of the small signal response of a typical stdcell output to a small noise pulse on an input. As one crosses the division between no output responses to output responses, there remains a region of the above graph that exhibits some amount of small-attenuated response. This can be calculated (after multiple SPICE measurement) and added to a vendor-supplied liberty (in order to increase the signal-integrity "safe" calculations within a vendor's stdcell offering). The reason that this is not as easy as that of the previous section is that there is a need to examine the output noise-pulse response that is present here, but not in the previous section, in order to see if there is sufficient attenuation input pulse to output pulse.

on the output. However, for the area of the upper right region that is closer to the curve, there is an ability to increase noise insensitivity within a liberty file. However, this region is beyond what a typical library vendor would supply.

As in the easier case given in the previous section where the possible solution space can be granulated up into a bounded number of test points (in terms of noise-pulse intensity and duration), one can do the same for the region between the two dashed lines in the illustration. Doing so and defining a SPICE characterization run for each of those points can allow for a decision to me made. That decision would either include or not include the noise-pulse intensity or duration as being "small enough" based on that particular point, allowing for sufficient attenuation (at least partially because of the inertial effects of the parasitic resistance and capacitance of the stdcell) of the input noise pulse to allow for a design-center customer-defined level of "noise immunity." That is to say, if the amount of resulting output noise pulse is sufficiently smaller or sufficiently shorter in duration (or both) than that of the incident input-noise pulse, then that particular test point (intensity and duration) can be added to the "good enough" range of noise nonsusceptibility. Otherwise, if the amount of the output noise pulse is beyond the design-center customer-defined limit of acceptable noise propagation, then that particular input range of incident noise means that that particular test point (pair of the intensity of a pulse and the duration of the pulse) is added to the "bad enough" range of noise susceptibility. For instance, this occurs when the output noise pulse is larger or longer (or both) than the input noise pulse. Once all such points are divided into acceptably good and unacceptably bad, then a polynomial equation for the division of the two regions can be mathematically defined. In most cases, this new division will be larger than the region defined for that particular input-to-output arc for that particular stdcell within the vendor-supplied liberty file.

Before we note that this is actually dangerous, which is the case for many margin-reducing techniques defined in this book, be aware that this technique can be used to intelligently increase the noise margin in the liberty files as well as intelligently reduce it. If a design center is suspicious of the noise susceptibility of a vendor-supplied liberty, then that noise susceptibility can be reduced. A wholesale reduction can be easily accomplished, but at the price of not knowing how successful such an activity has been in "making the liberty safer." In addition, this would be accomplished at the price of not knowing how much wasted margin was added in terms of reduced power and performance. However, instead of just wholesale offset or percentage reduction, one can define the input noise pulse intensity and duration data points and build SPICE around them. Once this is accomplished, one can observe the output of those runs in order to come up with a subset of otherwise marginally defined as vendor-acceptable points that, in the eyes of the design-center customer, might be better marginally defined as unacceptable. Those can be removed to the bad side of the region division, and the mathematical definition of the division between the two regions can be recalculated. As such, a more noise-immune liberty can be had without wholesale performance or power margin loss.

10.5 CONCEPTS FOR FURTHER STUDY

1. Worst-case pulse-train Verilog and SPICE experiment:
   - A series of EXOR (or EXNOR) stdcells can be strung together linearly. This can be done in such a way as to allow for concurrently applied input transitions to force the inter stdcell nets, especially at the end of the linear series of stdcells to toggle back and forth the number of times that corresponds to the number of such stdcells in the sequence. Inertial-based logical models will tend not to show this, whereas SPICE runs, assuming sufficiently small parasitic, can indeed show this. The student or interested engineer might be inclined to test this hypothesis by building such a network and running it in SPICE and in Verilog (once with inertial dampening turned on and once with it turned off).

2. POR brown-out reset experiment:
   - Some poorly designed POR analog circuits can suffer from power-line droop-allowed reset. The student or interested engineer who has access to a small-scale integration (SSI) level analog stdcell library might be interested in running a series of SPICE simulations of the various PORs in that library. This would be done such as to see just how far down voltage drain drain (VDD) has to droop (or how far up ground has to bounce) in order to allow the POR cell to think that it was powering up from the off state and sending a resulting reset signal.

3. AOI22 noise-analysis experiment:
   - The chapter illustrated an AOI22 having supposedly significant noise insensitivity between its four transistor pairs, each tied to its four inputs. The student or interested engineer should never take such a statement at face value. The student or interested engineer is encouraged to confirm this by means of SPICE simulations.

4. Subthreshold SPICE DC run analysis:
   - The student or interested engineer should set up a series of SPICE DC runs for a subset of stdcells in order to confirm there are noise pulses on one or more of the inputs to the cell in phase or out of phase. The purpose of this effort is such that those pulses, whatever their duration, do not affect the performance or power consumption of the cell when those pulses do not meet the various input thresholds of those particular stdcells.

5. Above-threshold SPICE AC run analysis:
   - The student or interested engineer should set up a series of SPICE AC runs for a subset of stdcells. This would be done in order to confirm that noise pulses on one or more of the inputs to the cell, even when those pulses go beyond threshold, do not affect the output of those particular stdcells as long as, for a given input pulse intensity, the duration of that pulse does not go beyond a certain limit. The duration as a function of intensity should be studied in order to determine a mathematical function describing the resulting curve.

6. Noise pulse "fall-off" (intensity and duration) algorithm:
   - The student or interested engineer might find it interesting to further the preceding input intensity and duration study in order to build an algorithm that indicates when such input noise-intensity and noise-duration combinations result in output-noise pulse intensity or duration that is sufficiently less as to cause significant noise attenuation through the subsequent stdcell.

CHAPTER 11

LOGICAL VIEWS

11.1 LESSON FROM THE REAL WORLD: THE MANAGER'S PERSPECTIVE AND THE ENGINEER'S PERSPECTIVE

Years ago, I was working as a library engineer for a data-communications division of a large well-known microprocessor company. At the time, our division was using an internally developed wrapper around an industry standard logical language. This was common practice for the company at the time, either with internally developed tools themselves or internally developed wrappers around most external tools that the company used. For instance, for one slightly amusing aside, they were seriously concerned that their engineers would have difficulty adjusting from English to metric measure, so they converted all English measures to "nearly" metric by assuming just 25 "nearly metric" millimeter units to the inch, which led to the development of the *mocron*, which was slightly bigger than the micron.

At any rate, the computer-aided design (CAD) engineer who was in charge of developing and supporting the logical wrapper had just completed a postgraduate degree in finite mathematics and decided that it made eminent sense to use these finite mathematic techniques to build the wrapper. This would have been fine except for the construct that was employed for building clock cycles within the wrapper. The algorithm that he used applied the clock at X/Y of the way through the cycle, with $X < Y$, and then ran the remaining cycle assuming a length of time of $1 - (X/Y)$. This was all fine with floating-point operations and it was all fine with assumed integer assignments for X and Y for integer (finite mathematics) operations. He tested the algorithm, and it always worked

Engineering the CMOS Library: Enhancing Digital Design Kits for Competitive Silicon,
First Edition. David Doman.
© 2012 John Wiley & Sons, Inc. Published 2012 by John Wiley & Sons, Inc.

exquisitely, yielding the correct operation, according to his test, and an order of magnitude increased cycle times as had previous versions of the software wrapper. That particular division even supported customers outside the company and designs for more than two years with this wrapper with no reported issues.

Unfortunately, he never tested it with a clock cycle that was not an integer. The division that I was in used that environment on a design that had such a noninteger clock. Within a couple of clock cycles, the combined resulting error between X/Y and $1 - (X/Y)$ started to noticeably add up. When we reported the issue in a bug report, it was assumed that we had been doing something wrong. After all, the code had been working on multiple internal and external designs for more than two years. The division kept telling us to investigate and fix our design methodology, which was obviously the issue. It took my division a couple of weeks to figure out the issue, and we called a meeting with the supporting division, including the engineer who had written the wrapper. When we conclusively showed that the cause was the inability of the integer math operations on the clock cycle, the engineer said "Don't do that." Ever after that, he was known as the Henny Youngman of the CAD division.

11.2 CONSISTENCY ACROSS SIMULATORS

There are two main logical view descriptive languages in common use across the industry: Verilog and VHSIC Hardware Descriptive Language (VHDL). (VHSIC stands for very-high-speed integrated circuits (ICs), a U.S. Department of Defense program meant to stimulate integrated circuit design in the early 1980s at technology nodes of 1.25 micron.) Verilog was developed and promoted by Cadence Design Systems, a leading engineering design automation (EDA) company for the last several decades, whereas VHDL grew out of a Texas Instruments internal logical descriptive language called hardware descriptive language (HDL) which was then accepted for development by the Institute of Electrical and Electronics Engineers (IEEE) starting in the mid-1980s. Over the last several years, the EDA open-source movement has accepted Verilog as an open-source logical view standard (just as they have accepted liberty as an open-source timing view standard). The reason that I mention this is that it is clear that these two versions of EDA logical view tools (Verilog and VHDL) will not disappear anytime soon.

In addition to the two main descriptive languages, there are several versions of interpreted or compiled EDA tools for running simulations with these languages (NC-Verilog, Verilog-XL, and even Verilog-A come to mind), and these tools all have similar pedigree and probably will be used by various design teams for years to come.

All of the above divisions, although attempting to maintain the general means by which logical constructs are handled, have slightly different ways of interpreting some of the more esoteric constructs within the logical languages. Tables 11.1 and 11.2 give examples of typical differences that may be found across such divergent logical simulator tools.

The first example is the difference between a model and simulator of an actual combinational function, a multiplex (MUX), which passes valid logic values whenever possible versus one that passes unknowns whenever possible. In the first table, notice the two patterns where the outputs differ (highlighted in yellow). This particular difference can cause issues with typical implementations of Joint Test Action Group (JTAG) modules given that the JATG module comes up in an unknown state, may not have an asynchronous reset, and relies on being able to synchronize itself over several clock

TABLE 11.1 Passing Logic Values Versus Passing Unknowns in Combinational Cells. Passing unknowns whenever possible ensures that designers write test vectors that force resolution, but passing logical values whenever possible ensures that self-synchronizing circuits can reach known states more easily. The shaded nodes once forced a redesign of a Joint Test Action Group (JTAG) TAP controller for a large IC design company.

A	B	S	Incorrect Output = $AS + BS\_$	Correct Output = $AS + BS\_ + AB$
0	0	0	0	0
0	0	X	X	0
0	0	1	0	0
0	X	0	X	X
0	X	X	X	X
0	X	1	0	0
0	1	0	1	1
0	1	X	X	X
0	1	1	0	0
X	0	0	0	0
X	0	X	X	X
X	0	1	X	X
X	X	0	X	X
X	X	X	X	X
X	X	1	X	X
X	1	0	1	1
X	1	X	X	X
X	1	1	X	X
1	0	0	0	0
1	0	X	X	X
1	0	1	1	1
1	X	0	X	X
1	X	X	X	X
1	X	1	1	1
1	1	0	1	1
1	1	X	X	1
1	1	1	1	1

cycles. There is a MUX usually implied inside of the JTAG RTL that passes the control-state machine or the data-stats machine in the JTAG controller long enough to force this synchronization no matter which way it resolves. However, that implied MUX in the JTAG controller has to resolve. If the gate-level implementation uses the first logical implementation given in the table, then it continually and incorrectly (different from reality) continues to pass unknowns and, in the design center will probably be better served.

The second example is an example of an actual sequential cell, a typical FLIP-FLOP, passing legal logical values whenever possible versus one passing unknowns whenever possible. The basic difference between the two columns is that if the clock goes unknown but the signal at the data input of the flop equals the data at the output, then the flop does not change state. Hence, it should not be simulated as an unknown. Here again there are probably no valid reasons to allow the overly aggressive passing of unknowns in the first column of that table as opposed to the second column. When such differences are found, it is beneficial to attempt to rewrite the logical view in such a way as to resolve the

TABLE 11.2 **Passing Logic Values Versus Passing Unknowns in Sequential Cells. Sequential functions are also susceptible to poor unknown handling techniques. Here, just because a clock goes unknown does not mean that the output of the FLIP-FLOP needs to go unknown.**

D	Previous Q	Previous CK	CK	Improper Q	Proper Q
0	0	0	0	0	0
0	0	0	X	X	0
0	0	0	1	0	0
0	0	X	0	X	0
0	0	X	X	X	0
0	0	X	1	X	0
0	0	1	0	0	0
0	0	1	X	0	0
0	0	1	1	0	0
0	X	0	0	X	X
0	X	0	X	X	X
0	X	0	1	0	0
0	X	X	0	X	X
0	X	X	X	X	X
0	X	X	1	X	X
0	X	1	0	X	X
0	X	1	X	X	X
0	X	1	1	X	X
0	1	0	0	1	1
0	1	0	X	X	X
0	1	0	1	0	0
0	1	X	0	X	X
0	1	X	X	X	X
0	1	X	1	X	X
0	1	1	0	1	1
0	1	1	X	1	1
0	1	1	1	1	1
X	0	0	0	0	0
X	0	0	X	X	X
X	0	0	1	1	1
X	0	X	0	X	X
X	0	X	X	X	X
X	0	X	1	X	X
X	0	1	0	0	0
X	0	1	X	0	0
X	0	1	1	0	0
X	X	0	0	X	X
X	X	0	X	X	X
X	X	0	1	0	0
X	X	X	0	X	X
X	X	X	X	X	X
X	X	X	1	X	X
X	X	1	0	X	X
X	X	1	X	X	X
X	X	1	1	X	X
X	1	0	0	1	1
X	1	0	X	X	X

(*Continued*)

TABLE 11.2 (Continued)

D	Previous Q	Previous CK	CK	Improper Q	Proper Q
X	1	0	1	X	X
X	1	X	0	X	X
X	1	X	X	X	X
X	1	X	1	X	X
X	1	1	0	1	1
X	1	1	X	1	1
X	1	1	1	1	1
1	0	0	0	0	0
1	0	0	X	X	X
1	0	0	1	1	1
1	0	X	0	X	X
1	0	X	X	X	X
1	0	X	1	X	X
1	0	1	0	0	0
1	0	1	X	0	0
1	0	1	1	0	0
1	X	0	0	X	X
1	X	0	X	X	X
1	X	0	1	1	1
1	X	X	0	X	X
1	X	X	X	X	X
1	X	X	1	X	X
1	X	1	0	X	X
1	X	1	X	X	X
1	X	1	1	X	X
1	1	0	0	1	1
1	1	0	X	X	1
1	1	0	1	1	1
1	1	X	0	X	1
1	1	X	X	X	1
1	1	X	1	X	1
1	1	1	0	1	1
1	1	1	X	1	1
1	1	1	1	1	1

difference. This may seem inconsequential, but it can be the cause for significant amounts of engineering debug time.

There are many other less esoteric examples in any library offering. Two less esoteric examples are AND functions passing X or suppressing X from one or more inputs when one or more other inputs are low, and OR functions doing similarly when one or more other inputs are high.

In addition, not all fabrication-house library developers know of these differences and hence, especially when that particular fabrication-house library only supports a few or even just one of the above logical languages and interpretive and/or compiled tools, can build the more esoteric parts of these languages into the stdcell functional descriptions.

However, if the logical tools that the design center uses differ from the ones on which the fabrication house develops the stdcell library offerings, then there is a serious possibility that such esoteric corners of the logical language environment can crop up and cause significant issue on a design . Similarly, if the design center has to merge various pieces of intellectual property (IP) from various sources onto a single design and those various sources use different logical tools and languages from the fabrication house, then there is a serious possibility that such esoteric corners of the logical language environment can crop up and cause significant issue on a design. It is the role of the internal library engineer to find and fix (or at least report the issue back to the library or IP supplier) these issues before they become critical to a design cycle.

This is not about how to write a Verilog or VHDL model so that it passes X whenever possible versus suppressing X whenever possible. Instead, this discussion is about how to test written extant models so that such differences, especially across simulators, will become apparent.

11.2.1 Efficient Testing

How do you go about efficiently testing models against the logical tools that a design center uses? It turns out to be more than just running all possible combinations of inputs against each stdcell and input–output (IO) model. The reason for this is that many issues arise on the transition from one state to another. As demonstrated in the previous section, this is especially true for improperly developed *user-defined primitives* (UDPs), sometimes known as *user-defined programs*, that are usually inherent inside of Verilog sequential functions. A UDP allows for extended definitions of what happens as certain signals such as the asynchronous inputs or the clocks go unknown. For example, a worst-case definition might say to have the output go unknown if the clock goes unknown, but a more realistic definition might allow it to hold the same value if the input has not changed from when the last valid clock edge occurred. When properly written, UDP will allow for such logical operation definition. However, this value-to-value transition means that there is more needed than just testing the logical view against all possible input values (while holding all other inputs sequentially through all possible legal values). Rather, it means that the logical view must be tested against all possible transitions from possible legal value to possible legal value (while holding all other inputs sequentially through all possible legal values).

Allowing all inputs to be tested against all possible values is easy. List each value sequentially to each input, changing just one input value on one input signal at a time, and the definition is complete. For N signals and X legal values, the preceding task can always be accomplished in X^N steps using extended gray codes, for instance. However, it is not so obvious for transitions from each state to each other state. However, there is a way out of the entanglement, and the following discussion demonstrates it.

Generally speaking, there are six legal logical values for inputs in most logical languages and tools:

1. driven low (L),
2. driven high (H),
3. undriven low (0),
4. undriven high (1),
5. driven Unknown (X), and
6. undriven unknown, also known as tristate or high Z (Z).

Some logical tools may allow more than this, but the preceding is sufficient for most activities and simulations. When viewed as a graph, with legal values being the nodes of

176   LOGICAL VIEWS

the graph (see Figure 11.1), and with all possible transitions between each legal node being the arcs between these nodes, there are two paths between each node: (1) the transition into the state and (2) the transition out of the state. That means that the figure is a doubly connected graph. Doubly connected graphs always have at least one Hamiltonian path, also known as a Hamiltonian circuits (and in actuality there are usually several), which is a minimum complete tracing of each path from node to node, crossing each path once and only once. Graph theory mathematics books are available that can easily demonstrate this.

Also note that this analysis holds for similar graphs representing simulators that do allow more than these six logic states. In all instances, because it is a doubly connect graph, a Hamiltonian path exists through the graph.

Hence, for Figure 11.1, several possible efficient minimal paths trace from each input legal value to each other input legal value. Just one of these possible Hamiltonian paths is

$$0 > Z > L > 1 > X > H > 0 > L > H > L > X > Z > X > 0 > X > L > 0 > 1 > 0 > H > 1 >$$
$$Z > 1 > H > Z > H > X > 1 > L > Z > 0$$

This is just one possible Hamiltonian paths, but it does demonstrate that a pattern of input transitions on a signal can be set up that efficiently tests all possible transitions without repeating any. For 6 legal input values, this becomes 30 required transitions (see the pattern just given). This is all that is needed when combined with setting all other inputs to all possible legal input levels. Hence, it now becomes possible to test the logical views for all possible transitions in a minimum amount of effort. The algorithm would be as follows.

- For each input pin:
  - Set each other input pin sequentially to all possible combinations of legal values.

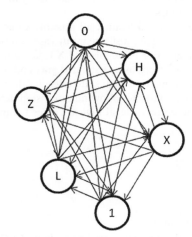

FIGURE 11.1 Doubly Connected Network of All Legal Signals Input Values. Doubly connected networks *always* have at least one (actually always more than one) Hamiltonian circuit. As a result, an absolute minimum spanning trace between all nodes with a return to the beginning node is always possible.

TABLE 11.3   **Required Vectors to Fully Test Logic Views of Wide Functions. Wide functions can cause extreme amounts of test time and resource usage.**

Pins	Vectors
1	30
2	360
3	3,240
4	25,920
5	194,400
6	1,399,680
7	9,797,760
8	67,184,640
9	453,496,320

- While holding these values, transition the given pin through the Hamiltonian path through the legal input values.
- Evaluate the model output for consistency of outputs with proper functionality or against all required logical tools and languages that the design center desires to use.

The given sequence will test all 6 legal transitions for X signals in $30*X*[6^{(X-1)}]$ vectors, which is found by applying 30 vectors to each of X input pins while setting the remaining $X - 1$ inputs sequentially to the 6 possible input states. For many stdcell functions, this is a small number, but this can still be a significant task for larger Booleans, where there are 7 or more inputs that require more nearly 10 million vectors to completely test (more than 453 million vectors for a 9-pin Boolean; see Table 11.3). Although this can be significant, it is not prohibitive, even for an entire stdcell library. So, setting up and running the required test across a stdcell library's worth of functions is a week's worth of effort in order to prevent significant design-center redesign in the future.

The preceding is an efficient algorithm for looking for differences between how a logic view is being addressed in different simulation tools. It is not a method for determining if those logic views are correct. That is dealt with in Chapter 14's section on consistency between views.

11.3   CONSISTENCY WITH TIMING, POWER & NOISE VIEWS

Consider the AOI21 Boolean circuit in Figure 11.2.

The worst-case timing arc across this cell is probably a falling edge on the B input while either A1 or A2 but not both (exclusive OR) are held low. This is because the charge transfer that goes through the transistor that is gated by the signal B that occurs after B goes low must go through either the P-channel transistor gated by the signal A1 or the P-channel transistor gated by the signal A2, whichever the transistor is active on. This charge transfer basically has a resistive path through the B-gated P-channel transistor and a second series resistive path through one or the other of the two P-channel transistors gated by A1 and A2, whichever is conducting, between the rail and the output. This second series resistance is twice what it could be if both of these P-channel transistors gated by the signals A1 and A2 happened to be held low (which would have the

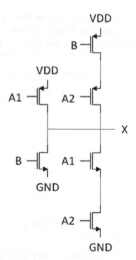

FIGURE 11.2 AOI22 Circuit. Best-case, worst-case, or even default power arcs will be different, in general, from their counterparts in the timing liberty conceptual world. In addition, similar to Figure 8.1 for delay calculation arcs, here different paths to rail can force the need for multiple power-consumption arcs.

two resistances in parallel, yielding half the resistance of either one separately). However, because only one is, this full resistance is added to the resistance in the B-gated P-channel transistor, which provides a large resistance to the charge path to the output capacitive load. The worst-case falling output edge probably occurs when the signal B is held low, and one of the signals A1 or A2 is held high, then occurs as the other of the two signals A1 or A2 is brought high, causing a falling edge on the output. This delay should be roughly twice that caused by a rising edge on input signal B because it allows the series resistance of the two N-channel resistors controlled by signals A1 and A2 to drain the charge on the capacitive output to the ground rail with twice the resistance as that through the N-channel transistor controlled by B.

On the other hand, the worst-case power arc across the AOI22 might be when the B input falls while both input signal A1 and input signal A2 are low, which allows for a parallel resistance across the two P-channel transistors gated by A1 and A2 before that resistance is added serially to the resistance of the P channel gated by B. What this means is that for timing, the path B falling to X rising, with A1 low and A2 high and B falling to X rising, with A1 high and A2 low, are more likely to be specific arcs in the liberty. Meanwhile, the path B falling to X rising, with A1 and A2 low, is more likely to be buried in the default timing arc information. For power, the exact opposite B falling to X rising with both A1 and A2 low is more likely to be a specific arc, whereas B falling to X rising with A1 low and A2 high and B falling to X rising, with A1 high and A2 low, is more likely to be buried in the default arc.

This situation is even more complicated when noise-view support is added to the liberty file. With all that, and realizing that the logical view has to be consistent with all three of these views, how does the *specparam* of a Verilog model (or its equivalent in a VHDL model) be developed such that it will get written to viably during back annotation? This sort of dilemma occurs on many cells in a library. The AOI21 is not a special case.

There are at least two ways that one can deal with the situation. The first is the minimalist approach, which builds a default conditional in the specparam section of the logical model that accepts just the default timing, power and noise arc. Along with that, ensure that the default timing arc, default power arc, and default noise arc are true worst cases for those three axes, although that means that they may not be corresponding for all three (or even for any two) of those axes. Although definitely viable and even having been the way models were written at technology nodes developed years ago on earlier technology nodes, this approach is unsatisfying and definitely wasteful of margin for the cell function inside of the Verilog logical file and the liberty file because of the clumping of arcs together. In addition, this approach is error prone because the default arc for a best-case logical file and liberty timing arc, power arc, or noise arc will probably not correspond to a worst-case timing, power, or noise arc. Further, neither will correspond to the typical case timing, power, or noise arc, hindering cross-validation efforts during logical and liberty build. The second solution, the exact opposite of the minimalist solution, is to ensure that all conditional arcs are enumerated both in the liberty and in the Verilog or VHDL logical model. In the Verilog or VHDL logical view, this approach corresponds to the extended or enumerated conditional case liberty-file approach detailed in Chapter 8. In this way, although the worst-case timing, power, and noise arcs corresponding may not be the case, they will all be extant in the model specparam section, allowing case-by-case correct back annotation.

This may seem effort intensive and potentially expensive, but it is not in actuality. All the simulations of these various conditional runs, in order to measure the various condition timing, power, and noise arcs, had to have occurred anyway (so that it could be verified that the correct worst-case timing, power, or noise arc was included in the worst-case liberty. However, the correct best-case timing default arc, power default arc, and noise default arc were included in the best-case liberty file, and the correct default typical case timing arc, typical case power arc, and typical case noise arc were included in the typical liberty file. Hence, the only additional effort is that these arcs are enumerated and included in the correct resulting liberty files, and the corresponding specparam tokens get included in the correct resulting Verilog (or VHDL) logical files. Yes, the resulting Verilog (or VHDL) logical files—just as in the case of the enumerated condition arc liberty files—can become large and cumbersome, even bordering on the incomprehensible or outright unreadable, with cell definitions continuing for pages of text. However, the effort to create these files can be automatable, is therefore of minimal added effort, and is not an added expense in any way.

One caveat for this extensive approach is that the worst-case delay, power, and noise default arcs that where proposed in the minimalist approach worst-case liberty files and similar for best-case liberty files and the typical case liberty files should nonetheless be appropriately included in the default sections of the each of the corner liberty views for each cell. Although this addition may seem to contradict the preceding note concerning this being error prone, that is only true for the minimalist version of specparam versus liberty verification. The reason is that in this expansive model approach the specparam section is completely defined for all cases and is thus even verifiable by construct. In such cases, the default case becomes just an automated scan of the appropriate field for the appropriate arc (best-case timing, power, and noise for best-case liberty; worst-case timing, power, and noise for worst-case liberty; and, by some measure, typical timing, power, and noise for typical liberty), the logical specparam not being an issue.

11.4 CONCEPTS FOR FURTHER STUDY

1. Verilog–VHDL comparison:
   - Verilog and VHDL are the two most common logical view-construction and simulation environments in use. The student or interested engineer might find it useful to attempt to build a series of stdcell models (using only those constructs commonly used within stdcell models as opposed to those that can be used in a wider and more-extensive RTL netlist). These models should be developed with all state-dependent arcs possible in the two languages in order to see, in their own view, which is easier to learn and use and which is more bulletproof.

2. Unknown handling in various simulators:
   - Many simulators are made to work with either or both of the previous two mentioned logical languages, including those that are compiled and those that are interpretive. It behooves the student or interested engineer to attempt to simulate the logical models of a stdcell library (either those that they built in the above exercise or those that they acquire from an extant library) in multiple different simulator environments in order to see how those various environments handle certain constructs within the logical models. Where there are differences, the student or interested engineer might be inclined to determine the offending constructs within the logical models and adjust them so that the models simulate correctly across all given simulators.

3. Complete unknown listing for the stdcells listed in Chapter 3:
   - The chapter gave the full logical truth tables, including unknown handling, for two stdcells. The student or interested engineer might be inclined to build the complete truth tables for the remaining combinational cells within the tables listed in Chapter 3.

4. Eight-value Hamiltonian path:
   - The chapter gave a complete Hamiltonian path for a six-value logic simulator. Some simulators allow for more than just six states. The student or interested engineer might like to build a similar eight-value Hamiltonian path.

5. Table 11.3 confirmation:
   - The chapter gave a brief description as to how the values were determined for Table 11.3. The student or interested engineer should attempt to reconstruct the table, using the guidelines given in the chapter, in order to confirm those numbers. In addition, the student or interested engineer might like to reconstruct such a table for an eight-value simulator.

CHAPTER 12

TEST VIEWS

12.1 LESSON FROM THE REAL WORLD: THE MANAGER'S PERSPECTIVE AND THE ENGINEER'S PERSPECTIVE

Some of my background has been in the IP development and support of communication-channel designs. One such particular effort more than 15 year ago was for a semicustom clock and data-recovery device implemented in 0.8-micron technology. The device's communication protocol was to function embedded in the clock within the data stream. The data algorithm ensured that a clock-induced data edge would be forced to present at least once every byte-worth of serial data. By monitoring when these edges occurred, one could surmise when the clock and adjust the internal to the device clock in order to align with the assumed clock of the generating device of the data stream.

There were three complicating factors in this design. One was that in order to achieve the bit error rate that was required, the data stream would have to oversample the data stream by a factor of at least 10. This pushed the technology node capabilities, at least for typical synthesized stdcells. Second, the data stream was noisy. Hence, the exact location of the edges as they crossed threshold from a logical low to a logical high and vice versa was jittery. Third, because of the otherwise unknown time of the guaranteed edge in the data stream, the sampling flops in general needed to be highly tolerant of metastability.

The eventual solution was a semicustom placement and hand routing of stdcells in that technology node. The doubled flops (for metastability) were arrayed 16 wide, each pair receiving its own continuously sampling clock of 10 times over the assumed system clock, from minus 3 oversample cycles early to 12 oversample cycles late, assuming that

Engineering the CMOS Library: Enhancing Digital Design Kits for Competitive Silicon,
First Edition. David Doman.
© 2012 John Wiley & Sons, Inc. Published 2012 by John Wiley & Sons, Inc.

the discovered edge would occur at some place between oversample cycle 0 and 9. Because of the noisy signal, the outputs of the doubled lops were best 2 of 3 majority voted. The output of these were compared in order to determine edges; these were then clocked into additional flops that were on the system clock, the output of which was used to adjust that clock, produced by a multiplying phase-locked loop on the device. Figure 12.1 gives an illustration of the previously described semicustom sampling circuit.

Because the oversampling rate of the clocks was 10 times the system clock rate, CK-3 was the same as CK07, CK-2 was the same as CK08, CK-1 was the same as CK09, CK00 was the same as CK10, CK01 was the same as CK11, and CK02 was the same as CK12. Similarly, the data in EDGE00 should be the same as EDGE10, EDGE01 was the same as EDGE11, and EDGE02 was the same as EDGE12, but this was left in order to accomplish implied clock cycle to implied clock-cycle continuity.

Notice that neither of the incoming pair of metastability flops clocked on any of the CK sample clock signals is a scan FLIP-FLOP. Hence, there is no means of inserting scan vectors into this section of the device. Also, note that because there is such a high performance requirement on the systematic sequential clocking of these sampling FLIP-FLOPs, there is no means of easily decreasing the time between each pair of successive sample FLIP-FLOPs. As a result, this circuit, which is critical to the functioning of the data stream, which in turn is critical to the adjustability of the internal multiplying

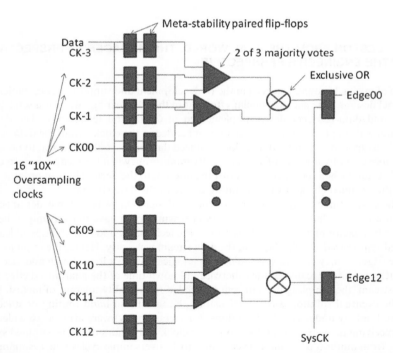

FIGURE 12.1   A 10× Oversampling Circuit Used for Clock and Data Recovery. The circuit had to ran as fast as the particular technology node, 0.8 micron, could achieve. As a result, the FLIP-FLOPs on the input side of the circuit could not be scan inserted (the scan MUX would have limited their performance and the metastability resistance). Further, because of the very nature of the clocking of the line of input FLIP-FLOPs, the structure could not be adequately covered with functional test vectors.

phase-locked loop (PLL), and hence the actual function of the device when taken in total is not testable in any manner. It cannot be scan inserted, and it cannot be functional tested.

The system clock can be slowed down sufficiently in test, and the length of time between the successive oversample clocks can be similarly delayed so that functional testing was possible. However, because of the critical need for tight "coupling" between what the internal multiplying PLL could see from the 13 EDGE signals back to the 10× oversampling clocks, there was no possibility of actually doing so. Even if that were possible, although such items as completely broken transistors and completely severed routes would be detected (as "stuck at" faults), poorly functioning (read slow) transistors and performance-limiting resistive nets could not.

The only way for this circuit to be adequately tested was to build a high-performance data input stream, emulating the actual communication channel into the input of the device, and to statistically determine how quickly the PLL could lock onto the data stream as it was adjusted faster or slower. This was *not* considered an optimal solution.

Although such instances do happen in design, it is the responsibility of the intellectual property (IP) development and support organization to adjust the stdcell logical models sufficiently to minimize such complications. One means of doing so is to develop a second set of logical models that directly map to the functional nodes of the individual stdcells [or hard blocks at either the medium-scale integration (MSI) or the large-scale integration (LSI) level] that will be used in the supported design center's efforts. This means that the Verilog or VHSIC hardware descriptive language (VHDL) logical views that come with the external vendor's IP, which may not support such an effort, may required replacing or at a minimum require being added to with a second set. This chapter will explain the issue.

A Venn diagram of the "two 2 of 3 majority vote into an exclusive OR" function will show that this is better implemented as a pair of exclusive NOR cells going into a two-bit NOR, as illustrated in Table 12.1. Two of three majority vote circuits require 12

TABLE 12.1 Simplified Majority-Vote Edge Detection. MAJ3 cells into EXOR cells are not the optimal solution.

sample N "A"	sample N+1 "B"	sample N+2 "C"	sample N+3 "D"	2 of 3 Majority Vote on 1st three samples	2 of 3 Majority Vote on 2nd three samples	XOR of resultant 2 Majority Votes
0	0	0	0	0	0	0
0	0	0	1	0	0	0
0	0	1	0	0	0	0
0	0	1	1	0	1	1
0	1	0	0	0	0	0
0	1	0	1	0	1	1
0	1	1	0	1	1	0
0	1	1	1	1	1	0
1	0	0	0	0	0	0
1	0	0	1	0	0	0
1	0	1	0	1	0	1
1	0	1	1	1	1	0
1	1	0	0	1	0	1
1	1	0	1	1	1	0
1	1	1	0	1	1	0
1	1	1	1	1	1	0

function = (A XOR D) and (B XOR C) = (A XNOR D) NOR (B XNOR C) transistors. An EXOR also requires at least 12 transistors. As a result, the "two 2 of 3 majority vote into an exclusive OR" function would require 36 transistors in three stages of logic. The suggested circuit as implied in the table requires two 12-transistor EXOR or EXNOR into a four-transistor NOR. It also requires just three stages of logic. Finally, all of the inter-stdcell signals implied in the circuit in the figure require two loads. The circuit implied in the table, on the other hand, has the outputs of the EXOR stdcells requiring just one load, with the inputs requiring no more than the two loads in the original circuit. Because both sets of circuit are completely made of two-stack logic, the difference is that the function implementation in the table is smaller with 28 transistors with the same number of stages and stack height; as a result, it should easily be faster. Even though I am a proponent of such cells in a stdcell library, they should only be used when they absolutely make sense.

12.2 SUPPORTING REACHABILITY

Before we can discuss supporting reachability and observability, it is important to note what is meant by *design for test* (DFT) in this environment. Design for test can mean many things, such as support by means of added views or added physical features, for the following:

- Formal verification
  - Requirement testing
  - Register-transfer language (RTL) code testing
  - State machine testing
- Functional testing
- Physical verification
  - Signal toggle testing
  - Logic and memory testing

In a later chapter, I will discuss the need for requirement testing as an aspect of requirement tracing. Such a requirement tracing is part of the QS-9000 specification for human-critical circuitry (such a devices designed for airplanes). In it, every given requirement in the specification must have either a signal or cone of logic (or circuit if more analog-like) or group of cones of logic (or circuits) that implement the requested function. Further, that cone of logic or circuit or set of cones of logic or circuits must have a single or small set of test vectors. Also, no additional circuit that has not been called out as complying with a specification requirement should be present, and no added test vectors should be added. That is to say, the specification completely maps to the circuit, which completely maps to the test bench, and these mappings are auditable. Such effort may become more and more of an aspect of future design as more and more human safety and health critical designs go into design. And, certainly, this aspect of testing been a too long overlooked aspect of test (except, possibly, in government programs). It is important to know what test vector maps to the testing of what part of a circuit that was developed in order to fulfill what specification requirement. However, that discussion will occur in the requirements section of the book. Further, it is not something that can be directly supported by logical view manipulation (other than the need for comment tagging within the logical view).

RTL code line-by-line testing is somewhat similar to the last step of the requirement testing aspect of DFT just mentioned. Such effort implies that a test vector or small set of test vectors usually set aside in a modular and easily insertable set of vectors can be generated so that they completely exercise a line (or block) of code within the RTL. Although such code testing is important to software test coverage in particular, and with the similarity of functional RTL to actual software, it might be assumed that such a technique in test-vector generation would be usefully applied to that RTL. But as the Venn diagram in the lessons learned section of this chapter might indicate, the gate-level postsynthesized netlist that is generated from the RTL might be vastly different from the actual RTL line of code from which it came. To illustrate this further, a blatant line of code of the assumed RTL for the Venn diagram in the previous section would look something like the following:

EDGE_D = (AB + AC + BC) XOR (BC + BD + CD)

However, the final line of code, as given in Table 12.1 is as follows:

EDGE_D = (A XNOR D) NOR (B XNOR C).

The amount of timing-performance pressure applied to the synthesis tool during the gate synthesis will determine the level of simplification that the synthesis tool does in order to find a shorter and higher-performance solution. The test vectors required to cover all of the signals in the first equation are different from those required to cover the signals in the second equation.

Another example is given in Appendix I. If one were to write a piece of RTL code for a four-bit by four-bit addition and were to synthesize it, then one might get a string of one HALF_ADD and three FULL_ADD stdcells, which would be a completely legitimate gate-level synthesis. Alternately, if one was to write the RTL codes better to emulate a flash add function, with the generation of propagates and generates as in the appendix, one would more likely get the circuit in the appendix. Both are valid solutions to a four-bit by four-bit add function. The string of HALF_ADD and FULL_ADD stdcells is smaller but slower. The circuit in the appendix is faster but larger. However, the test vectors required to cover all of the signals in the larger faster circuit are vastly different from the vectors required to test the signals in the HALF_ADD, FULL_ADD circuit.

In general, code coverage software testing is not sufficient for RTL testing and is insufficient to cover the postsynthesis netlist version of that RTL code. Beyond that, as discussed previously with requirements testing, this is not something that can be directly supported by manipulation of the logical views of the stdcells. There are formal verification tools available that will determine if a netlist in one format, perhaps with skewed clock delays on long paths, is a match for a netlist, cycle-by-cycle and function by function, with another, perhaps without such skewing. This is important to the design process, but it is not adequate to supporting the testing of the digital circuit.

The third subcategory listed earlier in this section is state-machine test coverage. State-machine test-vector coverage tools accomplish for state-machine RTL what software code coverage tools do for RTL code. However, they do this for a valid reason and in an environment different for straight code coverage. To adequately test a finite state machine, it is important to ensure that all of the paths through that state machine have been exercised. This is not adequate because one has also to ensure that all signals that might be generated within each state of each path of that state machine, but it is necessary. As a result, state-machine testing tools that ensure that all paths into and out of each state of a state machine are exercised is of great value even if the second requirement—to test for each signal generation for each state—suffers from the same problem as the RTL code coverage issue previously mentioned. However, once again, this is not such an effort that can be helped by logical view manipulation.

All of the preceding categories fall somewhat close to the realm of formal verification, although none completely encompasses that role. Formal verification can be thought of as the means to ensure that a gate-level implementation of a function is, in fact, the intended function. Engineering design automation (EDA) tools have been developed such that they can compare gate-level implementations with RTL (or other higher software description) for compliance with each other. The reason that this has become a necessary step in modern integrated circuit design is threefold:

1. Because higher (more abstract) levels of coding have become more and more prevalent, as compared to strict RTL coding, there is a need to ensure that all that is implied in such abstraction is present at the significantly lower-level gate implementation, not necessarily obviously.

2. With the increase in the implementation of performance-enhancing postsynthesis techniques, such as useful-skew cycle sharing, buffer insertions, and drive-strength substitutions, the otherwise demonstrably remote possibility of introducing functional variance is no longer such a remote occurrence.

3. The sheer size of modern integrated circuits, in terms of gate counts, makes visual inspection impractical.

However, the library view support of formal verification is not what is at issue here.

The next major subcategory is that of pure functional testing. Although this may seem just a minor step from the previous subcategories, being similar in whatever block is being functionally tested to what the state-machine testing accomplishes within a finite state machine, it actually is much closer to the real subject of this chapter. Here, an engineer familiar with the functionality of a block of circuitry within a chip writes test vectors to see it function "as it was designed to function." Note that this is not test-bench logical testing during development, where the goal is to write test vectors near the edge conditions of the specified operation of the circuit in order to see if the circuit can be made to fail at those edge conditions. The goal in that arena is to apply legitimate signals in unexpected manners and times, such as multiple simultaneous interrupts into an interrupt register in order to see if the register can adequately handle them. The goal of functional testing is to demonstrate functionality of an assumed logically complete and correct circuit. If the specification for the circuit requires that circuit to operate in some manner under some condition, then the test vector, with the condition satisfied, should show the circuit operating in that prescribed manner. Unfortunately, there is a fine middle ground between these two means of test bench development that is sometimes not explored by less than the most experienced test engineers, which is where the actual development of functional test vectors actually needs to be. Here one is attempting to find weak transistors and signals in order to get the assumed otherwise logically correct and functional circuit to demonstrate failure. The goal is to exercise the each signal within the circuit in an attempt to uncover each particular signal not operating as it is designed to operate. This can be extremely difficult test-vector development. In addition, at the size of the modern circuit, exhaustive coverage of each circuit block in the described manual manner is herculean. However, this effort is much closer to what the subject of this chapter. The reason for this is that adequate logical view manipulation can help detect logical failures within the stdcell in the same manner as adequate test vectors can detect failures between the cells.

The next subcategory is toggle testing. Here is a field of testing that can be helped with correct logical view manipulation. Toggle testing is the writing of test vectors with the goal of forcing every gate of every transistor to logical active and logical inactive states at least once each during the multiple-line test vector, perhaps many thousands of lines. However, many stdcells, especially those consisting of multiple stages, can have internal signals that are not necessarily shown within simplified logical views of those particular stdcells. The simplest case is the buffer (BUF) stdcell, with the output being defined as equal to a slightly delayed input as opposed to actually the more representative inverse of the inverse (double inversion) of the input signal. Later in this section it will become obvious that not adding such a complication in the case of the buffer stdcell is not too problematical, but there are other cells, such as a transmission-based mutiplex (MUX), where this type of extended logical modeling does become important.

Toggle testing has gotten a bad reputation over the years when it was routinely pointed out that doing such testing would only uncover approximately 20% of the possible faults in a circuit. This number is correct. However, this number is not a symptom of the inadequacy of toggle testing so much as it is a demonstration of the need to get the chip into a state so that such forcing of logical values onto the gates of each transistor can adequately occur. Further, it shows the need to then force the circuit into a state so that any resulting incorrect output states on the source or drain outputs of those transistors can be seen at the output pins of the chip. The addition of these two criteria, the ability to eventually *reach* a state where a particular failure on a particular signal can be demonstrated if it occurred, and the ability to eventually *observe* such a failure at an output pin is what test-vector development is about.

The last category is logic and memory testing. The key to such activity is the ability to identify those signals between transistors that can become "stuck at" faults, which is every such signal, when they do become stuck or otherwise shorted to another signal. For logic, this is either when a transistor does not function properly, leaving a resulting output either incapable of being adequately pulled to rail or to ground, leaving it stuck in the opposite state, or when another signal is be some means or another shorted to that signal, preventing it from properly functioning. Figure 12.2 shows a characteristic configuration for all CMOS stdcell circuits. Specifically, the output of every stage in every common CMOS stdcell consist of a connection from one or more bottom-most (those farthest from the rail and closest to that stage's output) P-channel transistors to one or more topmost (those farthest from the rail and closest to that stage's output) N-channel transistors. If the P-channel stack is not properly functioning, the N-channel stack is not properly functioning, or another signal is somehow shorted to this output stage, then there is a test vector that can be applied that can illustrate this failure. That vector specifically is the one that would normally force the output signal to the appropriate rail to which it cannot go because of that failure.

For memory testing, there is a pair of such P-channel to N-channel signals per bit (the outputs of each of the crossed inverters in the bit cell). In a failure situation, either one or the other of the P-channel transistors or one of the N-channel transistors cannot not properly function, preventing the bit cell from reach one of the other of the two logical states that it can be forced, or a nearby signal in a nearby bit cell is aggressively forcing an improper state. (This is assuming that there is adequate analog capability in the row and column drivers and the output MUX and sense amp in each bit-line pair.) Such symptoms are completely analogous with the stdcell logic situation just mentioned.

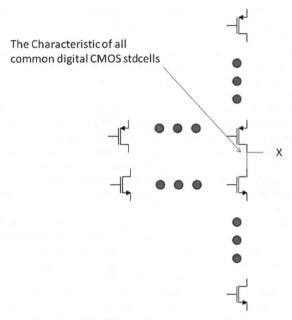

The Characteristic of all
common digital CMOS stdcells

X

FIGURE 12.2 Circuit Characteristic Shared by All Common CMOS Stdcells. Although this may not be true for some analog structures, all digital CMOS stdcells have, as an output of each inverting stage of logic within the stdcell, a signal running between the lowest P-channel transistors in the output stack and the highest N-channel transistors in that same output stack. Invariably, this signal is the output of that stage of logic. If a break occurs between this output and any one of the P-channel transistors or more than one of the P-channel transistors, then there will be at least one "stuck at zero" failure at the output. Similarly, a break between this output and any one of the N-channel transistors, or more than one of the N-channel transistors, will cause a minimum of at least one "stuck at one" failure.

EDA automated test-pattern generation (ATPG) test tools exist that allow for this identification. These tools determine the set of FLIP-FLOP states that must occur to allow a proper set of signals to be applied to the appropriate cone of logic in order to allow an incorrectly functioning signal to demonstrate that improper functionality. In an effort to minimize the identification effort, the tools use the logical views (Verilog and VHDL) as opposed to identifying the underlying transistors by some graphical display standard (GDS) extraction effort. it should be pointed out that these ATPG tools attempt to generate these vectors, which set the state of the machine not just one failure mode at a time but for as many as possible. That way, for a design with 200,000 FLIP-FLOPs, one does not need to scan in and scan out 200,000 vectors. The key to allowing for EDA tools to accomplish their task is to design the logical view of each stdcell not in a manner that is easiest to code or most optimal to simulate of synthesize, but one that most closely matches the stage output signal to stage output signal topology of that particular stdcell. In many instances, this is how one would normally code a logical view. Inverters, NANDs, and NORs as they come from the IP vendor are probably sufficient. However, buffers should be defined as sequential inverters, and AND functions and OR functions should be defined as inverted output NAND and inverted output NOR functions. Booleans, especially those with multiple stages, should be written as ANDs of minterms or ORs of minterms. This is typically not the case. However, not doing so will cause some loss of fault coverage, or at the least, an increase of effort in order to maintain a level of fault coverage. In addition, in such test versions of the logical views, a user-defined primitive (UDP) should not be used. This is acceptable because such UDPs are usually added to handle unknown level inputs, and all signals in ATPG are assumed to be driven. In addition, however, because UDPs are beneficial for logical simulation, this illustrates a need to allow separate logical and test views of the cells.

A second requirement that allows these ATPG tools to function properly is the inclusion of static scan-test–capable registers within the stdcell library. This allows the stitching together of such scan cells as a means to allow for the scanning in to the various registers containing the state of the machine (that is, every possible FLIP-FLOP in the design) the vector that allows for a symptom of a failure to occur. This stitching together of the scan flops facilitates the reachability of each failure symptom, but it also does the same for the observability of these failure symptoms, which is the subject of the next section.

You might have noticed that we have not mentioned the built-in self test (BIST). This is because BIST itself is a logical circuit. Although the methods for implementation, perhaps in a white paper, are certainly possible additions to any library release, there are no required changes to the various views of the stdcells or the memories beyond those normally required in order to allow any synthesis or simulation or place and route.

12.3 SUPPORTING OBSERVABILITY

The previous section discussed the changes to the logical views that allowed for ease of use of EDA-based ATPG tools. Most of the discussion concerned the need to develop signal-accurate Verilog or VHDL simulation models. These logical simulation models will never really be used for logical simulation because of the lack of support for unknowns in the resulting models. This is okay. The original Verilog and VHDL models, which do not need to be signal accurate, will be useful for simulation purposes. These signal accurate models are purely for ATPG test support.

However, one additional feature of the library was discussed: the need for scan-based sequential cells. Such scan-sequential cells allow ATPG tools to scan patterns into the controlling inputs of the cones of logic that are to be tested for failure. They also allow for the capture and scanning out of resulting outputs of those same cones of logic, allowing the individual and independent verification of them. Does a stdcell library require scan registers? The answer is no. Even in a design environment that requires scan insertion, the EDA tools are capable of adding the minterms required to "build" scan chains within the netlist. However, a stdcell library that has properly configured scan FLIP-FLOPs will greatly simplify the insertion task.

Scan insertion has been problematic because of the inconsistency of the timing of the scan and nonscan versions of the various FLIP-FLOPs. Any particular nonscan FLIP-FLOP will have a shorter setup constraint on the data line than the corresponding scan-based FLIP-FLOP of the same function. The reason for this is the obvious addition of a scan MUX between the data input of the scan FLIP-FLOP and what would correspond to the normal data input to the FLIP-FLOP without the scan. The circuitry of this scan MUX causes an extra delay as shown in Figure 12.3.

This extra delay causes additional timing closure issues on the long paths of the design. Specifically, a synthesis tool using just the nonscan FLIP-FLOPs of a library

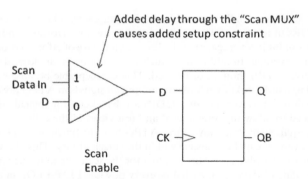

FIGURE 12.3 Why a Scan FLIP-FLOP Has a Longer Setup Than a Corresponding Nonscan FLIP-FLOP. Typically, a scan MUX more or less as illustrated here. It is literally a MUX connected to the regular data input of the nonscan FLIP-FLOP. This added MUX, as is the case for all circuits, has an associated delay that is directly additive to the nonscan FLIP-FLOP data input setup time.

(to prevent the accidental usage of scan FLIP-FLOPs perhaps by allowing the scan MUX to be used to add additional minterms during sais synthesis) will close on timing of the long paths of a design, if it is possible to do so, sometimes with great difficulty. It will do so using the regular nonscan FLIP-FLOPs of the library. However, it will not add any margin beyond the performance goal called for in the synthesis run. The tool does not continue to improve those paths that meet the performance goals as it attempts to finish those paths not yet meeting these goals. Once the performance goal is met by the path, it is considered completed even if more improvement were easily possible. Indeed, because of this, many more than "just" the longest paths tend to be at the limit of the performance goal. This means that when a scan FLIP-FLOP is substituted in for a regular FLIP-FLOP in order to allow scan insertion, many and perhaps most of the cones of logic that were barely meeting performance goals suddenly fail, causing perhaps significant cycles of required optimization in order to recover this lost performance on a large set of cones of logic. This seems at first glance to not be worth the effort just to allow for scanning in patterns to allow reachability and scanning out patterns in order to allow observability.

A shortcut can be developed for this effort by fooling the synthesis tool into believing that the timing for the nonscan FLIP-FLOPs is worse than they really are. Specifically, there need to be sets of liberty files for every pressure, temperature, and volume. The first set, which is used for timing analysis after the actual timing closure, has actual delay and constraint numbers for all cells subject to the adjustments given in the timing views chapter. The second set, which is to be used during synthesis and prescan insertion timing closure, has actual delays for all combinational cells but has had the scan FLIP-FLOP constraints (and clock-to-Q delays) substituted in for the corresponding nonscan FLIP-FLOPs. Figure 12.4 illustrates this. The reason for this substitution is so that the synthesis and timing closure tools think that the nonscan FLIP-FLOPs that are being used before scan insertion have the timing constraints of the scan FLIP-FLOPs that will replace them after scan insertion. Because the prescan insertion cones of logic now have to be closed to the longer scan FLIP-FLOP constraints, the synthesis tools will not stop optimization at what it would have otherwise thought was sufficient timing. In that manner, when the "real" scan FLIP-FLOPs are substituted in for the nonscan FLIP-FLOPs during ATPG, those long paths that were closed will remain closed. Significant numbers of otherwise long paths will more easily close without having to put additional

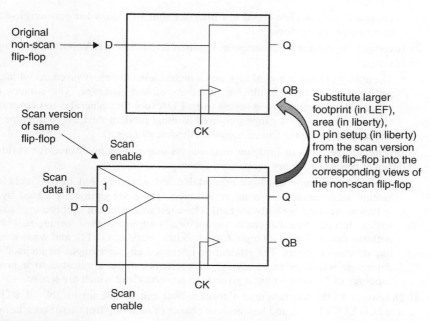

Original non-scan flip-flop

Scan version of same flip-flop

Scan enable

Scan data in

Substitute larger footprint (in LEF), area (in liberty), D pin setup (in liberty) from the scan version of the flip–flop into the corresponding views of the non-scan flip-flop

FIGURE 12.4 Illustration of the Substitution of Scan FLIP-FLOP Constraints for Nonscan FLIP-FLOPs. Such action in a synthesis liberty file better ensures that cones of logic timing that were closed before scan insertion can remain so after scan insertion.

strain on the synthesis tool by requiring all cones of logic to close to faster than otherwise required performance goals.

Due to the typical scan FLIP-FLOP being larger than the corresponding typical nonscan FLIP-FLOP, it probably makes sense to substitute the area and footprint numbers of the scan FLIP-FLOPs in for the corresponding tokens for the corresponding nonscan FLIP-FLOPs in these views, as stated in the illustration.

This method is better than just ramping up the performance requirements during synthesis because this method specifically targets the cones of logic that will most likely be affected by scan-insertion timing issues. Those paths that will not be affected will reach timing closure in the synthesis tools interpretation and stop further optimization. Effectively, this method allows for a surgical solution to a potential timing closure issue rather than the broader "turn up the crank" method. That method can then be specifically applied if desired to allow for performance margin.

Such an additional set of views allows for ease of ATPG scan insertion, thereby allowing for ease of both reachability and observability of failed cones of logic.

12.4 CONCEPTS FOR FURTHER STUDY

1. Design a low-speed test interface for the circuit given in Figure 12.1.
   - The circuit drawn in Figure 12.1 represents an actual circuit that was used in a real device. Unfortunately, the circuit suffers from the lack of ability to test at low speed. The student or interested engineer might be interested in developing a

minimally invasive low-speed test interface that will allow for per-sample-flop injection of test vectors.

2. Investigate signal-toggle coverage as compared to actual EDA-tool defined fault coverage:
   - The ability to force a logical high and a logical low onto every given net within a netlist is notoriously insufficient in terms of test coverage. The student or interested engineer who has access to an EDA tool that generates test coverage rates in the high 90% range (required in most modern devices) to attempt to understand where the forced toggle methodology fails.

3. Investigate the types of bridging mechanisms that can be uncovered for various different types of memory bit-pattern coverage:
   - Most memory suppliers have expandable test wrappers that can be included around their memories (in many instances, these are optionally added by a switch in the compiler software call). These test wrappers can supply a myriad of various test patterns that can be automatically applied to their memories. These patterns can be forced all high, forced all low, walking ONES, and what seems like dozens of others. The student or interested engineer might be inclined to determine which of these myriad of patterns can be applicable to a given topology of bit cells within a given memory array and which are less so.

4. Develop a ROM test coverage algorithm that can handle up to $2N - 1$ ROM words, is LFSR based, and has minimal chance of suffering from error bit aliasing:
   - Wrapper testing of ROM is slightly more complicated than that of testing RAM because for a RAM, a given bit in a given word can be written to a given value in a test vector and then it and the surrounding locations can be read to see if there are issues. A ROM is prewritten. A test wrapper can only read the data from the extent procreation. Ensuring test coverage by using such a method is problematical because it may be difficult to understand if a given bit in a given word is correct or incorrect without having a complete separate second copy of the program code against which to compare it. One way around this is to read the entire ROM array using a linear feedback scan register (LSFR), described elsewhere in the text. The student or interested engineer might choose to design such a ROM test wrapper, such that the verification pattern that rests the LFSR to a known final state, thus assuring testability, is preloaded into the last ROM word location, along with the program code being loaded into the rest of the ROM.

CHAPTER 13

CONSISTENCY

13.1 LESSON FROM THE REAL WORLD: THE MANAGER'S PERSPECTIVE AND THE ENGINEER'S PERSPECTIVE

The views of a library can be thought of as the projections of an item from various points of view. This is similar to the old puzzle-book question of describing what sort of object can cast a shadow that looks like a circular from one side, a square from another side and a triangle from yet another side. Figure 13.1 gives the answer. In terms of libraries, the views that are involved are those that are used by the various engineering design automation (EDA) tools. Each tool has a need for various parts of the description of the elements of the library, but no particular tool needs to care about all the aspects of the library. However, it is important to the success of any design that would use such a library that the various views tend to describe consistent aspects of the library. In the case of the puzzle-book question, the projections tend to be orthogonal of each other. As a result, the shape of one shadow tends to have little consequence to the shape of the other shadows but does have consequence for the actual shape of the object in the figure. In the case of the various views of a library element, the views may not be so orthogonal. A slight change in one view—say, the liberty file description—may require substantial change in one or more of the other views—for instance, the Verilog description.

Much of this book can serve as an illustration of that fact. The Verilog model must match the liberty model in terms of conditionals in the specparam section of the code. They both must match the functionality provided by the SPICE model. That must match the implied functionality of the pattern of polygons in the physical view. The physical

---

Engineering the CMOS Library: Enhancing Digital Design Kits for Competitive Silicon,
First Edition. David Doman.
© 2012 John Wiley & Sons, Inc. Published 2012 by John Wiley & Sons, Inc.

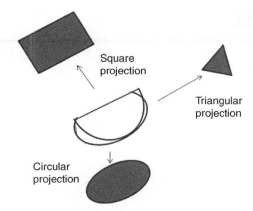

Square projection

Triangular projection

Circular projection

FIGURE 13.1 An Object That Casts Circular, Square, and Triangular Shadows. Because the three projections are orthogonal to each other, they can be dramatically different from each other (here a smoothly curving edge in pone projection is replaced by one with three angles in a second projection and one with four angles in a third projection). This orthogonal allows the three projections to have no correlation between them.

view must match the library exchange format (LEF) view used by the place and route engine. All of these views must play consistently with the verification decks in the physical design kit (PDK).

How these views are tested bears significance as well. Testing of the view is not an orthogonal event. The DV Club is an informal association of design engineers who are interested in monthly discussions on various aspects of design verification, which is what the initials in the name loosely represent. There are chapters in various cities around the country where integrated circuit design occurs.

At a recent monthly luncheon at the Austin chapter, a professor was giving a discussion on a tool that he and his graduate students were developing that could be used to better verify the completeness of the testing of state machines during formal verification.

One feature of the tool was that it first converted the original state machine based in register-transfer language (RTL) and the final gate-level state machine into a third format that it could then parse sufficiently in order to see if the two were equivalent. During the talk, one attendee was rather adamant about the lack of applicability of the tool. The professor fielded the question, but the questioning started me thinking on the subject. After the talk, I approached the professor and said that while I had no issue with the tool, having never used it to that point, I did think that the questioner had a valid point. I offered the following story to the professor in order to illustrate my point.

I need to verify the entire alphabet letter by letter that I see against the golden letter— for instance, the letter E.

- I cannot do that directly.
- I do have a tool that converts each letter into the ordinal number that represents the position of that letter in the alphabet.
- Unknown to me, the first time and every seventh succeeding time that the tool is used on every second Tuesday, it always converts all letters input to it into the number 26.

- On one such Tuesday, I convert the golden letter of the day the very first use of the tool that day, with E now being considered in the tool to be the 26th letter of the alphabet.
- On the seventh use of the tool that day, I input the letter M, and it converted to the number 26 as well.
- The tool now tells me that my tested letter M matches the golden letter E.

The preceding illustration is superficial, but it does illustrate the need for proper consistency checking of views in a library. Using a single tool to convert two different views and then to check these conversions for consistency between them is tantamount to checking the tool itself for consistency over usage, and it fails the stated purpose of using it as a view-consistency checker. To most engineers, this is a hidden fault. It tends to be "masked" to detection "in plain sight." Consistency checking by such means can lead to dramatic failures in devices that use inconsistent library views. Even when the underlying failure between two inconsistent views is detected, postmortem understanding of the correct cause can be easily missed, leading to repeated failure in future generated views on future technology nodes.

This chapter will discuss the various interconnected projections of one view onto another, and how those views can be legitimately checked for consistency.

13.2   VALIDATING VIEWS ACROSS A LIBRARY

View validation consists of an absolute minimum of two steps, the first of which is usually not explicitly thought about. Specifically, although it may be understood that views must match each other where those views overlap (are nonorthogonal), which is the second step, it is not regularly realized that there has to be a "first" or "golden" view. In the case of simple circuits, when the question of what is the proposed "golden view" is asked, it is common to get a reply that they all have to match the function. This reply is correct, but it does nothing to alleviate the situation of how the views need to play consistently together. It is too easy to write a Verilog model that accurately describes a given function and a liberty model that accurately describes the same function and have the two models not be consistent. As illustrated in Figure 13.2, the term *view* is appropriate. Yes, the function is the object that is being described by the various views, but those views are images of the function as projected onto the tools that requires them. Recall from the lessons learned section that these views are not orthogonal projections. They are at least partially dependent on each other. Therefore, changing an item in one view may necessitate changing one or more items in one or more other views. Once one view is defined, all other views, because they are not orthogonal to the first (or any other) view, must be defined in such a manner as to allow for consistent representation of the function in the remaining views without loss of correctness of or consistency with the original view. What needs to occur as an absolute very first step is a blatant definition of what view is the golden view. What this allows is the determination of which of two or more views that may overlap and do not agree in the region that they overlap, is the one to hold constant and which is the one to change in order to comply with the first. This layering of dependence of some views onto other views can be hierarchical, which is to say that an intermediate view needs to change in order to better coincide with the golden view; those views that are required to be consistent with that intermediate view will also have to be modified.

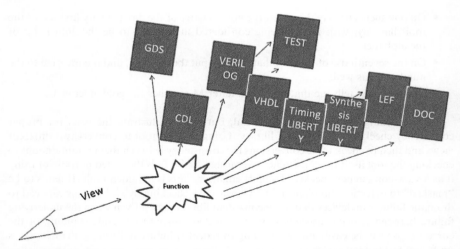

FIGURE 13.2 EDA View "Projections" Versus the Ideal Function. Unlike the case of the object that can cast a shadow that can be a circle, a square, or a triangle, the view projections here are not orthogonal. Because no particular view is orthogonal to any other given view, each has cause and effect relations with any and all other views. The changing of one aspect of one view might, and usually does, cause changes to multiple other aspects of multiple other views. This is the reason that a "golden view" concept is required. When a given view is found to be inconsistent with a second such view, knowing which is the golden view (or closer to the golden view) allows for determination of which inconsistent view to change in order to make it consistent.

Stdcell library developers may disagree with the preceding. They can point to instances where they have had to go back and change many or most (perhaps all) views when an issue arises, but this is actually confirmation of the argument. Starting with a bad choice for the initial view can lead the development into a corner as each succeeding view is hindered until at some point all preceding generated views have to be modified in order to correctly write a required portion of a view.

It is important to pick a good view to be the determinant golden view. Before we discuss how to determine which is best for a given design center, let us divide the views into subcategories. The classic approach is to divide the various views into several realms:

- physical views [including circuit description language (CDL)],
- logical views (the Verilogs or VHDLs),
- temporal views (usually the liberties, including power and noise),
- test views (which can be a subset of the logical views), and
- documentation.

We might assume that the physical views are good candidates for the golden views. The argument goes that they are the actual functions, but this is not the case. A function is almost a Platonic concept. A set of polygons that, if interpreted as the PDK layer list defines them, may represent one of many physical implementations of a given function, but it is not the only such implementation. In addition, a set of transistors interconnected appropriately also only represents a given function as opposed to being a function. The LEF, although a physical view, actually is a derived view from the GDS, and it does not directly affect the argument here.

The fact that the LEF is derived gives a possible means of prioritizing the three physical views. When the polygons of a cell are drawn, the layout engineer (or the automatic layout tool) starts with a circuit CDL and produces the GDS in accordance to the layout versus schematic (LVS) and design rule check (DRC) and architectural footprint criteria defined elsewhere. This is as opposed to the opposite flow of starting with a set of polygons and deriving a CDL from it, here CDL implies preextraction CDL (with no parasitic capacitances or resistances included). While this may seem ludicrous, it is not as far-fetched as, at first sight, it may seem. There have been innumerable cases, over the years where a design center has received a copy of the GDS only, and has had to extract the CDL from it and t to develop the Verilog from that. In general, CDL precedes GDS. Therefore, just as LEF (and postextraction CDL) is a derivation of GDS, GDS is a derivation of CDL. So, even though a given CDL is not a given function, just one of several representations of that function, it really amounts to the base view of the physical views and is the most likely candidate for the golden view if it is determined that the golden view should be physical in nature. The hierarchical prioritization of the physical views amount to the following order:

- CDL (preextraction)
- GDS
- CDL (postextraction)
- LEF

Next on the list are the logical views. As the case with the physical views, the logical are not a given function, but just a set of representations of that function. This should be obvious by the vast array of means to write a model. Not only is it possible to define a representation of a function in various lines of logical code, but also the very structure of the Verilog or VHDL language allows for varying code constructions that can be used to define the representation of the given function. What is more, there are legitimate reasons to define multiple logical representations of a given function using different code constructs. There may be a need to have a Verilog model that uses, for instance, logical coded lines for defining a view of a given function and have a second Verilog model that uses, for instance, primitives for defining a second view of the same given function. The first could be useful in accelerating compiled simulation whereas the second could be useful, assuming that it is written correctly, for test-coverage analysis. In such a case, determining the order of derivation can be difficult because writing models is an error-prone process. It is not just the case that one can look at the last modified date on the files and determine which came first. The dates usually only indicate which was approved last. On the good side of things, if there is a closest representative view to the concept of a Platonic function, then it might just be a logical view, aside from the various specparam additions to the code that are necessary to allow it to interplay with the other various views. The final piece of this hierarchy is that there exists a connection from the Verilog (or VHDL) to the temporal files sine they need to be able correctly to supply data back to the Verilog during both synthesis and timing closure. The hierarchical prioritization of the logical views then amounts to the following order:

- function Verilog or VHDL,
- gate-level (test-bench ready) Verilog or VHDL, and
- liberty (temporal view).

The next sets of views on the initial list are the temporal views. As is evident from the end of the previous paragraph, these have to be derivative. This is also true because they are derivative of the extracted CDL from the physical views. Therefore, these views are based on extractions from the physical and are consistent with the logical and cannot in any way be considered as a golden source.

After the temporal views are the test views, which have already been described as derivative of the logical views. This leaves just the documentation, which just describes the various features of the various other views together with the test and verification of those views and perhaps some application details for use of the various functions.

The best representative view of a function is the Verilog (or VHDL), whichever is used primarily in the design center being supported. From this golden "root," come pre-extraction CDL and test-bench–modified Verilog, from which come GDS then post-extraction CDL and LEF, then Liberty, then documentation. Any editing of a view lower in the hierarchy must be consistent with the upper view in the hierarchy and might cause adjustment to any view lower in the hierarchy. This hierarchy answers the issue of when two views disagree, which view should be fixed? The lower view, hierarchically, must be changed to match the higher view, hierarchically. The one caveat is that the Verilog must be hand reviewed to match the concept of the original Platonic function, even though the Verilog is considered the golden view.

How is this consistency to be checked? One possible flow is given in Figure 13.3. Over the last several years, the various EDA houses have added significantly to the view

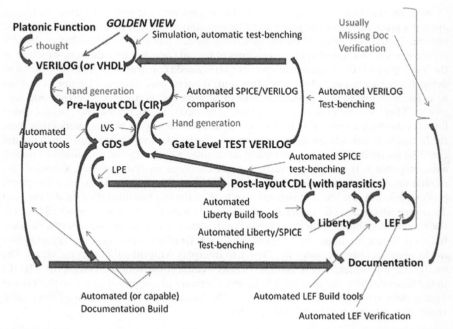

FIGURE 13.3   Representative View Generation and Verification Flow. Although much reduced for illustration purposes, the complexity of view consistency checking between views is represented here. Note, hand checking is possible but should always be backed up by automated checking through alternative paths.

verification side of the figure. This book will not endorse any given generation or verification tool, so the various obvious names of these tools are left off the figure. There are multiple entries capable of verifying in an automatic and verifiably complete fashion. Also, although several hand-generation steps on the view generation side of the flow diagram still exist, notice that in all cases these are matched to automatable verification steps.

The one missing step, documentation verification, is somewhat the result of the perceived lack of need by many a library support organization. Although this is perceived to be unnecessary, in actuality errors can creep in at this step just as easily as in any of the earlier view-generation steps. As it is currently, such as error would go unnoticed and could be in the released database for generations of technology before it is discovered (this is because of the level of perusal of the documentation views by the user community). When such an error is discovered, it may mean that the "error" has become "reality" and has long since been relied on by design teams for several devices and that the library support team must now concoct ways of supporting the originally erroneous issue. This is a value-added place for an EDA company to work on such tools.

Let me stress, as pointed out in the lessons learned section of this chapter, using the same tool to generate and to validate a view is a potentially bad decision.

13.3 VALIDATING STDCELLS ACROSS A TECHNOLOGY NODE

When a histogram of the number of different stdcells that are used in a given design is constructed, a surprising fact usually arises. Design after design, the same small number of cells is used repeatedly. There can be many possible reasons for this, ranging from the RTL coding style of the engineer submitting the RTL netlist to uncontrolled preferred assumptions within the synthesis tool being used. Both causes are testable. The first cause can be tested by using RTL code from a different engineer and see if the high-usage cell list changes. The second cause can be tested by using an alternate synthesis engineer from a different vendor and seeing if the high-usage cell list changes. When both of these assumed causes are tested and proven incorrect, and they more or less will be provably so, then a more likely reason needs to be assumed. That more likely reason is that the library being used is not conformal with itself. There are certain stdcells that are "better" choices, as far as any synthesis tool understands, across a wide range of applications and cones of logic. Certain cells, or combinations of cells, are always smaller, faster, or less power hungry over a large range of possible "standard cones of logic" than any other possible sets. In this condition, the vast majority of the cells in a library will be used a small amount of the time, and a very small portion of the cells in a library will be used all of the time. In many cases, the number of types of cells that are used 1% of the time across a device is less than a few dozen; sometimes it is less than a dozen. Figure 13.4 gives a typical histogram of the cells used in a typical embedded microprocessor design. In it, there are 17 cell types that are used 1% of the time or more, and a full 80% of the total cell count is made of just 72 cell types. Note that the chart is logarithmic. (This is for a real design at a deep technology node, however, the device name and the cell names have been removed in order to protect the design team that generated the data.)

This phenomenon occurs across multiple libraries over many generations of technology-node development. The fact that it does occur explains why there is an ongoing debate as to the optimal size of a library. If only a small fraction of cells ever get used, why waste the time and effort in developing the extra cells, and why waste the users'

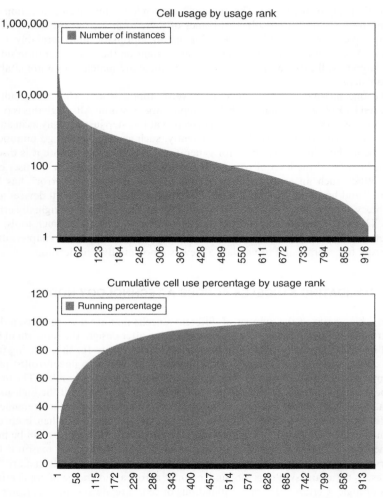

FIGURE 13.4 Typical Cell Usage Histogram for a Library in a Given Design. The so-called rule of 20 (20% of the resources complete 80% of the result) is taken to the extreme in most integrated circuit design. In many cases, the rule is enhanced to where is may become something like the rule of 10 (10% of the cells in a library are used 90% of the time in a device). This trend can be minimized if cells are properly tested against each other with those test-causing changes in the offending cells.

time in having to compile such large cumbersome library offerings? However, this argument is based on faulty assumptions. The paucity of extensive cell usage on a given design is not the result of a small set of cells being inherently better; it is based on the library developers not ensuring that the cells in the library are broadly spread out over the design landscape. The highly used cells are better choices for the synthesis effort but just are not inherently so. Invariably when a library that exhibits the type of behavior mentioned in this section is reviewed, it is generally easy to show where the high-usage cell types are not uniform with the trend of other cells in the same library.

Several axes of analysis that should be reviewed exist here.
First, for drive strengths for a given function:

- Across drive strengths, recall Figure 8.19, when a chart is constructed across output loading. For a given representative input edge rate, there should be defined regions where each of the drive strengths of a function that is in the library offering will produce the smallest delay. If this is not the case, then the offending drive strength will never get chosen and should be either banished from the library or reworked to resolve the issue (which gives the added benefit of defining some level of potential performance improvement for some potential cone of logic). The caveat exception to this rule is if the particular drive strength is specifically defined as a driver for a specific situation. Years ago, a microprocessor company had such cells in a 0.8-micron offering. A specific drive strength inverter, which would never be synthesized into a netlist on its own, was always edited back into any netlist a posteriori as the inverting input to a level-shifter cell. This editing was difficult because of the need to resolve the inversion of a driver signal that the addition of the inverter caused. Although scripts were developed that resolved the issue for all possible stdcell and input–output (IO) drivers of the signal in that technology, the second issue of the added delay on the path was never adequately resolved. Designers with such level-shifter cells in their designs had to synthesize those paths artificially fast in order to allow sufficient time for the postprocessed cone of logic. Based on the awkward manner in which the netlist were always postprocessed, starting at the very next technology node, that microprocessor company integrated the driver strength inversion into the level-shifter cell.
- In addition to the defined region driver strength rule, there should be no increase in area per transistor (inverse of density) for higher driver strength versions of functions over lower driver strength versions of functions. Indeed, the opposite should be nearly true. Density should either remain constant (at worst) or improve as drive strengths of functions increase. At very deep submicron technology nodes, where lithographic patterning of polysilicon gates forces discrete distances between transistors, the trend should be toward constant density. Here again those cells that fall outside of this trend should be either banished or reworked.
- In addition to the preceding two, there should be defined output loading ranges for given input edge rates where each drive strength consumes a minimum amount of power as compared to the other drive strengths for that function. In addition, again banish or rework any offending cells.
- Another issue is routability of higher drive strength cells. It is true that the higher drive strength cell has more area with the same number of pins and should have improved routability capability. It is also true that there is an increased demand on internal routing in order to connect all of the appropriate sources, gates, and drains on the increased number of folded higher driver strength transistors. This increase in internal routing demand can cause some cells to have internal routing on higher metals usually reserved for pure automated place and route (APR) route. This may seem a minimal hazard at first glance. However, realize that if every cell in a library had just one "router-level" metal track taken in order to connect internal-to-the-cell connections, then for an N-track tall footprint, 1/Nth of the router resource would be gone. This resource will be on the lowest router level, which should be the most critical to APR connection of signals to the cell. If the library in question were, for

instance, an eight-track architecture, this would mean that 1/8 (12.5%) of the bottom router metal was unavailable to the router. Care should be given to rework any cell that has such increase in order to reduce or eliminate it. Otherwise, physical synthesis tools might limit the amount of times that the particular drive strength is chosen, and whenever it is chosen the result is detrimental to the route.

Second, for signal inversions within the various functions in general, first enumerate the cells of a library so that they can be defined by both the number of transistors in a design (note that for multistage functions, the last stages are the only ones that get multiplied up for higher drive strength cells) and by number of inversions. For the enumeration of the transistors in a cell, some fascinating differences in the following trend analysis can be uncovered in comparing the number of "functional" transistors (those based on an ideal circuit) versus the number of "actual" transistors (those based on the circuit as used in the cell). This will be refined shortly. In most cases, where functional transistor counts are calculated, it is assumed that the entire transistor count, even in multistage cells, increases with drive strength, but feel free to calculate it either way. Also, for the enumeration based on the number of inversions in a cell, the basic operation of a CMOS stdcell always involves some level of signal inversion so that AND functions and OR functions are actually inverted output NAND functions and inverted output NOR functions.

- The density of a cell should always increase as the number of transistors in it increases. It is okay for the size to increase, but as transistor count grows, they should be able to be packed closer together such that the cell will generally grow either at the same rate as the transistor count increases or as the inversion count within the cell increases or at a slower rate than at least one of these. This general rule does eventually break for deepest submicron design technologies, where lithographic requirements force fixed polysilicon gate spacing, but even in that instance, the density never decreases as the transistor count increases. Just as before, cells that fall outside of this rule should be removed from the library (or reworked to bring them better aligned to the rule). This falls into one of the categories that should be viewed across both the *actual* transistor counts and the *functional* transistor counts. Although the density of the actual transistors might be okay, realize that the marketing number that will be used by the design center to advertise the density of the design, and that will be advertised by an EDA company to demonstrate the routing capability of the APR tool, is that of functional transistors per unit area.
- Delay through multiple inversion functions should be slower than delays through single inversion functions but faster than through combinations of single inversion functions followed by inverters. As an example, a buffer (BUF) cell should be slower than a similar drive strength inverter (INV) cell but faster than two INV cells, the second of which is the same drive strength as the BUF cell and the first at a size capable of driving the second. Also, the AND should be slower than a similar drive strength NAND cell but faster than a NAND driving a similar drive strength (to the AND) INV cell. As was mentioned in the characterization section of this book, this is where the sheer number of possible cell combinations becomes nearly intractable. Luckily, several small EDA companies do have such library analysis tools that can facilitate such an effort on the market. Many of these analysis tools

can do so across multiple dimensions (for instance, reviewing output delay as a function of both input edge rate and output load simultaneously). Although this is the difficult validation task, it is critical that cells that do not follow the trend are identified. Without this task, designs that use the resulting library will exhibit the "small list of high-usage cell" phenomenon. This task, combined with either banishing or fixing the cells that are shown to be outliers, will significantly minimize this phenomenon.

- Dynamic power consumption, as defined in the liberty files, at a representative loading per cell and across all cells with a similar drive strength, should follow the same trend that the more the total amount of transistor width (assuming actual transistors) in the final drive strength scaled transistors, the larger the dynamic power. Cells that fall outside this trend should be reviewed. They may indicate significant design flaws.

- Static leakage power should always track total transistor width (again, assuming actual transistors). Here also is an indication of serious design flaws on those cells that do not track with this.

- Output edge rates for all cells with similar drive strengths and with similar numbers of stacking N-channel (or P-channel) transistors should have similar falling (or rising for the P-channel) output edge rates. Review and possible adjustment of cells that do not follow this trend should be undertaken.

- Finally, input capacitance should be always nearly consistently minimal. Realize that this has to be the case. Nearly regardless of the function, an input signal will connect from the router level through a small amount of interior to the cell metal, through a contact (or local interconnect, depending on the technology node) to the gate of a P-channel and an N-channel. Very few functions require an input signal to connect to much more than this, perhaps a MAJ3 or MAJ5 cell (see Chapter 2 on stdcell design) will require driving more than this. Any cell with a pin that does not exhibit such nearly constant "across the library input pin capacitance" range and that cannot be otherwise explained such as with the MAJ3/MAJ5 example, should be reviewed and fixed. The one caveat here is that single-stage functions input pin capacitance should grow linearly with drive strength. Here, any cell family that has an input signal that does not exhibit this inverse trend should be reviewed.

- Clock signals on sequential cells will have even less loading (in all technology nodes except the deepest submicron because of the unit widths and lengths of most polysilicon gates at such technology nodes) because they usually drive a small inversion of a double buffering inside the sequential cell FLIP-FLOP.

With the previously described review and adjustments of the stdcell library against various trend lines, more cells will get used more often across any given design. However, the synthesis engines will not force synthesis of better cells into not otherwise optimized cones of logic as opposed to forcing the cones of logic toward usage of a scant few falsely preferred cells and drive strengths. This will allow for smaller, faster, and less power-hungry designs and simultaneously allow for easier (i.e., quicker) synthesis runs (the engine not needing the extra effort to force adjustment to the nonoptimized cones of logic as often).

It is possible to adjust these trends in order to force certain behaviors in a synthesis. Although the resulting netlist will not be optimized in a general sense of the word, it will be optimized in the skewed view caused by the trend adjustment. An example of this adjusting might be to attempt to force a flash adder not to use full-adder and half-adder functions that may exist in a library offering. It could be possible to force this by placing the proper "Don't touch" and "Don't use" tokens in the liberty file that is to be used during synthesis, but this forces absolute banishment of the functions from the netlist. It might be desirable to allow the synthesis engine to possibility to use the full-adder and half-adder cells "as a last resort" but to attempt to use other cells in order to accomplish the synthesis "if at all possible." The reasons to do this are myriad, including the fact that such cells, with two outputs (SUM and CARRY), will be forcing a nonoptimal place-ment of the cell (for either output independently) during APR at some point partway between the two optimal locations that would otherwise be chosen if the function was accomplished by two independent stdcells. At any rate, in order to accomplish such a skewing of the synthesis tool, it might be sufficient to make the offensive stdcell (or stdcells) appear to be nonoptimized with regard to one or more of the previously men-tioned trend lines. Artificially, in the liberty that is to be used during synthesis (but not during actual timing closure), increase the delay or power consumption or input cap-acitance beyond the rest of the library.

In such a manner, the data in the library is being made to work for the design center as opposed to making the design center subject to the whims of the data.

13.4   VALIDATING LIBRARIES ACROSS MULTIPLE TECHNOLOGY NODES

A common technique employed across the integrated circuit design industry is to "shrink" a working design from one technology node to another. This is done at any one of many possible stages.

For instance, a "hard" design (with "frozen" polygons) can be shrunk with any one of several EDA tools available on the market that will adjust the polygons from one set of DRC rules to another. This technique is sometimes employed to shrink analog structures, such a phase-locked loop (PLL) or analog-to-digital (ADC) and digital-to-analog (DAC) converter structures. In general, and especially in the case of analog shrinkage, it is important to resimulate the design after the shrink, just to ensure that some important aspect of the design wasn't shrunk out of the assumed operating range for the rest of the circuit to continue to function. In the digital world, this type of shrinkage is generally possible; in the analog world, because some structures would become unusable for with a direct shrink, consistent across the entire design, multiple independent adjustments of pieces of the design, followed by reconnection, might be necessary.

On the other end of the "hard" to "soft" continuum, the RTL that was used to syn-thesis a previously working design in one technology node can be reused to synthesize the same design in a new technology node. Another term for this is just *design reuse*. As long as proper formal verification practices on the resulting synthesized netlist are followed, this is of minimal concern.

In between these two extremes, it is sometimes viable to take a gate-level netlist synthesized in one technology and just perform APR again and retime it in another technology node. Although this third, intermediate shrink methodology works best when there is a one-for-one direct naming of cells, including drive strengths, between the two technologies, this is not a real requirement. The actual requirement is that there just exists the simpler one-for-one direct mapping of any cell in one of the technologies to a given cell in the other technology. It is only really required that this mapping be in the direction of shrink. If the new technology node library has more cells and drive strengths and options than the older technology node library, then this is not an issue. This technique can be used in mapping from one library vendor's offering to another vendor's offering in the same technology node as well (something that might be needed for a second source requirement on a design). As long as a direct mapping exists between the initial vendor's library and the second vendor's library, and as long as the design can be run through APR a second time using the new library, and that both resulting devices are capable of passing validation, then this is viable. This is true even if for any reason, it is impossible to "just" resynthesize the original RTL in the new vendor's library. See Chapter 5 for a more extensive discussion on "hard, firm, and soft IP."

The issues that are discussed in this section deal mainly with the last two of these shrink techniques. For actual technology shrinks, it is important to understand how just how much improvement in area, power, and performance can be expected in an ideal shrink versus reality. In a process shift at a given shrink, either because of the fabrication house giving out modified models or in the moving of a design between fabrication houses, it is important to be able to determine how such shifting can affect area, power, and performance.

In the first case, how one determines if a technology shrink is appropriate involves several steps.

- Ideally, raw area improvement should improve with the square of the shrink. As technology allows smaller and smaller geometries to be printed on silicon in such a manner as to allow these geometries to be electrically connected in order to form transistors, then the length and width of transistors, and the length and width of the interconnect between them all shrink linearly with the technology. The area of such will then shrink with the square. The problem with this is that as technologies have shrunk below the light wave of the stepper engines that print them, some critical geometries cannot shrink in pace with the rest of the technology. As discussed in the library architecture section of this book, many creative minds have found (and continue to find) ways of keeping the library shrink level, however, consistent overall with the shrink in the technology. With this in mind, a raw density of a library should double for each major technology step. Recall Table 6.1, which gives representative raw-gate density numbers for major technology-node levels.

- In addition to this, interconnect capacitance will tend to shrink at this same square scale factor. The reasoning there is that even though oxide thickness is decreasing with scale, so have routed metal thickness, routed metal length, and routed metal width. The routed metal thickness shrink will at least partially cancel out the oxide thickness shrink, and the combined width and length shrink will then allow capacitance to decrease with the square of the shrink.

- The interconnect resistance will grow approximately linear with shrink. The reasoning here is that even though the thickness and width is decreasing both linearly with shrink, so is the length, which will cancel out the effect of one or the other.

- Concerning interconnection itself, typically router utilization tends downward (less efficiency) as shrink increases. This is because of the need to isolate signals more so that they do not become either aggressor or victim nets, which will cause noise coupling between them and will force a decrease in the performance of the design, awaiting the noise on such victim nets to settle. Luckily, router technology has been improving over the years, and this trend is rapidly being fought.

- Dynamic power should be a function of capacitance, voltage, and frequency, none of which is a direct function of shrink (each of which admittedly can be a driver for the need for a shrink).
- Static leakage power will grow explosively with shrink. Do not be surprised to see an order of magnitude increase per major technology shrink from 180-nanometer technology nodes downward. Design centers have had to learn to live with this fact. Power-consumption reduction design techniques as voltage frequency scaling, well isolation, and state-retention sequential circuits have been developed for such reasons as this.

Table 13.1 summarizes the per-shrink trends that are to be expected across a library.

The preceding trends can allow for an extensive study of the area and power comparisons between technology nodes, allowing validation of one library against a previously approved higher-technology node. Notice, however, that beyond the effects on parasitic capacitance and resistance, no mention has been made yet about performance.

As technology shrinks from technology node to technology node, certain features do not shrink evenly, as has been mentioned. Most of these are the result of litho-graphic issues such as minimum printable area of features or optical proximity between features. These issues do not affect all cells the same way. Some functions are notoriously loose; NAND2 are good examples of these. Functions such as these tend to be able to absorb the inefficient shrinks of some features without being hindered by them. Some functions, especially those that are used frequently, the earlier discussion on how to spread usage across a library notwithstanding, have been hand packed. Functions such as these tend to not shrink at anywhere near the theoretical. Well, aside from hand exploration of each individual cell in a library in order to see "improve-ment" possibilities, how can the "theoretical" raw density improvement be calculated? One possible way is to bin cells into categories, develop rough shrink percentages per bin, and arrive at an area improvement number with a little more weight than a simple across-the-board guess. This effort can be somewhat loose and nonmethodical, but it does allow for rough trend analysis. One possible binning categorization is as follows in Table 13.2.

With a little practice, the cells of a library can be easily dropped into one of the preceding four categories. Once done, and assuming a representative shrink of one or two example cells within each category, a rough "estimated shrink per category" would allow for a scaled library-wide estimated shrink with little more than a weighted average of the estimated shrink for each category. Admittedly, this is a loose definition,

TABLE 13.1 Summary of "per Shrink" Trends. These should *all* be viewed with some suspicion. At any given technology node, any one can be adjusted for a myriad of reasons. However, the above general guidelines is approximately correct.

Feature	Trend
raw area:	decrease with shrink^2
router parasitic capacitance	decrease with shrink^2
router parasitic resistance	increase with shrink
router resoure utilization	tends toward constant to slightly decrease
dynamic power	orthogonal to shrink
static power	increase by order of magnitude with shrink

TABLE 13.2   Categories Sufficient to allow Rough Library Versus Technology Shrink Analysis. Different categories tend to shrink at different rates. Splitting functions across such categories allows for greater control over the thought process of estimating cell shrink factors without having to build entire libraries for benchmark studies.

Category	Shrink Effect
Loose functions (large numbers of pins per number of transistors with few stages)	Contracts Consistent with Technology shrink^2
Loose functions but with significant amounts of interconnect	Contracts Consistent with Technology shrink^2, but uses router metal resource
Complex multiple stage cells	lowest derive strengths do not contract consistent with shrink^2, but higher drive strength cells do so
Tight cells (few pins per transistor)	does not contract with Technology shrink^2

but describing a tighter definition is difficult without using an exact example library. Although the category that each cell falls into will differ from technology to technology, for these categories the cells might arrange as follows (your arrangement may differ):

- Category 1—single-stage, non-Booleans;
- Category 2—Booleans and MUX;
- Category 3—all reaming cells, except FLIP-FLOPs; and
- Category 4—FLIP-FLOPs.

It is possible to describe estimated performance benefits for a technology shrink. On the first order, things are direct. Transistors are hotter, parasitic capacitances are less with the square of the shrink, and parasitic resistances are no more than linear increased with the shrink, the performance should increase by more than the shrink. However, as the binning of the cells that has just been described occurs, some types of cells must suffer smaller transistors in order to allow for proper shrink. Depending on the technology node in question, the number of types of cells for which this may be an issue could be the majority of the library. This can mean that the drive strength across the library may not keep pace with shrink. This is another reason why the performance numbers in Table 6.2 were loosely defined and, in general, undefined. The bottom line here is that performance benefit resulting from a technology node shrink is not an assured thing. The only viable solution is to take representative data paths from previous legitimate designs in earlier technologies, scale the APR routing accordingly, and run SPICE analysis. One cannot validly compare performance across a technology otherwise.

The second discussion previously mentioned is the need to validate new SPICE models at the same technology node. This is but another aspect of the previous discussion about performance across technology node. Minor comparisons of transistor characteristics can be made whenever new SPICE models are released by a fabrication house. However, proper validation of how the new SPICE model will affect actual design performance will require a full set of characterizations of the pertinent pressure, volume, and temperature liberty files, followed by timing analysis, based on these new liberty files, on previously timing closed designs.

13.5 CONCEPTS FOR FURTHER STUDY

1. Inconsistent reordering of state-dependent arcs within a logical model (Verilog) and a temporal model (liberty):

   - An interesting experiment is to rearrange the state-dependent timing or power or noise arcs within a Verilog model of a cell, do the same (but in a different order) for the timing or power or noise arcs within a liberty file, and to see what back annotation will do with the result. The student or interested engineer could do the previously described changes to a subset of stdcells, use those cells within a design, and then use the modified Verilog and liberty files to see if the back annotation is consistent or if suddenly the default timing arcs appear to influence the dynamic simulations of the circuit.

2. Gate count study:

   - The student or interested engineer with access to one or more system-on-a-chip designs might be interested in building a histogram of the usage of the stdcells in those designs. This would be done to determine whether they follow something close to an 80–20 rule where the vast majority of the actual used instances within the netlist are the result of a relatively small percentage of the stdcell functions and drive strengths within the used library. More so, the student or interested engineer might find it fascinating to determine, assuming he or she has access to more than one design that used the same stdcell library (even if those designs were made in different design centers), if the resulting usage histograms were consistent between the various designs.

3. SPICE to logical test vector and logical test vector to SPICE conversion script:

   - A student or interested engineer that has an interest in scripting could write a two-way conversion script that can be used to apply consistent test vectors between SPICE netlist of stdcells and logical Verilog views of those same cells. This would be done in order to see if there is a consistency between what the SPICE netlist function actually does versus what the Verilog logical view thinks that it does.

4. Trend lines within a stdcell library consistent with those in the chapter:

   - The student or interested engineer with access to a stdcell library might be interested in making a complete study of the stdcells of that library in order to see if they follow the trend lines as described within the chapter.

5. Trend lines between technology nodes consistent with those in the chapter:

   - The student or interested engineer with access to a series of two or more stdcell libraries from the same vendor but from various technology nodes might be interested in making a complete study of the stdcells of those libraries in order to see if they follow the trend lines as described within the chapter.

CHAPTER 14

DESIGN FOR MANUFACTURABILITY

14.1 LESSON FROM THE REAL WORLD: THE MANAGER'S PERSPECTIVE AND THE ENGINEER'S PERSPECTIVE

It has been claimed that the best design for manufacturability (DFM) for any given structure to be processed at a given technology node is to draw that structure using the design rules for the next higher technology node. Actually, this is a slight overstatement. It is commonly held that a rough estimate would be that 90% of the benefit, in terms of yield improvement, that can be accomplished for deep submicron design can be had by increasing the width or space or overlap rule in question by one manufacturing grid beyond the design rule check (DRC) rule. This yield-improvement trend continues for each succeeding manufacturing-grid increase—that is, a second manufacturing-grid increase would gain an additional 90% of the remaining 10% yield loss for 99% total gain in yield improvement. A third grid increase captures another 90% of the remaining 1% yield loss not otherwise captured, or a total of 99.9%, which is beyond the normal 3-sigma range that typical modern integrated circuit (IC) devices are designed toward. Table 14.1 roughly enumerates this.

Assuming a manufacturing grid of 5 nanometers, three manufacturing grids beyond a 65-nanometer technology node is 80 nanometer (one-half a technology node step below 90 nanometer). Three manufacturing grids beyond a 90-nanometer technology node is 105 nanometer (approximately one-half technology-node step below 130 nanometer). Therefore, as these two examples indicate, designing at a *half-step* technology move backward, while not a best-solution technique, certainly is a good enough technique for

Engineering the CMOS Library: Enhancing Digital Design Kits for Competitive Silicon,
First Edition. David Doman.
© 2012 John Wiley & Sons, Inc. Published 2012 by John Wiley & Sons, Inc.

TABLE 14.1 Yield Benefit as a Function of Design for Manufacturability. Drawing polygons bigger than minimum increases yield but at a loss of density or performance or both.

Margin	Result
Match DRC	Set Yield Loss of X%
DRC + 1 Manufacturing Grid	Yield Loss of .1*X%
DRC + 2 Manufacturing Grid	Yield Loss of .01*X%
DRC + 2 Manufacturing Grid	Yield Loss of .001*X%

improved DFM yields. The one additional caveat here is that it is usually not advisable (and most often not allowed) to grow a hole. So VIA and contact layers are quantized. DFM advantage is gained for these layers by *doubling up*—placing two or more VIA and two or more contact when connecting between layers wherever possible.

Several years ago, I worked at a company that took the preceding manufacturing-grid addition concepts one step further by binning the space and width and overlap design rules into high-critical, medium-critical, and low-critical categories and then writing a DFM-guideline rule-check file that could be run on the design-center released designs. High-critical guidelines tended to be the polysilicon overlap of active rules (where if the polysilicon did not completely cross active, from end to end, there would be no actual transistor formed). Low-critical guidelines dealt with the self-aligning structures (where the structure was guaranteed to exist although it might be slightly closer or farther from the ideal drawn location). Medium-critical guidelines tended to be all of the rest of the DFM-modified DRC rules. In that manner, it became possible for a library designer to judge during library development and under consideration of the goal of the particular library being designed whether a library was a dense, a performance, or a low-power library. The designer could choose to grow a space or width or overlap by judging the cost (against the library goal) of doing so versus the cost of not doing so (against the weighted DFM guideline).

This was a disaster for the library support team. The library used at that design center was a priority dense library. Although performance, power, and manufacturability were important, library density was the most important feature. DFM was added with the caveat of "all that was possible without growing the cell." However, because of this policy, just the high-critical DFM rules tended to be adjusted, and they were usually only grown by one or two manufacturing grids as opposed to three. Did this mean that devices that used this library would fail 10% of the time? No, it meant that it was acceptable to have a yield loss on devices using this library that was 10% higher than ideal so that these devices were as small as possible. However, the DFM-guideline deck was released to the design center so that it could DFM adjust the chip-integration level issues as well. As a result, any design-center engineer who ran the DFM-guideline deck, which was viewed in the design center as an additional DRC-rule deck as opposed to a DFM-guideline deck, against a chip that used the library, found hundreds of thousands of "suggested improvements" to the cells of the library. All of these DFM adjustments would have forced growth of those particular cells of the library, violating the density requirement. In addition, it was possible for the user to scale the weights of the suggested DFM improvements for the various bins separately. Some chose exponentially increasing weights from three grid improvements to two grid improvements to one grid improvement while choosing linear increasing weights from low critical to medium critical to high critical, whereas others chose linear in both directions and still others chose exponential in both directions. Table 14.2 gives three examples.

TABLE 14.2   Three Possible Bin-Weighting Schemes. Such weightings, among others, effectively determine which DFM additions are instantiated and which are ignored.

	1 Grid Growth	2 Grid Growth	3 Grid Growth		1 Grid Growth	2 Grid Growth	3 Grid Growth		1 Grid Growth	2 Grid Growth	3 Grid Growth
High Critical Rules	300	30	3	High Critical Rules	10000	1000	100	High Critical Rules	9	6	3
Medium Critical Rules	200	20	2	Medium Critical Rules	1000	100	10	Medium Critical Rules	8	5	2
Low Critical Rules	100	10	1	Low Critical Rules	100	10	1	Low Critical Rules	7	4	1

The weightings given in the leftmost of the three tables imply that just the one-grid improvements across all levels of DFM criticality would probably be made. The middle of the three tables implies that just the one-grid improvements of just the high-critical DFM issues would be made. The weightings in the table on the right would make it more likely that all levels of grid-based DFM improvement across all levels of DFM criticality would be judged against each other. These are just three of the weighting factors that were used that could be set even as equal weighting across the entire array. It was pointed out to company management that such even weighting effectively removes that binning process: each bin of DFM guidelines is just the same as any other bin, so why do the binning in the first place? However, because of allowing the individual engineer the luxury of adjusting the weights of each bin, some engineers scaled these dramatically different from others. This resulted in e-mails into the library support team along the lines of "please reduce that *high-critical* DFM improvement in favor of these two *medium-critical* improvements (or these *three low-critical* improvements)," many of which were inconsistent with the request by other designs either on the same device or on a different one for which different weightings had been chosen. These "suggested improvements" from the design team became so overwhelming that we had to eventually pull the DFM-guideline deck back and prevent the design-center engineers from using it.

DFM decks are a useful tool when handled internally by the organization responsible for the DFM of a particular phase of a design (for libraries that would be the library support team). However, allowing the deck to be used by an entire organization just enables those who do not have to accomplish the work to chance to second-guess the design choices otherwise made.

14.2   WHAT IS DFM?

14.2.1   Design for Manufacturability or Design for Mediocrity?

To understand what DFM is "in reality," it is important to understand what it is "in definition." Thus, the difference between theory and practice can more easily be explored.

Figure 14.1 gives a general plot of geometric printing requirements as a function of technology node versus the capabilities of steppers used to produce those technology nodes.

The chart shows the minimum geometries to be printed at a given technology node (dashed line) versus the wavelength of the stepper technology that was dominant during

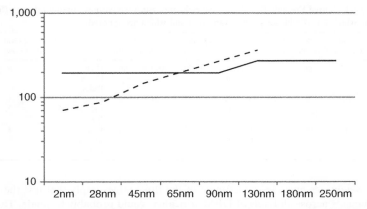

FIGURE 14.1 Technology Node Geometric Requirements Versus Stepper Capabilities. Required drawn feature size is now well beyond the ability to deterministically reproduce it, leading to heuristic approaches (which are mostly successful up through the current technology nodes). However, the nondeterministic means leads to the loss of either yield or density or performance caused by DFM.

the development of that particular technology node (solid line). Below a certain point, the ability to print a pattern moves from the certain mechanistic realm to the heuristic practice-based realm. This occurs when the wavelength of light that is used to print the pattern is larger than the pattern. As can be seen in the figure, this currently occurs at around 65 nanometer. The lithographers in the industry have been remarkably creative in figuring ways to do this using serifs, crenellations, and other optical proximity correct (OPC) effects (see Figure 6.1). However, when the pattern is smaller than the wavelength of light, what is printed may not be close to what is on the mask. Even with these OPC and other lithographic imaging features added to the polygons, the resulting patterns on silicon are nowhere near rectangular. As a result, the concept of DRC has morphed over time. In the past, DRC was a guarantee that anything drawn to the particular width or space or overlap would show on silicon "as such." DRC has come to mean *minimum possible* width, space, or overlap and is no longer a guarantee. Design for manufacturability has been added to push polygons a little wider or farther apart or with more overlap. This is done to help ensure (but still not guarantee) that what should exist on silicon actually is wide enough to exist, what shouldn't connect on silicon actually is far enough apart to not connect, and what should extend beyond a structure on silicon actually is far enough beyond that structure. DRC, in the meantime, has been happily shrunk along with technology node. Even DFM, although closer than DRC to this earlier standard, is not the guarantee that DRC used to be. See the discussion in the lessons learned section of this chapter. Three-manufacturing-grid DFM growth is a 99.9% guarantee, not the 100% guarantee.

This has left an ever-widening gap between what the DRC rules say can be done versus what the DFM guidelines say should be done. "You can draw it that wide, but you should draw it this much wider. You can space it that far apart, but you should space it this much farther apart. You can overlap it that much, but you should overlap it this much farther."

A problem with this reasoning is that the DRC rules, because they are following the technology shrink, remain on the trend lines in terms of the amount of space a transistor should take and the amount of parasitic capacitance it should see (as defined by the trend-line scaling of previous technology nodes). The DFM guidelines, on the other hand, are falling behind. Recall the story in the lessons learned section of this chapter. A "good enough" DFM stdcell looks as if it had been drawn one-half technology-node step backward. Figure 14.1 begins to look more like Figure 14.2.

The two curves do not look that far apart in the figure because it is a logarithmic scale. The percentage difference, as little as 4% at 130 nanometers, is now more than 20% in the deepest technology nodes. Every extra manufacturing grid longer or wider than a polygon, as the DFM guideline suggests, contributes parasitic capacitance and robs performance. In addition, every extra manufacturing-grid wider space (or width or overlap) robs a polygon of density. This last point might seem provocative; after all, many of the one manufacturing-grid DFM improvements, even some of the two-manufacturing-grid and three-manufacturing-grid improvements, might be added without causing any particular cell to grow. However, any particular manufacturing-grid increase in a width or space will eventually affect some structure, perhaps a priori unknown, to grow that given cell. If this is not this case, then the original cell layout for every given cell in the library is "loose" against the original DRC rules, and your library supplier should be fired. Therefore, it does remain fair to claim that every extra manufacturing grid really does rob density somewhere in your library.

As a result, circuits that follow any amount of DFM-guideline growth (from the minimum one-manufacturing grid to the three-manufacturing grids that assure more than 3-sigma yield improvement) does so at the expense of performance and density. One possible reality-based definition of DFM becomes, as the section title indicates, a design for mediocrity.

This is not necessarily a bad thing, although the very term may sound as if it is. It may be true in some cases that it is better to have a circuit that meets a design density or performance goal slightly less well than it could but is more consistent in yield across a wider range of processing. Conversely, it may be the case that it is much better to have one that matches design goals exceptionally well although for a fairly small range of processing (or for a noticeably smaller yield). In some situations, this will be the case, and sometimes it will not.

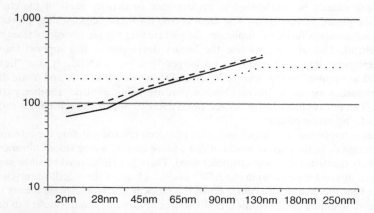

FIGURE 14.2 DFM-Adjusted Technology Requirements. The difference between DRC and DFM will continue to grow in the future.

How do you decide what level of DFM-guideline growth versus DFM-guideline yield improvement is sufficient? The answer is easy enough to calculate, but for a given integrated circuit design it comes down to "good devices per wafer." Is increasing the size of the device because of the growth of some of the stdcells (and reducing the number of possible devices per wafer, increasing their potential cost) offset by the increase in yield (or actually decrease in yield loss) or is the opposite true?

For an IO-limited device or one that the worldwide lifetime supply will come off a single lot (or even a single wafer)—and there are such devices, especially in military applications—the answer is simple. Adding whatever DFM is possible will increase the number of good die without growing the size of the device (limited by the size of the IO ring). Even the increase in the number of extra cells in the long paths of the device, because of the decrease in performance of the DFM-adjusted cells, should not increase the cost of such a device. Although the addition of extra cells will mean that, there is more of a chance that a fault on silicon will find a critical geometry, the number of added circuits should be relatively small. The relative increase in faulty die per wafer resulting from the increase in amount of circuitry per device should be offset by the increase in yield by the DFM-improved stdcells.

For core-limited devices, the story may be substantially different. Although it is unlikely, any added performance degradation in a critical high-speed section of a design may make that design suboptimal. The goal should be to reduce the parasitic that could slow a critical path. A core should be laid out as tight as possible. The opposite conclusion from the preceding is reached. For a core-limited design, it is probably better not to add the DFM improvements. For other circuits that fall between these two extremes, partial DFM-improved stdcell libraries could be optimal.

In a given technology, because of design-divergent goals on various designs, using the same original stdcell library source, it is altogether possible for a design team to require multiple DFM-adjusted libraries. Indeed, because some core designs are reused on other new or derived devices, a design may need multiple versions of a library on the same final IC device. This means that the design-center methodology needs to comprehend this. Cells that are to be used in a performance-critical part of the circuit cannot be synthesized in DFM-critical or density-critical parts of the design and cannot be placed among them, although cells that are to used in a DFM-critical part of the circuit cannot be synthesized in performance or density parts of the circuit and cannot be placed among them. The same is true for dense cells. Adding power-critical library adjustments further complicates the picture that the methodology team needs to comprehend. This also means that the library development and support team needs to comprehend this. A NAND2 is no longer "just" a NAND2. It is a "library X" NAND2 as opposed to the "library Y" NAND2, where libraries X and Y are divergent goal-oriented versions of stdcell libraries that are being used on another part of the design. The release mechanism for the library has to adjust the cell names sufficiently for this to be accomplished.

What is happening is the *customization* of stdcell libraries. Before the advent of true ASIC design or in the current world of full custom design, where stdcell libraries do not exist, every transistor is or was customer sized. There is a viable need to allow something similar to this to happen also in the ASIC world, at least at the stdcell-library level. The need to beat the next design center down the block in the marketplace means that it is imperative to adjust the fabrication-house source library on the same design to divergent goals in order to meet potentially divergent density, power, performance, and yield goals across various parts of the integrated circuit device.

Having redefined what DFM is in actuality as opposed to in theory, there are a few more topics that will bend the definition even more. These will be covered next.

14.2.2 Design for Methodology and Design for Mobility (Between Fabrication Houses)?

In the last section, it became obvious that it is no longer true that "a NAND2 is a NAND2 is a NAND2," with apologies to Gertrude Stein. The need to support multiple versions of a stdcell library, because of reuse goals, on a given design is a real requirement, now and for the near future. However, the mere requirement to allow divergent names of cells, each referring to divergent optimized layouts of the same function, is not enough.

Engineering design automation (EDA) tool versions change over time. A core that is to be reused on the next device but was originally designed for the previous device may have been designed with outmoded versions of tools, some of which may no longer be available to the design center and features of which might not be currently available from the versions of the tools currently in place. One way that this could happen is the well-known fact that not all new versions of EDA tools are backward compatible. Another possible reason for this is that the design center has decided to reduce cost by cutting funding for ongoing licenses in the particular EDA tool that was previously used. Perhaps another possibility is that the design center, as marketing pressure forces adjustment to the various entries in the design centers' understanding of their version of the classic "Frankenstein methodology flow," has moved parts of that methodology flow from one EDA tool to a competing tool. For whatever reason, it is likely that a design that is required to be reused will, at some point, require EDA capability (i.e., library capability) that is no longer available.

It is important to extend the concept of DFM from "just" the physical view to other views as well. Here DFM comes to mean "design for methodology." Preparing and adjusting the various logical, test, placement, or timing views ahead of time to allow for similar capability in alternate EDA tools and the testing of these various capabilities, may pay off in the future by allowing for future reuse that would otherwise be impossible (or severely limited). In addition, perhaps "jumping onto the band wagon" of a new feature—for instance, scalable polynomial-based delay models of a few years ago—is not necessarily a good idea.

What is needed is compliance with the longer-term trends of classes of EDA tools. Identification of these longer-term trends for the market-driven design center might be easy, but for certain companies, mapping to industry-identified technology trends might suffice. Many fabrication houses publish prescribed platform flows. Extraction of some of the viable features of a typical one might look like the list in Table 14.3.

TABLE 14.3 Representative Fabrication-House Platform Flow. Each new technology node requires more and more specialty handling to successfully close designs and produce working silicon.

Features inside of Liberty that mesh with Modern Timing Closure APR tools
Features inside of Liberty that mesh with Power Measurement and ClosureAPR tools
Including capability to use Well Isolation, Multi-Vt, Voltage/Frequency Scaling & State-Retention
    Storage Elements
Pre and post APR DFM feature insertion
Statistical Timing Tools

How to ensure that the previously used but still required libraries remain compliant on these and other design-specific features is design-center specific. But ensuring that each version of the library that is used on each design-center device to be compliant with a set platform flow will go a long way to ensuring that archived devices remain reusable, even as EDA tools change (because of version changes or vendor changes). Regular validation of these and any design-center specific capabilities against new versions of EDA tools is of high importance and should be a regular and scheduled event before licenses are given to old tools.

However, the ability to maintain compliance across new versions of EDA tools is just half of the topic. Just as it is possible that EDA tools (and versions of those tools) can change, the appeal of fabrication houses can change. This may be the result of either contracts (a different fabrication house offers a more cost-effective solution to management) or by customer requirement (the customer of the design center wanting some or all of the fabrication done in one local or another) can change. Here the definition of DFM changes again to "design for mobility (between fabrication houses)."

As long as the design-reuse level is limited to soft register-transfer language (RTL), this is not as much of a problem. However, if the design that may need to be reused is either firm gate-level or hard-timing or power-closed graphical display standard (GDS) blocks, this can be a major issue. Fabrication houses tend not to want to share their intellectual property (IP), which they would consider the stdcell library to be, with competing fabrication houses, justifiably so. This is one reason why it is important, whenever possible, to use fabrication-house–neutral libraries (that is, either third-party GDS that has been validated in multiple fabrication houses or design-center–specific stdcell libraries).

This possible future requirement should always be under consideration when a design is in automated place and route (APR). If any special hard blocks are specifically designed to be compatible with a specific fabrication house—for instance, a nonvolatile memory block—that block should be isolated, with some additional guard-banding space placed around the structure. In and of itself, this is a good design methodology for power and substrate noise-isolation reasons among others, but it more easily allows replacement on future reuses of the design in which it appears just in case that substitution become necessary. This is a safe process for any third-party–acquired piece of intellectual property. Not only does adding such spaces allow for the safe substitution of alternate third-party IP, but also it allows for safe growth of the given third-party circuit should this become necessary. For all of the preceding reasons, how much room should be allowed? This is a choice for the design-center manager. Certainly, for design centers with long-standing relationships with fabrication houses and with third-party IP vendors, this "extra space" can and should be small. After all, every extra bit of space added will grow a device, increasing the cost of production. On the other hand, any decrease in security of the relationship with either the fabrication house or the IP vendor can be partially minimized by added space in such areas.

One last reuse issue that needs to be considered during design-center device closure are the differences between fabrication houses. A worst-case scenario is that the process is sufficiently different that the previous design is completely unusable. This *will* happen occasionally. As mentioned, using a third-party stdcell library will help significantly in such instances. A much more likely event, however, is that some levels of the processes will have different physical parameters. A classic case is that of a layer having a significantly different sheet resistance, making resistors that have been designed for that process layer subject to redesign. Such sheet-resistance changes (or, similarly, potential interlayer-capacitance changes) can and will have significant influence on performance of the more analog aspects of previously closed designs. Do not be surprised that circuits that rely heavily on such features—for instance, electrostatic-discharge structures, analog-to-digital converter (ADC) or digital-to-analog converter (DAC), and phase-locked loops—might be in need of significant redesign.

14.2.3 Design for Models and Design for Measurement

The next time you review the process verification report of a fabrication house, no matter the technology node, note the widths and lengths of the transistors. The vast majority of the test structures use geometries, especially transistor widths, which are up to two orders of magnitude larger than the transistors in the stdcell libraries. The smallest such test structures have geometries of nearly an order of magnitude greater than the stdcell library (and on the same order of magnitude as those transistors in the IOs). Figure 14.3 is a representation of the actual such sizes of a real but anonymous fabrication house for a 90-nanometer technology node. The structure on the left is the size of the most common N channel in the 90-nanometer stdcell library that was being supported in that fabrication house. The structure on the right, which was 60 times the stdcell library N-channel width, represents that size of one of the original N-channel test structure that was used to

FIGURE 14.3 Comparison of Representative Stdcell Library Versus Test Structure Channel Widths. Test silicon test structures are notoriously larger that the type of circuits they are meant to match. As a result, all SPICE models need to be questioned as to their reflection of reality.

build the initial SPICE model. The structure in the middle, which is still 10 times the stdcell library N-channel width, represents the compromise N-channel transistor width that the fabrication house agreed to use for SPICE model generation, at 90 nanometer only, from the point that this was identified as a problem onward. Note that this was true across multiple test structures.

The reason for this seeming incongruity is that it is easier to generate statistically valid measures of such parameters as leakage currents, measured in milliamps per micron, for wider geometries as opposed to narrower geometries. It is common, when asking for an explanation, to hear that things such as narrow channel effects contaminate parameter measures of narrower devices. This argument seriously occurs. Unfortunately, stdcell library users need to understand how those very same narrow channel effects reflect on the transistors in the library and how that affects the liberty model characterization. Such effects as hot-carrier injection (HCI), which slowly causes performance degradation of transistors over the life of the device because of the accumulation of permanent charge in the transistor channels, were first identified not by the fabrication houses but by the design community. Whenever the design community identifies a new effect, the fabrication-house community does a good job of adding test structures to verify and validate from that point forward. However, by definition, this is after the fact.

Here is another definition of DFM: design for models and measurement. Specifically, this definition implies the need to adjust design as a function of future adjustments to the SPICE models. Understand that the fact that models are adjusted after the fact means that is fundamentally impossible to predictably adjust the margin for such adjustments to SPICE models and measurements. As previously noted, the models tend to become adjusted to the performance of the library and to the issues identified by the user community a posteriori to the use of the library. This is why fabrication houses tend to issue new SPICE models over time that tend to grow in range (especially the worst-case models). This is in the exact opposite direction that one would expect. As a fabrication house learns to control a process over time, the statistical range of the best-case and worst-case limits should decrease. The process does not drift over long periods as much as the issues tend to be identified over time.

To be fair, part of the reason for this increase in margin is because of the creativity of the layout engineer. Fabrication houses are very good at closing in on the actual performance of transistors as they learn the process. However, the transistor that they learn how to model is a fabrication-house–assumed layout. However, layout engineers can regularly demonstrate novel variations on how a transistor and its surrounding topology are arranged. As a result, the fabrication-house SPICE is somewhat limited by having to relearn the effects of these new novel arrangements. Standardization in layout would result in a quicker process learning because of the cleaner processing, and this cleaner processing will result in tighter SPICE models. The reason that this is brought up here is that as technology nodes continue to shrink, the creative variability that characterized topology design in earlier technology nodes will be lessened, thereby allowing such accelerated process learning and consequently tighter SPICE models.

How does one take into effect that the models used in designs made with one use of a library, shifting before that design is reused? The answer lies in the question of just how much power and influence the design center has on the fabrication house.

- If you happen to have an internal fabrication house, simply refuse to accept any SPICE model change of more than, say, 5%. The internal fabrication house is not a profit center. It exists to support the design center's profit center. Note: This also works if you are a large enough of a customer for an external fabrication house. Threatening to take a significant amount of the design material monthly run from a fabrication-house vendor can do wondrous things.

- Most of the design world does not fall into either of these two scenarios. What do you do? Most of this book describes ways to reduce fabrication-house margin so that the design center can compete on performance or added feature as opposed to cost. Adding some of this margin back in order to cover worst-case SPICE model performance decrease, specifically between design uses, is intelligent design technique.

Although it might seem that I am suggesting that we are substituting one margin for another, this is not exactly the case. Proper design technique would have added this design-reuse margin anyway. Adding design-reuse margin on a reduced fabrication-house added margin liberty still should produce faster devices (and smaller final integrated circuits) than adding design-reuse margin on a straight fabrication-house liberty file used library. This is because it is a multiplicative effect, and a 110% design-reuse margin of a 95%-reduced fabrication model is smaller than 110% of a 100% fabrication model. Probably more important, the amount of design-reuse margin added is controlled by the design center as opposed to the fabrication house. This adds a degree of freedom to the design center.

14.2.4 Design for Management and Design for Metrics

This section could be easily summarized in two statements. First, company management of companies doing military or other governmental contracts does not want to be called up before a congressional committee to be asked why they haven't been able to fulfill some such aspect of the contract that they have been awarded. Second, managers of companies doing consumer products do not want to be hauled in front of civil courts by their clients or customers for breach of contract. With that in mind, it is always important in engineering to do exactly what the contract asks for and, assuming that it prescribed, in the exact prescribed manner. One definition of design for management says "Do it right." This is always an important, indeed central, requirement of proper design work—and, indeed, common sense. Every ethical engineer follows this practice.

Although engineers do take shortcuts, they usually only do so after having tested them so they do not unknowingly break the chain between initial assumption and final result. As an example of such, a valid shortcut might be building a 50 or 100-cell "library" in order to test a new architecture against a current alternative. Fifty to 100 representative cells will give a decent estimate of the density or performance of a library. The architecture under review will end up giving the opposite density might certainly be the case. Perhaps performance might be the opposite of that shown in the 50–100 cell test case. This effect may only occur in the remaining several hundred cells, but this is unlikely enough that "doing the correct thing" and building a complete library for the test "just to be safe" is wasted effort. Completing the entire task is the incorrect alternative. The "shortcut" of building just the critical representative cell subset of an entire is the correct alternative.

Another instance of proper shortcutting is the educated guess. On a recent 45-nanometer technology low-yield debug of a customer's custom design, on listening to the failure mode and reviewing the layout of the failing latch, an associate of mine pointed to one of the transistors in the sequential cell design that was connected to the failing net. This connection was through a lone VIA and asked to get a microphotograph of that VIA in a good part and in a bad part. Other engineers said that this was too critical to be grasping at straws and insisted on a multiweek review involving significant time and resources. At the end of the effort, after multiple false paths of analysis, it was determined to be the particular VIA having not been drawn correctly on the mask, which caused a resistive failure between the failing node and the transistor. A microphotograph showed the issue. The fact was that my associate's original insight was not grasping at straws but rather an informed belief by a senior engineer fully capable of narrowing the field of possible issues significantly—and it should have been followed. The bottom line is that shortcuts, as long as they don't allow for breaking the linkage from incoming require-ment to outgoing product, are value-increasing, effort-multiplying techniques.

An additional aspect of design for management (see Chapter 20) is reporting to management. To many engineers, managers can sometimes be viewed as the bad people. They are not. The role of the manager is to handle the problems that the team reports that it cannot. This is true at all levels of hierarchy. The manager's managers handle issues that managers cannot, just as managers handle issues that engineers cannot. At any rate, as soon as an issue is identified, communicate it up the chain of command, pref-erably in writing. Issues tend to be much easier to resolve as early as possible as opposed to when they become critical. As for the written part of this issue of communication, make sure that the manager actually hears (or reads) what you think that he or she should hear (or read) and not just what he or she wants to hear (or read). Written reporting of issues forces this in a slightly better manner.

There is a second aspect of design for management to be reviewed: the treatment of engineering as a commodity. This aspect of DFM might just be described as "design for metrics," and it really should be treated—and this is unfortunate—as a falsehood. Moore's law goes something like the ability to put transistors on a piece of silicon doubles every 18 months to 2 years, the time span depending on which expert "who was there when he said it" you ask. Gordon Moore actually gave the comment in a 1965 paper, not in a speech. Over the past 40 years that this "law" has been valid, the industry has moved from designing single transistor integrated circuits to designing hundreds of millions of transistor-integrated circuits to now more than 2 billion transistor circuits. During that same time, the number of engineers in the industry has not been keeping pace with this exponential growth. Typically, the number of engineering jobs in the world historically grows at around 4% per year. With these two figures in mind, Figure 14.4 gives the comparative trends.

The numbers of transistors per device per worldwide engineering population does not really mean much for the years 1970 through 1990. After all, the vast majority of engineers, both then and now are not actively or directly involved in IC design, so the very fact that this number has increased by a factor of nearly 100 between 2000 and 2010 is rather alarming. If the trends continue, the number of transistors or devices per engineer will be nearly 40,000 by 2020. The general engineering population, even when combined with engineering productivity per engineer, is not keeping pace with the demand. Although EDA-based design-productivity–boosting tools plus design-reuse techniques will help alleviate some of the stress here, a 200–400% increase in engineering output per engineer every decade is unlikely to be sustainable. If there is going to be a break in Moore's law, it is not the physical limits of the technology that will cause it but the human inability to produce the ever-increasing amount of design required.

Year	Gates/device	World wide engineer population	Gates/device per engineer
1970	2,000	750,000	0.002666667
1980	40,000	1,100,000	0.036363636
1990	1,000,000	1,620,000	0.617283951
2000	20,000,000	2,400,000	8.333333333
2010	2,000,000,000	3,550,000	563.3802817

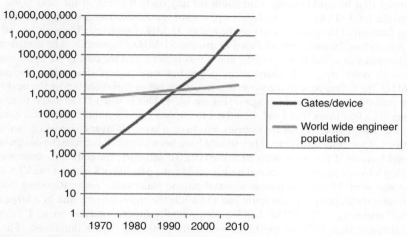

FIGURE 14.4 Moore's Law Versus Integrated Circuit Design Engineer Population. The wall at which Moore's law will fail is one due to the lack of human capacity to design transistors rather than the lack of physics to produce valid transistors.

14.2.5 Design for Market

The preceding sections have discussed some aspects of added yield improvement, however yield improvement is defined. However, there needs to be an exploration of when and even if it is appropriate to add yield improvement. Sometimes, the amount of added engineering that should be applied to a device should be limited. The old caricature of an engineer never wanting to stop improving a design while the management is in need of something to actually sell may be a caricature, but there is some truth to it. At some point, diminishing returns on the effort and time involved in yield improvement will make that improvement uneconomical.

I was once involved in the embedding of a microprocessor core in a device to be used in a pager, actually debugging the device after the fact. The version of the microprocessor core chosen always had poor electromagnetic interference (EMI) performance, and that EMI was always directional out of the right side of the core. The issue was quickly identified as resulting from the large devicewide clocking circuit with rails that formed an efficient directional antennae pointing from the left side to the right side of the core. Another design center, in an attempt to reduce the EMI of the core, removed all of the VIA connecting this particular core ground rail to the rest of the ground plane of the core. Unfortunately, the center neglected to either replace them or tell the rest of the company that it had, in fact, removed the VIA. Unfortunately, or fortunately, depending on the point of view, during the custom layout of the core some years earlier, a layout engineer had the foresight to snake a minimum metal route across the device to the ground rail of the clock circuit. Because this metal connection existed, the microprocessor passes layout versus schematic (LVS) verification with no issues. The design center that did the actual embedding of the core in the pager device, did not know that the clock ground plane was only connected through this minimum metal route. When the part came out, it performed exactly to the performance spec (25 MHz). The business manager of that design center, knowing that he could charge a premium for any parts that ran at the next higher performance level (33 MHz), asked for some end of the fabrication line binning, expecting some fraction of the parts to work at the higher 33-MHz frequency. Through several lots, no part did so. Indeed, they all failed at around 31-MHz frequencies. The design-center engineering was asked to investigate, and it was noticed that the core clock appeared as a sawtooth waveform, with sharp rising edges but very poor falling edges. Indeed, at 31 MHz, the falling edge of the waveform never got below threshold before the next rising edge pulled the waveform high again (Figure 14.5). This is when I was asked to join the debug. The first thing that I did, because I knew that I had a similar embedded core in a part that ran at much higher frequencies, was to do a layer-by-layer comparison of the two versions of the embedded core. They should have been identical, because the design center should have used the same source for the core. In actuality, the only difference was the missing VIA construct. I was asked how it could have passed LVS without the VIA array. That was when I found the minimum metal ground plane snake route. Realizing that this minimum metal ground plane route had to handle the entire current sink of a large clock circuit operating at 25 MHz, I did a mean-time-to-failure analysis on it. I informed management that 50% of the cores would fail within a 3-year timeframe. The vice president of the division said that most people replace their pagers every one and one-half years. Although the design center never used that particular version of the core in any other design, it also never made any changes to "fix" that particular device. For the life of that pager product line, we never had any returns.

Many device specifications call for a device to have a mean time to failure of at least 11.4 years (or 100,000 hours) of cumulative operation. The pager episode illustrates,

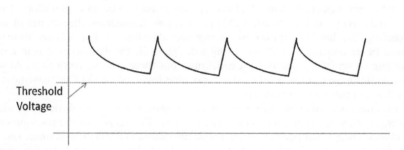

FIGURE 14.5 Poor Ground Plane Clock Output. A frequency-limiting waveform is represented on an actual microprocessor.

however, that this is not necessarily a valid requirement. Similarly, off-the-cuff boiler-plate compliance to other requirements needs to be called into question. The need to operate over an extended temperature range for a set-top box that will always be used with a fan in the box, or the need to operate over an extended voltage range for a wall-plug–connected box might be places to look for requirements that can be reviewed and modified. Of course, doing so, especially for temperature and voltage, requires approval by the potential customer base. One specification that does not require customer approval (just management approval) is the need to operate over a given process range. As discussed in the SPICE chapter, it is sometimes financially viable to limit the number of sigma worst cases (or best cases) that a device is required to operate if, by doing so, the device can be qualified for an extended operating range or a reduced power range. For some processor applications, this is not an unfamiliar design practice. Although for certain communication-channel specifications, if the produced silicon device does not meet a minimum performance standard, then it is sand. For some processor design, the fraction of the product that does not meet a certain standard can be bin sorted at test and sold into a lower-priced, lower-performance market. As a result, although a communi-cation design tends to require timing closure to some definition of best-case and worst-case processing (liberty files), it is common for processor cores to be designed to typical case processing (liberty files) only.

14.2.6   Design for Money

Since the start of this chapter, the definition of DFM has changed several times. Admittedly, these definition changes may be slightly unfair, but they have been made for a point. DFM is not and should not just be "design for manufacturability." However, what is the real (or best) definition? Each of the previous definitions emphasizes certain aspects and benefits.

- **Manufacturability**: This "book" definition of DFM, at the cost of increased para-sitic and power consumption and potentially decreased cell density, emphasizes adding extra geometric features, usually added width of, or space or overhang between, polygons used in the design in order to ensure better processing yield but that also forces the need for multiple truly customized versions of original libraries.

- **Mediocrity**: Engineering analysis is required on the benefit trade-offs of higher yield versus larger (or more power-hungry) devices than otherwise necessary for meeting performance goals and the decision about which can legitimately go in either direction.

- **Methodology**: This forces either long-term continuity in the EDA tool selections and versions used by the design center (and supported, therefore, by the library support engineer) or regular verification of older releases of libraries and designs against upgraded (or replaced) EDA tools, thus allowing design reuse.

- **Mobility**: This similarly forces either longer-term relationships with fabrication houses or the need for "same technology node" design-morphing tools and engineer-ing resources to ensure designs can be alternately sourced into other fabrication houses, but thereby also allowing potential design reuse.

- **Models**: These force familiarity with fabrication-house validation techniques and materials and their resulting SPICE models, as opposed to integrated circuits. but they could justify the use of reduced process range design.

- **Measurement**: This forces added (admittedly, design-center owned) margining of designs against performance, power, and noise goals.
- **Management**: This forces active compliance to standards and the communication of same in documented means, even if just for legal protection.
- **Metrics**: These forces constant innovation and productivity improvement, although as a whole the industry is currently failing at this.
- **Market**: This forces the active review and justification of the design goals and standards to which a design is made.

None of these bears directly on the reason for a company to be in business. The ultimate definition of DFM should be "design for money." Like all activity, DFM is not free no matter which definition is used. It requires active resourcing, takes time, and costs money. The only way that DFM thus is a benefit is if it returns more than it cost. Does the added effort produce a part that is better in some noticeable manner (in terms of feature, performance, power, or cost)?

Because so many design centers and fabrication houses actively pursue DFM and so many EDA vendors actively support this effort, the apparent answer to this is that, yes, it does return, generally speaking, on the resource, time, and other costs involved (at the least, it is perceived to be doing such). However, all things need to be pursued in moderation. The old adage of the design engineer never wanting to stop refining a design to release it for production can be true. It is easy to lose track of this. Adding too much yield improvement, in whatever form, can be counterproductive. Doing the incorrect type or amount of DFM improvement in whatever form can be detrimental. Missing a market window, causing the product to fail in the market as a result, counteracts any benefit from the little extra add yield.

In addition, we need to address the preceding parenthetical comment. Yield improvement is often perceived as a good thing in and of itself. It is not—or at least no more so than any other feature that can be added to a design. True cost-versus-benefit analysis is required and should be undertaken. Adding a polygon—just because the fabrication house advises it—is just the same as adding a line of code to the RTL just because the vendor asks for it. If there is a real benefit for doing so and there is time and resource for doing so, most assuredly do so. If this is not the case, then the correct action would be not to waste the effort (cost) of adding the polygon.

In conclusion, another ultimate definition of this DFM is "time to market." Do everything that makes sense and that can be accomplished in order to improve the yield of a device, without adversely affecting the size, performance, or power of that device beyond a reasonable (and a priori defined) limit right up to the required release. However, do no more beyond that.

14.3 CONCEPTS FOR FURTHER STUDY

1. DRC-to-DFM deck-builder script:
   - The chapter described a viable DFM-deck possibility with the idea of 90% of the theoretical yield improvement being achieved by increasing (growing) the various polygon widths and spaces and overhangs by one manufacturing grid beyond (larger) their DRC limits. It went on to say the saving of another 9% of

the yield improvement (for a total of 99%) by growing these widths, spaces, and overhangs by a second manufacturing grid. In addition, it went on to say that the another 0.9% could be saved (for 99.9% of the theoretical yield-improvement limit) by growing by a third manufacturing grid. A student or interested engineer might develop a script that will automatically build a 90% or 99% or 99.9% DFM deck from a given DRC deck.

2. DFM-automated weight-adjustment script:

   • The chapter described a weighting system that would scale the value of making the various high-priority, medium-priority, and low-priority DFM adjustments of one, two, or three manufacturing grids each. A student or interested engineer might be inclined to provide an automated script that could be used to adjust the weights of these various regions of the DFM solution space in order to allow a "best overall" weighting as determined by the student or interested engineer.

3. Running a DFM deck against a library and making the density trade-offs in layout:

   • For the student or interested engineer who has access to a DFM deck (or series of decks), there might be an interest in running the same against a subset of stdcells. The student or interested engineer can then make the adjustments required by the deck or decks in order to improve the yield for those stdcells and to see how much can and cannot be accomplished without growing the size of the stdcells.

4. User acceptance:

   • If a DFM deck or decks is available, and if the student or interested engineer has used the deck (or decks), then it can be used to accomplish as much DFM-yield improvement as seems reasonable to the student or interested engineer in such a manner as to not grow the stdcells. Then it might be interesting to deliver these adjusted stdcells for use by a design center with and without the DFM deck. Once accomplished, judge the satisfaction by the two design centers to determine whether there is a need to improve DFM yields on those adjusted stdcells.

CHAPTER 15

VALIDATION

15.1 LESSON FROM THE REAL WORLD: THE MANAGER'S PERSPECTIVE AND THE ENGINEER'S PERSPECTIVE

In integrated circuit (IC) engineering, otherwise interchangeable terms such as *verification* and *validation* are used to identify divergent activities. Verification is the testing and assurance that the various views of the library represent different aspects (see Figures 13.1 and 13.2) of the same platonic function. The various views play together well. Validation is the testing and assurance that these views map directly to postprocessing physical reality—that is, they meet physical goals.

For the library development and support engineering team, this is a two-step process:

1. The views map to the desired models and decks that are represented in the physical design kit (PDK), and
2. The density, power, performance, noise, and yield goals of the libraries are actually achieved through testing.

Both steps are or should be continuous processes.

Several years ago, I was working at a company that believed that the library design effort, by the library design and support team, was completely out of control. Invariably, whenever a new SPICE model would be released from the PDK development and support team (the company had internal fabrication facilities), the customer base, all of them internal design centers, would be shocked to find that the liberty files were not

Engineering the CMOS Library: Enhancing Digital Design Kits for Competitive Silicon,
First Edition. David Doman.
© 2012 John Wiley & Sons, Inc. Published 2012 by John Wiley & Sons, Inc.

concurrently released. Whenever these new liberty characterization files were released, they were often slower or more power hungry or more noise susceptible than previous characterization releases. Whenever a new design rule check (DRC) deck would be released, these internal design-center customers would be shocked to find that either the library physical views for multiple cells would fail multiple adjusted rules or the automated place and route (APR) built from APR decks generated by library support teams. The generated graphical display standard (GDS) would no longer correctly route without failing many of the adjusted rules. When the new library physical views were released, they would often lead to significantly less-dense routes. Whenever a design-center engineer would telephone a library development engineer and ask for an adjustment of some manner on one or more views of one or more library cells, they would be dismayed to find that more often than not the request would go unfulfilled. Whenever they were fulfilled, they usually broke the density, performance, or power or noise immunity of previous releases. Several successive library development and support team managers were either reassigned or removed from their positions as a result.

The real reason for this failure on the part of the library development and support team was not lack of diligence or experience by library development engineers nor lack of skill or leadership by management teams. The real reason was failure to control the two opposing extremes of the communication continuum. One of these failures was uncontrolled under communication, and the other was uncontrolled over communication.

- First, because of the lack of communication or undercommunication between the process engineers in the PDK team and the library development and support engineers of the digital design kit (DDK) team, there was no foreknowledge of upcoming SPICE models releases. Because of the limited number of computer nodes available, it would take as many as 6 weeks to deliver a set of three pressure, volume, and temperature (PVT) characterizations. With multiple design centers requiring multiple PVT characterization corners, a new SPICE model could cause far more than a 3- to 6-month cycle of effort by the library development team. In addition to that, the design centers had enough power to force the process engineering PDK team to regenerate and rerelease SPICE that they felt was too different from previously released SPICE models. As a result, DDK liberty release delays could cause more than a quarter of a year's worth of delay of multiple design-center projects. In addition, when the PDK team released modified DRC, lack of communication of these upcoming releases caused the library team to learn of them coincidentally with the design-center teams. No time was thus allowed either to bring the library into compliance with the rules or to negotiate the extent of the rules in order to allow easier such compliance. As far as the effect of these releases on previous design goal requirements of the various libraries, this was not addressed.

- Second, design-center engineers felt they could directly communicate (overcommunication) their request to the library team, and doing so resulted in their requests sometimes being completed, despite those requested changes possibly breaking others by competing design-center teams. This very communication would cause rounds of gyrations of effort by the library team plus rounds of reengineering effort by the various design-center engineers.

The solution to both of these issues and to bringing the library development and support team under control was to formalize the communication flow. All future PDK process model development and DRC deck development needed to be communicated with the

DDK library team (and to key design centers) several months ahead of time in order to allow sufficient development of consistent PDK and DDK views and decks. In addition, no future communication by the design-center engineer would be allowed with the library development engineer. All requests for changes of the cells and views would be communicated and authorized through a regularly scheduled prioritization meeting with all design centers.

However, the final follow-up requirement that was put into place and the reason that this illustration is the subject of the lessons learned section of the validation chapter was the need continuously to validate all new library development. This requirement dealt with both the individual cells and the individual views of the cells as measured against the particular library goals using previously completed "known good" design-center circuits.

Continuous testing of cells and views was instituted. The ability to evaluate the libraries became so automated that it became possible to use the *validation* flow as a design-development *verification* flow. New cells were daily submitted to the validation cycle in order to have them automatically evaluated against DRC and layout versus schematic (LVS) decks and against other similar cells and functions of competing drive strengths. How these cells fit into the trends of the other cells in the library and how they affected the density or power consumption or performance or noise immunity of the evaluation and test netlist, was daily reviewed by automated scripts. Cells and views are graduated across levels (see Figure 15.1). Before a cell was clean to DRC and LVS, it remained in a daily development local on the disk. As a cell passed initial DRC and LVS, it was moved to areas of the disk that were more open to the various design centers, even as it was refined in order to follow density, performance, power, noise, yield, and whatever trends it was deemed to require. When it was determined sufficiently clean for open usage, it was finally placed into the current queue for the next scheduled release of that particular DDK, concurrent with the next scheduled release of that particular PDK.

No combined PDK–DDK release was allowed without validating it against these previous results and against each other. Going forth, it was always known how such changed views would affect any future design-center device effort. Since that time, the

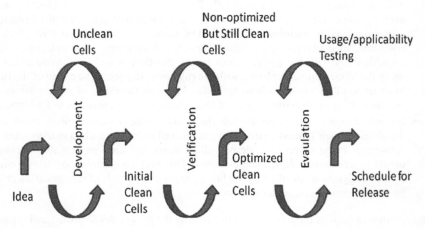

FIGURE 15.1 Graduated Development and Validation Cycles. If implemented properly, the three-cycle method can daily and continuously produce upgrades to design centers of better libraries.

library support organization of that particular company has expanded this effort to continuous testing against new versions of EDA tools and versions as well.

15.2   QUALITY LEVELS

15.2.1   Tin: Engineering Work in Progress

Library development usually occurs in one of two variations. First, there is the clean unadulterated development that is accomplished separated from the design community. Engineers in library development organizations, just as in any organization operating in such environs, love doing development in this manner. The library is designed in order to be optimal all on its own merits. Messy inefficient requirements are not placed on the resulting library by the user community. The resulting library is thus extremely clean and efficient (and, usually, useless). The second mode of development is one that either starting from scratch with a new library or starting from the result of the first variation, an initial design-center project (or entire design center) is defined as the driving customer. This type of development effort is messy. Proposals are made by the design center, counterproposals are made by the library development organization. Evaluation effort remains ongoing for days to weeks to even months. Eventually, a library is developed that is optimized to the needs of the driving customer (project or design center). If the initial driving customer is representative of the entire company, then the library is optimized to those same needs.

The second variation on the development model is the more useful and the more common of the two, and justifiably so. However, it is also the messier of the two. At any given moment, the library resembles a "work in progress." Because there is an ongoing need driving customer-defined evaluation cells (and certain views of those cells), not all cells and not all views are developed or maintained consistently. Daily updates and piecemeal handoffs of these cells and views are continuously made. Consistency between views is impossible to maintain as different views are optimized to divergent requirements. Many views that would be required in a fully released version of the library are not even attempted at this point (or, if they are, then they are "borrowed" from older more stable libraries and partially morphed to suffice). If this development is on a new (to the organization) technology node (assuming internal fabrication) then the PDK is likewise in a continuous state of flux. SPICE models, DRC, LVS verification, and layout parasitic extraction (LPE) decks are either outright missing or are of questionable validity. The questioned SPICE makes the timing and power views, when they are developed, suspect. The questioned DRC makes the assumed density of the cells suspect. Route technology files may go either missing or underdeveloped for long stretches of development time. The cells' routability, therefore, is suspect. Logical views—indeed, any particular view—may be missing or based on earlier libraries of suspect applicability.

This period in the DDK development cycle usually occurs in the very early stages of a technology-node development. No design center is going to release product based on the DDK. Most design centers will not even release test silicon based on such DDK. However, it is a proper and desirable required stage of the DDK development cycle. Within the library development organization and within the driver design center, piecemeal development and evaluation will be taking place. Significant amounts of broken cells and views will be found. However, over time, as these views and cells are refined, the number of new issues will trend toward fewer and fewer (although, by no

230   VALIDATION

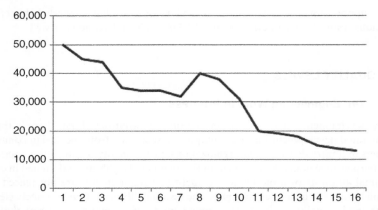

FIGURE 15.2   Rough Approximation of the Issues and Errors Uncovered in a Tin-Quality (Use at Your Own Risk) DDK. This is illustrative and representative of the general trend of issue resolution as a function of time spent in the tin-quality range. The vertical range could be any number (here they represent a weighted factor representing severity times the number of known issues), the horizontal range could represent days or weeks after initiation of effort (here they represent weeks), but the key aspect is the general features and trend of the graph. Consider this type of trend symptomatic of a "use at your own risk" tin-quality-level library.

means monotonically and by no means all the way to zero issues, at least within this phase of the development cycle). Based on an actual library development, Figure 15.2 gives a rough approximation of the trend lines.

Note that in the figure were times of little apparent progress (issues that were resolved during these times were not systemic, but rather individual point problems) and times of dramatic progress (during week 10, for instance); issues resolved during these times were more systematic endemic concerns. Also, note that there was at least one week where the number of raised concerns increased. Such times as these correspond to times when additional usage of the tin-quality library is occurring within the driving design center, and more engineers handling items tends to uncover more issues.

How does a library development effort reach this phase? It is at a tin-level (work in progress) quality:

- once a architecture is settled on in a technology node, together with the required goals of the library (in terms of density, performance, power, noise immunity, yield, and cost); and
- once test development is instigated as in the three-phase development effort given in Figure 15.1 and initial releases start.

15.2.2   Silver: Expert Use Only

At some point during the library development cycle (but not necessarily readily identi-fiable at the moment and place that it occurs), tin-quality, under-construction libraries morph into silver-quality, *expert-use-only* libraries. Although the moment of this occurrence may be difficult to define, especially a priori, the resulting library (and the resulting usage model for it) may have remarkably common traits.

- There is morphing from a driving customer (design center) into a driving project, possibly but not necessarily typically a test chip (the need for workable solutions leading to a test device does wondrous things). As mentioned, this is not always a test chip—in fact, usually it is not. The decision to develop libraries is usually forced on a company because of a performance, density, power, or cost issue that prevents a design center from developing an IC device in a previously used technology. As such, the design center has a time-based stake in starting development of the integrated circuit with the new library. The library development and support team is required to release the library—"let the library out," as it were—in a timely manner. As a result, the first design, which can be thought of as a *paying-customer-based test chip*, is in actuality, a "real design." Design-center synthesis engineers will be synthesizing gates from RTL using the library the day it is released. Test-vector development will start occurring just as quickly. Floor planning of blocks will begin almost as quickly. As a result, the design center becomes the ultimate test facility for the testing of the output of the library development team. Because of the need for quick turn-on issue resolution, the number of issues identified over a given time period and the number that are resolved over a given time period can both greatly increase. The continuation of the issues (Figure 15.2) therefore can resemble Figure 15.3.

- The morphing of the appearance of the library effort from "thrashing" to one of "resolving" occurs. Early library development, similar to other engineering efforts, is characterized by multiple false starts down developmental dead ends. Many times, these dead ends are not apparent for several rounds of staged attempted effort. Finally, though, the direction of development is chosen and blazed far enough that issues will be either resolved or overcome as opposed to causing dead

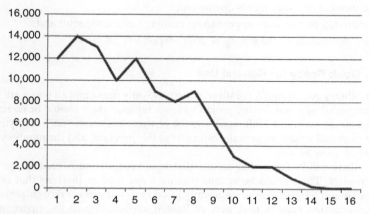

FIGURE 15.3 Rough Approximation of the Issues and Errors Uncovered in a Silver Quality (Expert-Use-Only) DDK. Note that there are wide swings, upward and downward, as the driving projects work toward identifying issues with libraries and both the library development organization and the driving project work toward resolving them. Just as in Figure 15.2, the vertical axis could represent anything (here it represents an issue-weighting severity factor times the number of issues), and the horizontal could represent any time scale (weeks here), but the key feature is the increased frequency of gyrations in the curve plus the downward trend of it. Consider this type of trend symptomatic of an expert-use-only, silver-quality library.

stop and redirection. Because the "expert use" phase of design has a driving project, together with a required schedule, management (at both the design center and at the library development center) usually enforces this discipline;

- There is recognition that learning occurs both from the library development team to the driving customer design team and from the driving customer design team back to the library development team. The designers of the driving project are the experts in the needs of the project, whereas the designers of the library are the experts of the abilities of the technology. As you can imagine, knowledge of the requirements and the abilities, closely connected, is a prime means of convergence on solutions to the design project needs.

- The knowledge that strong regular discussions between any customer using such libraries needing continuous communication and interface with the library development team (which is also sometimes known as *hand-holding*) allows for quicker development (for both the design team and the library team).

The bottom line is that this *expert-use-only* phase of library quality is a second necessary phase of development. That second phase is limited to single driving projects. The goal is to allow the driving project to work out all the known bugs and issues of a new library so that second users and all succeeding users benefit from cleaner released versions of the library.

How does a library reach such a level? Here again, it is by definition. When a first driving project decides:

- to spend resource in order to bring a new library up to such standards as to allow that project to reach tapeout, and actually does expend such resource and time;
- the quality of the library increases;
- the number of issues trends downward;
- the driving project does appear to be heading toward tapeout; and
- then, by definition, the library is *silver* (expert use only).

15.2.3 Gold: Ready for General Use

Once the library goes through the silver, expert-use-only phase and all issues are resolved sufficiently to allow a first driving project to reach tapeout, there should be no remaining bugs and issues—correct? Unfortunately, the obvious answer is "no." If anything, the types of bugs and issues can actually bifurcate at this junction, and the numbers of them may shortly grow as well.

First, there will be issues that remain that have yet to be uncovered by the first-use driving project. These could be in cells that were not used or in views that dealt with design goals and tools that were sufficiently loose as to not cause a recognizable issue or in characterization corners that were of limited applicability to the environment for which the driving project was being produced. Perhaps resolving them could have produced a smaller, faster, or less power hungry or higher yield result for the driving project, but with the project reaching its design goal, that extra benefit was not required and was, therefore, not searched for and the limit caused by the issue went unfound.

A second new source and type of issue is the class found by general-use design engineers who are using the new library as they deem correct and perhaps not with the constant hand-holding that was characteristic of the expert-use driving project phase of development. Constant interfacing between the library development team and the driving project team is a beneficial aspect of the expert-use phase because it allows knowledge to pass in each direction, as mentioned in the last section, but it also unfortunately allows library designers to influence how the library is assumed to be used. That last point is the issue. Although it is good to indicate correct usage, as assumed by the library development organization, it is bad to allow perhaps unknowing communication of incorrectly assumed ideas about how the library is to be used. However, because the library developer is a library developer as opposed to a chip designer, it is common for said library developer to have incorrect ideas on "real" library usage. With the library passing into this general-use phase of its life, the direct influence by the library development engineer on usage is minimized and previously unrecognized issues will arise. This is, in fact, how such issues as minimum-data setup-plus-hold window concerns first become known. Actual users determine issues and concerns that are not assumed such in the library team (because they were not issues in earlier technology nodes). Actual users identify the previous second-order effects that become first-order effects at each technology node better than library developers (or at least as well as library developers).

The library user in the design center is a great potential source of added library features in a new technology node as well as a source of library requirements and goals in that same technology node. It is why the driving project during the expert-use phase of library quality, with its constant interaction between the library user and the library developer, is of such value.

With the successful release of a first project to tapeout; the initial full release of the library to the general design-center population begins the gold, general-use phase of library life. This is finally a well-defined transition point, unlike the development of the tin, work-in-progress and silver, expert-use-only phases with their de facto "the release is out there" transition points. It is characterized by two events:

1. the release of the library through the company-defined mechanisms (and the assumed extant company-wide revision-control system); and
2. the identification of a new bin in the company-defined system of bug reporting and tracking.

Design-release flows and mechanisms in general (and library-release flows and mechanisms in particular) have caused significant pain across the electrical engineering IC design world for years. As a result, each company has gone through painful revision after painful revision until it is sure that its method is the best method. Unfortunately, IT engineers, or whoever is viewed by the company as the owner of the release flow, will move from company to company, bringing their previous companies' ideas with them. As a result, no release flow, even those that are viewed as successful and stable, remains so for long. For the sake of not causing arguments, as far as library release mechanisms go, yours is better than any that I can describe. That having been said, several key features of your successful release flow should be described.

- First, the files, including documentation and verification and validation reports, are tagged and time stamped appropriately.
- All metadata concerning the data structure of the library have been developed and are present, those pieces that cannot be developed until after installation with the design center are identified, and automated scripts to generate them after installation are included in the release.

- An automatically generated "current" pointer scripting is added in all levels of the hierarchy (which runs on installation on the design-center disc) to allow for multiple revisions of any particular view to be present. This allows multiple projects to work from multiple versions of the various library views concurrently, with the "currently active" projects working from the versions of the views pointed at by the "current" pointer with other older projects pointing at older revisions within the same data structure.

- Typically, the revision number contains three levels of identification that allow for major release, minor release, and "bug or hot fix" number, something along the lines of "XXX:YYY:ZZZ" (although your numbering will be different).

- All issues and bugs that are resolved in this release are documented and e-mails are sent to the instigator of the bug report and all others listed on the original bug report as desiring to be notified, asking for timely feedback on the applicability of the solution to the original bug and approval to close the bug as resolved, with automatic closure if no response is received within a certain length of time.

- The release (be it major, minor, or hot) is broadcast to all design centers across the company and external customers (worldwide if needed) as fast as possible.

All bugs that are applicable to the previous release of the library that have not been resolved by this release should be cloned automatically into the new releases bin (including automatic e-mail notification of the instigators of these issues stating that the new release does not resolve the particular issue). Also, all bugs in the previous version of the library that are resolved in this version should be automatically moved to a tentative resolution area, and e-mail should be sent to all pertinent instigators of those bugs asking for their validation of closing of same.

The number of bugs and issues for the library should average zero. Yes, previously unknown issues will arise, but proper attention should be able to resolve most, possibly with the release of hot-fix revisions of the library or minor or major revisions of it over time. Unfortunately, as the technology node becomes "old" in the eyes of the engineers of the company (especially those in the library development organization) and as these individuals move on to newer technology nodes, there is usually a general trend toward not resolving newly reported issues. This leads to the counterintuitive and unfortunate increase in the number of open issues for a library over time. As a result, Figure 15.4 better represents the number of open issues for a library.

When should a library reach this level of quality? By this, I mean the good aspects of an open release of the library for general use as opposed to the creeping loss of quality as unresolved issues tend to pile up. As experience with the design centers continues, reaching a gold, general-use quality level will be reached sooner and sooner. Certainly, though, a general release that is sufficient for general use should occur quickly after a first driving design reaches tapeout. If this does not occur, then it indicates incorrect resolution of the issues identified during the driving project phase. Rather than resolving the underlying issue, treating the symptom is being used as the method of issue resolution. Doing so, in some instances, is sufficient, but a more correct means of resolving issues is to identify the underlying problem and resolve it at that level.

How long a library stays in this phase depends entirely on how long enough resources can be spent on issue resolution. Eventually, however, library support resources will be moved to the library development of the next library.

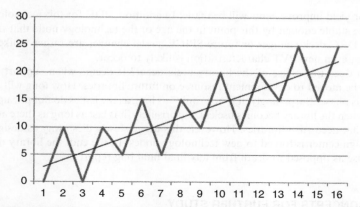

FIGURE 15.4   Representative Approximation of the Number of Open Issues in a Gold (General-Use) Library. Here the horizontal axis represents quarters since last major release, with minor releases and hot fixes taking care of whatever resolution of issues occurs, whereas the vertical axis can be represent any number. A trend line has been added to ensure that the declining quality of the library is made clear. Consider this type of trend symptomatic of a general-use, gold-quality library.

15.2.4   Platinum: Long-Standing and Stable

At first glance, a library (or other piece of IP) achieving a platinum quality level may sound impressive. It may be perceived that all IP should be driven to this point. However, in reality, it is not only difficult for a piece of IP to reach this level, it also indicates the decline in support of the IP when it does. Eventually, as mentioned at the end of the last chapter, resources have to be moved from support of previous libraries and onto the development of next-generation libraries (at either the same technology node or another, usually deeper submicron technology node). When it is determined that this transfer of resource becomes necessary, it is important to set the soon to be unsupported library on as solid a footing as possible before it is set adrift—because it will be.

If a library is to remain useful during this unsupported phase, then it is important to do the following before this loss of support occurs.

- Scrub he current list of unresolved issues and split the list into several subcategories.
  - Any issue that can be resolved and is of sufficient weight as to cause loss of usefulness of the library if it is not resolved should have resources addressed to it and attempts to resolve it.
  - Any issue that is too costly to resolve but can still cause loss of usefulness of the library, if it is possible to be isolated by removal of some views and or cells, then that effort should be made.
  - Any remaining unresolved issue needs to be well documented as an unresolved issue as blatantly as possible so that the future user group of the library can attempt to avoid the problem through design techniques.
- Finally, a remaining stable design should be run through the library in order to ensure that what remains is still a viable library.

What remains is a hands-off stable and clean but unsupported release. The remaining user's group should understand that it is unlikely that future density, performance, power, or yield improvements will be made. In addition, PDK for this technology node should be stable enough by this point in the age of the technology node that the likelihood of DRC deck changes or future SPICE model changes are rather unlikely. As a result, not even new PVT characterization is likely to occur.

When does this phase of the library cycle occur? It occurs when support resource needs to be moved to development resource on future libraries. How long will a library last in this phase? Surprisingly, it is not "only" as long as the issues do not pile up beyond a point when the library becomes useless. Surprisingly, it is last as long as there are design centers that need such libraries. However, this usage slowly dies away as design center after design center is forced to new technology nodes. In the end, the library dies from lack of use as opposed to death from any final fatal bug report.

15.3 CONCEPTS FOR FURTHER STUDY

1. Ensuring that the views of a stdcell library map to the desired models and decks that are represented in the PDK:
   - As discussed in the chapter, one goal of validation and verification is to test the various views of the various stdcells of a given library against each other (and against the models and decks of a PDK). The student or interested engineer who has access to a stdcell library that may or may not yet be self-consistent might benefit from developing a method of comparing various stdcells of that library against each other and of comparing the various views of those various stdcells against each other.
2. Ensuring through testing that the density, power, performance, noise, and yield goals of a stdcell library are actually achieved:
   - Again, as discussed in the chapter, the other major goal of validation and verification is to test the various attributes of a given stdcell library against the density, power, performance, noise insensitivity, and yield goals for that library. The student or interested engineer who has access to a stdcell library might benefit from actually measuring these attributes for that particular library and then comparing them against industry standards for stdcell libraries built in that particular technology node.
3. Setting up a multicycle development, verification, release mechanism, and flow:
   - An outline for a multitiered development, verification, and release cycle flow has been described within the chapter. A student or interested engineer might be inclined to attempt to work out a similar automatable environment where elements are regularly and routinely checked in and are automatically checked and, if incorrect or insufficient, are cycled back to the developer or, if correct and sufficient, are promoted to a release queue for use by the design center.
4. Measuring quality levels:
   - The student or interested engineer who has access to the bug and issue reports for one or more stdcell libraries might find it enlightening to review those records in order to see if they follow the trend lines as given in the chapter and, thus, give the corresponding stdcell library a quality judgment.

CHAPTER 16

PLAYING WITH THE PHYSICAL DESIGN KIT: USUALLY "AT YOUR OWN RISK"

16.1 LESSON FROM THE REAL WORLD: THE MANAGER'S PERSPECTIVE AND THE ENGINEER'S PERSPECTIVE

At the microprocessor company where I was employed when I built the 90-nanometer stdcell library, the manufacturing grid that was chosen for development of that library was 5 nanometers. However, every minimum design rule check (DRC)—in terms of spacing, overhang, and width—was a multiple of 10 nanometers. This may, at first, seem slightly strange. Why build a manufacturing grid that is twice the refinement (half the granularity) of any geometry that can be built with it? Actually, that is a slight over-statement. The *minimum* DRC rules were all multiples of 10 nanometers, but there was no restriction on building features that were an odd number of multiples of the manu-facturing grid larger longer or wider than the minimum. In addition, for design for manufacturing (DFM), this is exactly what we did. (Please refer to the discussion in Chapter 14 about lessons learned on such DFM extensions, and what they mean in terms of yield.) However, 5 nanometers is a very tiny level of granulation. At such a level, a physical layout engineer is viewing a layout at a level of magnification only slightly larger than what would be sufficient so that full DRC rules (in terms of widths or spaces or overlaps) can be seen. The slight change in a shape because one corner of the polygon being shifted one 5-nanometer grid becomes nearly impossible to detect. This is especially

Engineering the CMOS Library: Enhancing Digital Design Kits for Competitive Silicon,
First Edition. David Doman.
© 2012 John Wiley & Sons, Inc. Published 2012 by John Wiley & Sons, Inc.

true if the graphical display standard (GDS) view that the layout engineer is viewing is cluttered with multiple other polygons on multiple other GDS layers. This caused the typical layout engineer to have to zoom in (to work on a polygon edge) and zoom out (to see correct DRC measures to surrounding geometries) repeatedly when doing typical stdcell layout design. In addition to this, if the layout engineer was working over an Internet connection, perhaps from home, then the amount of data transfer for objects drawn on such a grid as compared to the bandwidth limitations of the connection would make it nearly impossible for effective work to be accomplished. Recall that the development of the 90-nanometer library technology node was the mid- to late 1990s and Internet access through the modems available at the time was not near what it is today.

Why was an awkward manufacturing grid chosen? A clue can be found in the very level of technology that was being developed. The 90-nanometer technology meant that at least some geometry (specifically, the 90-nanometer-wide polysilicon transistor gates) was required to be drawn in the minimum allowed dimension as an odd-number multiple of a 10-nanometer manufacturing grid. If a layout engineer wanted to draw a minimum-width polysilicon figure, then he or she might be apt to draw it as a *geometric path* (as opposed to a *geometric polygon figure*) 90 nanometers wide. In such a case, the turn points that define the centerline of the figure would be on the 10-nanometer grid, whereas the sides of the geometric path, being 45 nanometers on either side of the centerline, would be halfway between the grid lines. Because the 45-nanometer distances to the sides of the geometric path are not multiples of 10 nanometers, the GDS layout tools used during drawing (and the mask-fracturing tools used during mask preparation) would snap them, more often than not in a somewhat random manner, to a 10-nanometer grid. The result would be randomly crooked, randomly DRC-violating geometries. Similarly, certain VIAs in that technology were defined as 130 nanometers by 130 nanometers drawn. It was (and still is common) to predefine such VIAs as their own subcell that can be placed as needed (as opposed to forcing the layout engineer to redraw the VIA every time that such a construct is needed). Because it is easiest for the layout engineer to place such constructs by clicking on the exact center of where such construct is required to be, as opposed to clicking on some corner that is offset from the exact location, it is common practice to make the center of the VIA as the origin of such VIA constructs, just as in the 90-nanometer-wide geometric path forcing sides that want to be on a 5-nanometer grid to be semirandomly snapped to a 10-nanometer grid, the 130 nanometers by 130 nanometers VIA constructs force sides that want to be on a 5-nanometer grid to be semirandomly snapped to a 10-nanometer grid. If, on the other hand, the grid was 5-nanometers, then the same geometric paths, and the same VIA constructs, with the need for sides on a 5-nanometer grid, would not force such semirandom snapping of figure edges. Figure 16.1 illustrates this condition for both aspects. The 5-nanometer manufacturing grid was somewhat counterintuitively chosen.

However, even with the preceding analysis being a legitimate reason for such a fine manufacturing grid, making that choice did not resolve the time and bandwidth issue on being able to accurately work with such a grid. The solution I implemented during the library-development phase of 90-nanometer stdcell library was to force the drawing of all geometries on an even-number multiple of a 10-nanometer grid. This was relatively easy because, for that particular technology node for that particular company (which had its own fabrication capability), the only geometries that had such odd-number minimum DRC widths were the aforementioned polysilicon and VIAs. So, polysilicon for the 90-nanometer library was originally drawn at 100 nanometers; and VIAs constructs, which were both allowed and required by official DRC manuals to be 130 nanometers by

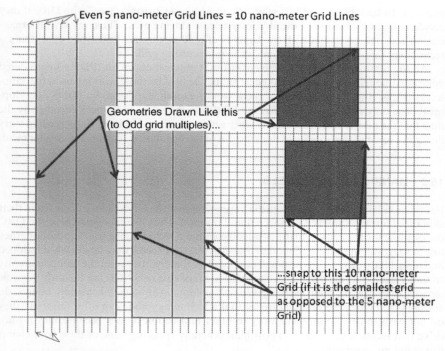

FIGURE 16.1 The Need for a 5-Nanometer Grid for Drawing Objects That Are Multiples of 10 Nanometers. Objects that are drawn with odd multiples of 10 nanometers force such fine granularity.

130 nanometers, were drawn at 140 nanometers by 140 nanometers. This allowed the library to be developed on a 10-nanometer grid, which both eased the need for constant zooming in and out by the layout engineer and reduced the bandwidth requirements (by a factor of 4 because the grid is 2× in both directions) for the layout engineers working at home. In addition, I built a special "in-house-use–only" development-usage DRC deck in which not only the polysilicon and VIA sizes were adjusted upward, but also the various spacing (for instance, polysilicon to polysilicon, VIA to VIA, and contact to polysilicon) and overhang (for instance, active extension beyond polysilicon, and metal overhang of VIA) were adjusted downward. This adjustment would ensure, if the polysilicon and VIA constructs were somehow sized down by 10 nanometers after the fact, that the original DRC would be valid. Once a cell was deemed "correct" against the cell specification, it was processed through a GDS sizing tool that shrank all polysilicon and VIA sizing by 5 nanometers per side. The polysilicon overhang of active was upsized by 5 nanometers to recover that lost by the 5 nanometers under sizing and the adjusted GDS was written back into a 5-nanometer grid database. The original 10-nanometer grid cells were kept around so that any redevelopment effort could continue using the easier-to-use version of the cell, which was always reprocessed after such redevelopment through the 5-nanometer adjustment script. The chance of improperly using the script was even minimal because it was, in a sense, self-policing. This was so because running the officially released DRC deck just before stdcell library release to the design centers would

cause the 10-nanometer gridded cells to "light up like a Christmas tree" with DRC if they accidentally made it to the official library-release staging area. Thus, my layout organization was able to design with the benefits of a 10-nanometer grid while developing a technology that required a 5-nanometer manufacturing grid.

All too often, in library-development effort, a physical design kit (PDK) is considered golden. Although it is true that this belief should be held by those working on actual chip-level designs, it is of less benefit for those working on digital design kit (DDK) development. The only real necessity is that the PDK and the DDK mesh back together by the time that the design-center user community gets its hands on the two. The preceding story illustrates some of the benefits of thinking out of the box during such library development.

16.2 MANIPULATING MODELS

Fabrication houses like to promote the impression that the GDS shapes that exist in their supported libraries are well engineered and optimal. They like to suggest that the DRC decks allow for the densest features possible within a technology node. In addition, they like to suggest that the SPICE models that describe the process were intricately developed and model the transistors as optimally as possible. Finally, they want the user community to understand that the various devices that are modeled in SPICE are the only possible devices that can be built in the given technology. Although fabrication houses do endeavor to deliver the best feature set that a technology node can offer (after all, they do not wish to drive customers away to other fabrication houses because they do not drive that technology as hard as possible), generally speaking none of these four claims can stand up to scrutiny.

First, on GDS optimization, there are too many design goals that a customer may require that are different from other customers' requirements for the fabrication house to produce an optimal library for every customer "across the board." It is not just that the three design corners of density, power, and performance need to be addressed, thus building a maximum of three libraries; one for each corner ultimately resolves the dilemma. In Chapter 19, I discuss a special-purpose 90-nanometer silicon-on-insulator (SOI) library that I produced for a high-performance design center several years back. In it, I had produced multiple height cells that fit into separate placement rows. A methodology was developed that allowed for this to be used. (See Chapter 19 for a more in depth discussion). The design center needed neither a density-optimized library nor a performance-optimized library. During the construction of that particular library, the standing joke was that we were trading between "fortnights of time and acres of silicon" because the customer needed a combination of the two. Similarly, it is easy to construct reasons for a library to be built that must match a combined density and power goal that would sacrifice some reduced power consumption for some density and some density for some reduced power consumption. It is far too easily conceivable that other combinations could be possibly requested by some design center or another. What is more, there are going to be customers who require these trade-offs in varying degrees. It is impossible for a fabrication house to produce a unique library to meet a unique customer's unique requirements. The best of such fabrication offerings will be three libraries: one for density, one for performance, and one for reduced power consumption. This is just not enough for any of the various fabrication houses to claim an optimal GDS offering.

Second, the fabrication houses may claim optimality in another manner: that the various polygons that represent the various functions have been optimized against the various DRC and optical proximity correct (OPC) requirements of the technology node in which they are produced. This is also verifiably incorrect. Placing objects in a two-dimensional array in an optimal fashion is a known NP-complete problem. It is impossible to demonstrate optimality in an automated place and route (APR) of a netlist at the block or chip level (without somehow placing the N objects in a netlist in the N! different orders and seeing which one is best in some measure or another). It is also impossible to demonstrate that same optimality in the arrangement of the polygons in each cell of the library (again, without arranging the total number of such polygons in a similar manner as previously mentioned). No library-design organization can afford the time required for such optimization, especially because that optimization will differ for each of the design corner goals as mentioned earlier in this section.

The only way that the fabrication house can viably claim optimality is in terms of ensured passage of the various stdcells against the various required DRC and OPC verification decks. However, ensured passage of other polygons produced by other library sources, such as internally developed cells, can be guaranteed through validation of these alternate stdcells against the same DRC and OPC decks in combination with using them in the fabrication-house–approved versions of the supported engineering design automation tools.

The very nature of DFM improvements, which can be added to a library offering for yield improvement at the (admittedly slight) cost of density or performance, shows that fabrication-house optimized GDS is an illusion. The bottom line is that the argument that it is impossible to improve further, for the specific requirements of the specific customer and design, is unsupportable.

DRC rules tend to be produced purely by process and lithography engineers. The generation flow is for such engineers, with a rough feel for what did and did not work at the previous technology node, to guess at a first-pass set of DRC guidelines describing prohibitions for design. These DRC rules come from the negative space. They do not tell what can be produced by layout engineers that will be allowed to be processed in the fabrication facility; they prohibit what can be produced by those layout engineers that would otherwise be allowed to be produced in that fabrication facility. However, layout engineers need to produce workable design that meets some required goal. As a result, a game of cat and mouse can easily develop between the DRC rule-generation authorities and the layout engineers. The DRC rule generator informs the layout designer that he or she cannot do anything. The layout engineer then develops some sort of work-around that meets the letter of the rules but not necessarily the intent. The DRC generation engineers then refine the rule, and the layout engineer then attempts to work around that. Sometimes, the DRC rule is eased, but sometimes it is hardened. As a result, many DRC rules can require dozens of extraction steps. The bottom line is that the ability of the process to produce features at a given technology node is at best a guess. What is more, the preceding argument can be just as easily applied to OPC rules. A layout engineer who wants to produce some feature should question any DRC and OPC rule that appears to be too restrictive for that feature. The worst that a fabrication house can say is, "No, follow the rules." Figure 16.2 illustrates a simple one-layer DRC case that occurred to me several years ago. I had produced a medium-scale integration level bloke that was so dense in one section that I could not fit some 2 × 2 arrays of VIA onto two nearly adjacent nets without violating the spacing rule on the metal layer in question. However, the violation was at a pair of opposite corners of the two metal areas in question. The DRC rule writers for the fabrication facility that was to receive the design had not thought about that eventuality. They had only assumed that minimum spacing could occur side to

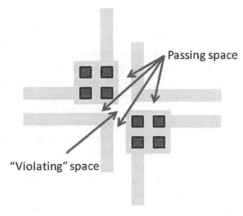

FIGURE 16.2   How to Obfuscate a Metal Spacing Rule. Layout limits forced reconsideration of this rule on a case-by-case basis, leaving those who did not realize the limits of the written rule to produce inefficient design.

side. They approved the variance for the design but never changed the rule in the DRC deck. The design was processed and made 100% functional. However, who knows how many other designs at that same technology node that did not question the rule were forced to produce larger (and less-profitable) design. Figure 16.2 shows an actual DRC rule at a 65-nanometer technology. The fabrication house will remain anonymous.

Please question the DRC and OPC rules. You will occasionally be rewarded.

Third, concerning SPICE modeling by the various fabrication houses, too many circuit designers accept the models as golden. They should never be accepted as such. The SPICE models that are received from a fabrication house are just models that appear to fit the observed functionality of the various transistors, within their limited and sometimes highly defined width and length ranges, and the associated other devices that the fabrication house has decided to support. However, in many cases, the actual "fitting" of the observed data can be questioned. As an illustrative example, a little more than a decade ago, I was asked to advise a SPICE modeling team on improvements to its methodology. At the time, the particular modeling team wanted to model a given parameter that it desired to shift upward by, say, approximately 6%, and that would eventually become part of the transistor model on a long-ago discredited process. Figure 16.3 shows the actual observed measures of that particular parameter. I was actively involved in the taking of the measures and can verify them as real. The data were described as a Gaussian distribution exhibiting some negative kurtosis (that is, the histogram exhibited "squatness" as compared to a true Gaussian distribution with an otherwise same standard deviation). It is my strong belief that this is not a Gaussian distribution at all, but a bimodal distribution, probably because of unexpected and untracked (at the time) process variation. Still, the particular parameter that was put in the nominal-case SPICE model fit the Gaussian average (even though no actual material was observed that exhibited at that average value), and the standard deviation that was used to fit the best-case and worst-case SPICE models fit the assumed Gaussian standard deviation. Admittedly, on subsequent and better-controlled material out of that factory, the observed histogram did indeed better fit a Gaussian model. On those subsequent test cases, the range of the distribution of this particular parameter was also reduced from the

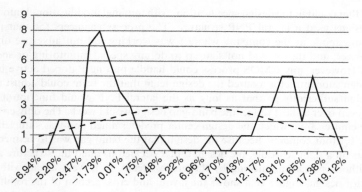

FIGURE 16.3 This Is Not a Gaussian. The solid line represents the particular observed measures of a parameter that was to be incorporated into a SPICE model. The dashed line represents the equal-energy Gaussian distribution that modeled the particular observations.

here illustrated 26% (–7% to +19%). However, with such questionable data passing as Gaussian, it is not difficult to question all parameters in any given SPICE model. This is why, as I mentioned in the chapter on SPICE, I have built a model comparison tools and suggest that such a tool be part of every library-development-and-support organization as well as every design center.

The preceding is not a rare case. I have other examples from different fabrication houses and at different technologies (although the preceding is the most extreme that I have seen). In addition, certainly fabrication houses are not in the business of randomly generating absurd data and passing them along as legitimate. I have included the preceding example not to make fun of the absurdity of it. Rather, it is important to realize that of such data, especially early in a technology node's existence, are some models made. Models are not reality. Sometimes, the reality that they reflect is to be questioned.

The fourth issue, adequate coverage of the possible devices capable of being built in a technology, will be covered in the next section.

16.3 ADDED UNSUPPORTED DEVICES

Fabrication houses publish documentation as to the various parameters for each device they monitor during processing. They deliver SPICE models (and symbols for use in schematic capture) for each device monitored. However, it is relatively impossible to determine the entire set of possible features that can be built in a process. It might sometimes be capable of building a device, admittedly, that is unmonitored in the particular fabrication house but nonetheless useful to a function. The design center may be the actual people involved in the effort of the design, including its required interface with the fabrication house. But the library team will have to be involved in the development of the required internally supporting views and, if for no other reason, to ensure that any future design team that is considering subsequent usage of the resulting intellectual property (IP) block is aware of the fabrication-house unsupported device model in it.

For example, several years ago, the design center where I was working had assigned a band-gap design to a subcontractor. That subcontractor had several designs on its own

IP shelf that could fit the design specifications with minimal modification. Unfortunately, each designs used a lateral PNP structure. The process engineers for that particular technology mode at the company where I was working did not advertise or monitor such devices, and they would not sign up to do any special monitoring or special processing. However, any CMOS process could handle putting P-doped active in N-well. The design subcontractor assured the design team that the team could adjust the band-gap design sufficiently so that as long as the lateral PNP would function as a linear device, even over a short range, the band-gap design would work. The only critical issue was the emitter area, which the design center predetermined and was sufficient, coincidentally for four contacts. Because the design that the band gap was to be incorporated onto was on a tight schedule, the design center decided to build the structure and views even without special monitoring by the process team, so it went ahead with the subcontractor design. Figure 16.4 shows the layout and cross section of the resulting PNP. As a precaution, because of the lack of certainty of the current flow in the process engineering unmodeled lateral PNP, several copies of the design were placed adjacent to each other. This was accomplished so that they could be metal connected in either parallel or series with a single mask change on top metal (and that connection was planned out so that it was directly over sacrificial metal stacks with nothing around them). This was done so that any adjustments could be first tested by means of focused ion beam (FIB) stitching of the structures as deemed necessary. A second complete set of lateral PNP with twice the emitter area was included. Again, this was done with multiple versions of the structure placed adjacent to each other so that they could also be stitched together either parallel or serial fashion at the top-level metal and substituted into the

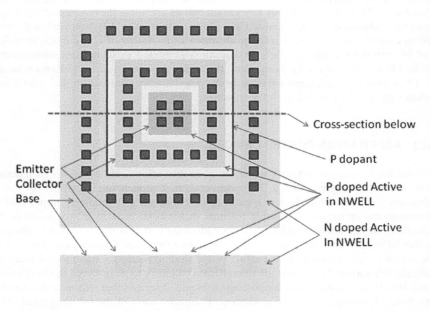

FIGURE 16.4   A Lateral PNP in CMOS Technology. Even though this structure was not approved by process engineering, it functioned properly for a band-gap design sufficient for a high-speed data-recovery communication device.

original design using that same top-metal stitching. Finally, isolated copies of both structures were included on the chip, with either needle-probeable or wire-bondable pins to the emitter, base, and collector of each.

As far as SPICE modeling goes, the first design was completed with just the best-guess assumptions concerning what it would look like. However, because of isolated test structures, which are discussed shortly, we were able to probe and pin out the isolated structure in the test floor and develop actual SPICE models for it.

The design was completed and processed. When it came out, the device current of the "lateral PNP" was not what it had been estimated to be beforehand. It was roughly a third. With that in mind, on several of the initial parts, we stitched together three of the adjacent structures together (well, two additional ones to the original). Each design was sufficient to pass the communication-channel data-recovery specifications. With that, we adjusted the top-level metal mask to include the three lateral PNP in parallel, with all of the others connected to ground. The design was a success. What is more, the band-gap design, as well as other such designs from the subcontractor, was used in several more devices at that technology node, the only difference was the removal of the extensive redundant lateral PNP structures in order to reduce device area.

Just because a fabrication house does not support a device and just because process engineering says "No, we cannot monitor such a construction," this does not mean that it is unable to produce that structure. Do not be afraid to put such structures on a test design, evaluate it, and use it as needed.

16.4 CONCEPTS FOR FURTHER STUDY

1. Half-rule DRC deck:
   - One way to check to see if any two cells of a given stdcell library can be placed next to each other in an APR environment is to place every stdcell next to every stdcell (including themselves) and in every possible flip and abut arrangement. Although for most stdcell libraries this remains tractable for left–right abutment, it becomes practically intractable for top–bottom abutment (as stated in the chapter, because of the large number of possible adjustments of the top–bottom placement by a grid at a time left or right from a true north–south alignment). One means to get around this issue and allow a priori analysis of each cell being placed adjacent to another cell in any arrangement and in any legal direction is to build a half-rule DRC deck. A half-rule DRC deck checks that for any given polygon width or space rule, no width or space of half that much occurs for that particular polygon shape to the cell boundary. In practicality, this deck is slightly more complicated than that because of the need to test for some features that do not have specific half-rule width or space restrictions. For instance, if there is a rule stating that no three VIA can be placed at minimum space between them, then every cell has to have no more than *one* VIA a half-minimum space from the cell boundary. The student or interested engineer might find it useful to build such a deck and, if possible, run it against each stdcell in an available stdcell library.

2. Double-grid layout and GDS adjustment script:
   - As discussed in the chapter, some deep submicron technology nodes require manufacturing grids that force layout engineering to move consistently into and

out of high levels of magnification in order to ensure that polygons stay plumb. One way around this is to do the actual build of the stdcells at a larger manufacturing grid and then to adjust the polygons that are required to be on the smaller manufacturing grid by means of a script after the fact. The student or interested engineer might be inclined to attempt this. Perhaps this can be accomplished by doing a 90-nanometer stdcell (or set of stdcells) on a larger grid, writing the GDS out of the layout tool, then reading it back in on the finer manufacturing grid, and then building a GDS adjustment script (perhaps in SKILL) that adjusts the VIA polygon coordinates inward to the finer manufacturing grid.

3. Lateral PNP and lateral NPN structures and layout:
   - The student or interested engineer might be interested in developing lateral NPN or lateral PNP structures in a given technology node, especially if those devices are not supported in the vendor's PDK. Once accomplished, these can be placed in the layout, together with bond pads to each of the terminals that would be required to characterize the structure. The resulting combined circuit can be placed on a test device mask for processing, packaging, and analysis of the performance of the structures across many parts across many wafers across many lots once they come out of processing.

4. High-yield improvement, medium-yield improvement, low-yield improvement DRC adjustments:
   - In Chapter 14, I suggested that the student or interested engineer might be inclined to build a weighting factor into a DFM deck that would allow for weighting of some DFM rules over others. Such a deck implies increasing the likelihood of those issues with DFM adjustments being made as opposed to the down-weighted DFM issues. The concern is that the binning of the DFM was rather arbitrary at that point. One way to make them less arbitrary would be to base the high-impact DFM rules on the high-impact DRC, the medium-impact DFM on medium-impact DRC, and the low-impact DFM on low-impact DRC. Tor this to occur, it is important to understand which DRC fall into which of the three categories. The student or interested engineer might find it beneficial to discuss with various process engineers familiar with a given technology which DRC are nearest to the edge of failure and to thus bin the DRC appropriately.

5. Gaussian parameter study:
   - The student or interested engineer who has access to various transistor and backend stack parametric across several test devices across several wafers across several lots and possibly across significant physical timeframes for a process might find it enlightening to attempt to do Gaussian analysis studies on those parametric measures.

CHAPTER 17

TAGGING AND REVISIONING

17.1 LESSON FROM THE REAL WORLD: THE MANAGER'S PERSPECTIVE AND THE ENGINEER'S PERSPECTIVE

Several years ago, I was supporting a 90-nanometer stdcell library that I had created. That particular library had been given to several companies because of some consortiums. The technology node that the library was based on had a minimum metal area rule in it. This was the first technology node where such design rule check (DRC) issues had existed. The automatic place and route tools available at the time could not handle such issues; the library exchange format (LEF) had to ensure that such minimum metal area issues could not occur during place and route "by construction." That particular minimum metal area rule was sufficiently large that the middle metals in any stacked VIA construct would have to encroach on one or more of the surrounding router tracks on whatever particular metal happened to be the sandwiched layer of the stacked VIA. If the area was grown sufficiently large for the rule and done so in a symmetrical fashion, then it would encroach on all four surrounding router grids.

Because stacked VIA were desirable by the design teams that were using the library, I experimented with drawing four separate versions of oblong versions of the minimum area and placing them in the part of the physical LEF file that was used to define VIA constructs. The hope was that the router that was then used would choose from the four stacked VIA constructs such that the direction encroached was minimally disruptive to the surrounding routes placed there by the router. Unfortunately, router technology at that time was insufficient to the task. The router always choose the first of the four constructs

Engineering the CMOS Library: Enhancing Digital Design Kits for Competitive Silicon, First Edition. David Doman.

listed, even when this particular choice hindered the next route over and even if one or more of the remaining directions were empty and could have accommodated the encroachment with ease. Because letting the router make the encroachment choice was not working, and because the router could make a choice of how to route between nonstacked VIA, I worked with the various design centers to remove the stacked VIA construct altogether. That way the router would have to route over at least one router grid from a VIA from one layer onto the particular layer that would have been the intermediate one in a stacked VIA construct and then to the next layer. Doing so legislated the minimum area rule away. All one-grid routes were large enough to pass the rule.

I happened to put a digital checksum in a comment in the LEF file and a second hidden one in the UNIX directory that contained the release, and then I sent the file out to the design centers of all the companies in the consortium. The next morning there was a bug report on the new release saying that the LEF produced minimum metal area DRC violations. Just after I read the report, and although I was checking to see if my earlier calculations had been in error and that a one-grid route was not sufficient to pass the minimum metal area rule, my boss and his boss came in and asked why I was releasing buggy databases. While they were both in my office, the vice president in charge of the design center that had reported the bug called me and asked why I was releasing buggy databases. While he was on the phone, the designer that had submitted the "bug report" called with the very same question. But then, in front of everybody as we all were listening and while I was trying to come up with some defense of the situation, the designer explained that he had seen the DRC in the stacked VIA constructs produced by the LEF. I knew that I had removed all of such constructs in the release but wondered if it had somehow gotten back into the release after I had removed it, perhaps by a download of some backed-up early database. I quickly downloaded the attachment that was part of the bug report, which was the LEF that the designer had used. I reviewed the checksum. It was different from the one that I had sent out. I had the entire released database downloaded by that time and ran a checksum on the version of the LEF that I had released. That version matched what I thought it had to be. I asked the designer where he had gotten this particular LEF. He said that he had gotten it from me. I informed him that he had not and that I had the checksum proof. He then told everybody listening, my boss, my boss's boss, and his vice president that the LEF in question was virtually the same as the one that I had released; the only difference was that he had just added stacked VIA constructs into it before he ran the place and route.

That was the last time that I had to worry about that particular bug report.

17.2 TAGGING AND TIME STAMPS

17.2.1 ASCII

There are two legitimate reasons for tagging. First is for a form of data control (being able to tell when a file might have been modified), which will be discussed shortly, and second is as an aid to chip design quality and revision control, which will be discussed toward the end of this section.

With a couple of noticeable exceptions—the view of the polygon database, known as graphical display standard (GDS), and that of various compiled versions of otherwise readable precompiled views—almost all engineering design automation (EDA) tool-compatible views tend to be human-readable (ASCII) format. In many ways, this makes the job of the library-support engineer easy. First, everything *is* human readable. This means that these views can be edited, compared across versions, adjusted for additional capability (in new cells to extend the library functionality; new views to extend the library usability; new pressures, volumes, and temperatures to extend the library applicability; or new tokens in given views in order to extend the library portability across multiple methodologies). In addition, the Backus-Naur format (BNF) for several of the major tools is open source, giving the library-support engineer the capability of valid full verification against the tool (as opposed to the "does it compile" version of verification otherwise capable without such access). The downside to all of this is that it gives the design-center engineer casual access to all of the preceding and in equal measure as with the library-support engineer. Therein lays the issue. It has often been observed that engineers in general, and design-center engineers in particular, have a natural tendency to tinker. This is a valued attribute for design-center engineers to have and design-center managers should look for such attributes in their staff. The buzzwords "constant improvement" are a given in the design-center engineering community, and searching constantly for better, easier, and cheaper ways to make something happen is one path toward the goal of this constant improvement. However, library-support engineers have serious headaches because of the often good or even excellent but sometimes deluded or even dangerous adjustments that these well-meaning design-center engineers may have.

The opening case in this chapter is a perfect example. The design-center engineer was interested in improving the routing capability of the library by adding something that the library-support engineer had left out "probably because he was unaware of the capability within the tool." The design-center engineer added it to improve the release and was probably going to send in some sort of an "improvement for next time" suggestion once he saw that it worked.

In a sense, because of this ability of design engineers to "see inside" the views that library-support engineers develop and support, these library-development-and-support engineers "wear targets." Library users always know better ways to improve the library. In many instances, those improvements are either explicitly incorrect or only specific to certain instances and detrimental to alternate projects that might use the same release.

Please note that some "changes" are accidental and apparently inconsequential but ultimately disastrous. A case in point: every UNIX engineer has heard of the "put a space here" answer. Developed more than 40 years ago, UNIX is fairly syntax forgiving but has the occasional strict formatting of commands. In those specific strict interpretation parts of UNIX, an added or a missing space on a comment can cause improper operation. Most UNIX engineers know that they are experts in the UNIX world when they can look over the shoulder of a less-experienced UNIX engineer who is flummoxed by a command in a script that is not working correctly and point at a particular section of the non-working line and say "Put a space here." Things like that can happen in some library views as well. For instance, when a design-center engineer gets a new release, he or she may just open a file in edit mode, because of laziness, just to read it and inadvertently add or subtract such a dangerous space as the preceding story relates, thus making the file potentially uncompilable.

When the "improvements" are inevitably added and eventually break something, how does a library-support engineer prove what is a "good" enhancement (and owned by the library-support team) and what is an "ad hoc" enhancement (and owned by the design team)? After all, perhaps a library was sent out with broken issues. One cannot just say, "It didn't work, so it is something you did." It may not be possible to just do a UNIX DIFF on the database because of the potentially hundreds of files involved. Time stamps

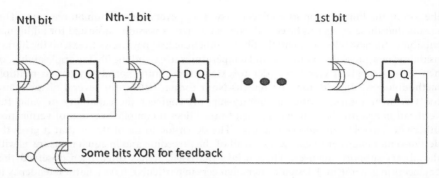

Nth bit   Nth-1 bit   1st bit

Some bits XOR for feedback

FIGURE 17.1 A Cyclic-Redundancy Checksum Tool (Usually Incorporated in Software). Linear feedback scan register (LFSR) taps (the nodes that are used to do the actual feedback) determine the rate of possible error "mismatch" aliasing. For this reason, be careful to use taps that produce maximum nonrecurrence.

can be useful, and most library-delivery systems look at file modification times to determine the need for updating with newer versions of files. However, depending on how a library release is instantiated within a design-center directory structure, it can be sufficient to just inspect the last time a file was modified. For instance, perhaps as part of the design center's project "MAKEFILE" a library file is "UNIX TOUCHED" so that the MAKEFILE will force a build from the library onward for that given project. One way to do so without having to do such file compares over vast quantities of "before release" and "after release" files is to place checksums, both visible and invisible, in the various ASCII files and UNIX directories. Many such checksum software tools exist. They all do similar operations. Specifically, as illustrated in Figure 17.1, they scan the data of a file serially through some version of a linear feedback shift register (LFSR) (in software) to develop a unique key for that particular file, which is then incorporated into a comment line within the file. This incorporation effort is done in such a manner as to be transparent for future rerunning of the tool. Any other change to a file after the checksum is incorporated, however, will produce an alternate checksum. Without proper access to the tool, however, this new checksum will not be incorporatable into the file. A fingerprint has been established within each and every file in which such a checksum has been inserted that indicates if it has been "touched" in a meaningful (and perhaps detrimental) manner since the original checksum had been established. A second checksum on each directory of a library release and placed in a UNIX DOTFILE (a file usually hidden and not usually touched) that incorporates the size of each file within the directory can localize on inspection the directory in which the modified file or files may reside.

Design-center engineers are creative, not malicious. Their adjustments to a library release are meant to be improvements, not detriments. As a result, simple checksum algorithms that are not meant to be secure are needed. Identifying files that have been changed is the goal as opposed to identifying malicious and potentially hidden changes. Checksum algorithms, such as CRC-32, a 32-bit, error-detecting, cyclic-redundancy checker, are probably sufficient.

The previously illustrated algorithm is a true error-correcting algorithm and can be built in hardware, making it of great importance and usage. It is often used in chip design to validate data inside memories and across communication channels. Simpler versions of this algorithm that are used include parity (equivalent of a 1-bit CRC), universal serial bus (USB) (which uses a 5-bit CRC internally when generating packets), integrated services digital network (ISDN) (which uses an 8-bit CRC), and code division multiple access (CDMA) (which uses a 30-bit CRC). An open-source pretty good privacy (PGP) algorithm that incorporates a 24-bit CRC for the known published feedback pattern even exists. The IEEE CRC-32 for IEEE-802.3 encoding feedback bits are: "32; 26; 23; 22; 16; 12; 11; 10; 8; 7; 5; 4; 2; and 1." The IEEE CRC-32 feedback bits have been shown to be good (that is, they have little chance of allowing aliasing or hiding errors) in such areas as Ethernet packets, and they might be excellent for checksum efforts for library files. However, many typical checksum tools allow the user to define the size and the feedback bits.

The second major reason for this type of tagging of files is to allow a design center the ability to measure the level of quality of the various pieces of intellectual property (IP) that are present on a design. Many modern design-cycle times can run into months and involve hundreds of engineers across multiple locations with regularly (or irregularly) scheduled design builds during that extensive effort. There is a series need to understand that the IP (down to and including stdcells and IOs) is up to date (or at least at the level of quality it was when it was included in the particular design build). Knowledge that a particular functional cell is new versus one that has seen silicon versus one that is partway between those two extremes might be very useful for a design manager to foretell a design quality level. Sometimes a released design needs to be resurrected for any one of a number of reasons, and it becomes paramount to track the quality level of the various pieces of IP from the stdcell and previously. For instance, a bug report may come from a customer, and the archived design needs to be put back onto the system to allow some testing in an attempt to re-create the issue. Another case might be that the design is to be redone in a smaller technology node. Another reason might be the decision to include part of an earlier design as a subcircuit on a new design. Adding a tag inside a comment field in each view of a stdcell library (or IO library or compiler release) allows this level of tracking to take place (at the least for the stdcell and IO IP). In that case, it is important to align the comments to what the design-center manager is expecting and to determine what the rest of the IP (system level, for instance) is that is being incorporated on the design. This allows for easier auditing of these comments on design builds. Important items might include date of design, revision of design, and a level of quality (from inception or creation to known good silicon).

17.2.2 Binary

From the previous section, recall that there are two legitimate reasons to include tagging in released library views: (1) to ensure that the view has not been changed, either inadvertently or otherwise, after the release from the library-development cycle and (2) to allow for quality-level auditing of designs in progress or released from a design center that happens to use the released library. Both functions become difficult when dealing with binary (i.e., non-ASCII) data.

Two main classes of non-ASCII tool views are the compiled versions of the aforementioned ASCII views, the main one being the compiled liberty view (usually known as .DB files) and the actual graphical representation of the physical polygons, also known as GDS.

Coverage in this section can be perfunctory. Because they are just compiled versions of extant ASCII views that can be properly tagged and have checksums placed inside comments in the tools, it is possible to handle data validity in one of two ways. Both means might be considered slightly suspect because of their reliance on the design-center engineer rather than reliance on an engineering resource within the library-support team responsible for data validity. From the previous section, recall that the time stamp is not a realizable trace tool because of the common design-center practice of using UNIX TOUCH on some foundation views (such as compiled liberty) in order to force UNIX MAKEFILES to build from the base foundation views. The first way is to compile the liberty files within the library-support organization and deliver them extant. This means, however, that although there is a trace capability from the delivery of the view backward up the supply chain, there is none downward for these particular views, to the design. If an update to a liberty occurs, then there can be issues with existing designs. If the design center decides to accept a liberty update, but a particular design being done in that design center is approaching timing closure with an older version of the compiled liberty, then there is no way to review the a posteriori database and determine whether that design was closed with the older compiled liberty or the newer compiled liberty. The only alternative is rerunning the actual timing check with the new compile, which is redundant and wasteful and undesirable. The second way shares this fault and pushes reliance onto the user community as opposed to the developers' community and is even less desirable. Because it is impossible to determine if the compiled view is up to date without opening a compiled liberty within an EDA tool and extracting it to review the comments, some library-support teams do not deliver the compiled view. Instead, they rely on the design center to do its own compiling of the liberty files. Both ways are problematic and weak, but currently, when combined with one additional admittedly weak validity check, they are the best scenarios for whatever amount of traceability is possible. The weak validity check is that the liberty-compiling algorithm is deterministic. It will always produce the same-sized output given the same input liberty and input switch settings. It is possible to compile the liberty within the library-support organization, determine and record the size of the compiled version, and add a check to the design-release checklist that the size of the compiled liberty file be equal to it.

The graphical display standard is a non-ASCII database as well. The ability to place readable checksums and tags in the view so they are readable from outside an EDA tool is as somewhat limited as they are in the case of compiled liberty files. It is true that it is possible to place text strings that are visible from inside a GDS viewing tool, but that is different from what is possible for most of the other views described so far. For each ASCII view, a simple parsing of the file could display the pertinent information and that information could then be checked in some manner without regard to the EDA tool for which that particular view was developed. However, unlike compiled liberty, GDS is stored as a *heap* in a computer memory. In general, heaps remain very useful for many database manipulations problems and are a somewhat problematical computer memory-storage technique for validating library releases (where some of the data in the release—in this case, the GDS—are stored as a heap). Specifically, heap-reading algorithms literally rearrange the heap database during a read. The typical heap features that force this are shown in Figure 17.2.

Unless you are a software engineer or enjoy doing such things in your spare time, it is not necessary to understand a specific HEAP_READ algorithm, just the basic heap-access mechanism. The HEAP_READ inherently sorts the data in a heap according to some key, pops (reads) the top element (smallest key), and then promotes the next keyed element (next smallest key) to the top of the heap. Writes onto heap are similar, but the order of the elements in a heap is *not* preserved. Can it happen coincidentally? Yes. Does it have to happen? No, it cannot be counted on to happen.

- **HEAP Features:**
  - Tree format
  - Every node has a "Key" Hash value whose value is less than the value of the key of the parent node (or more than for reverse HEAP trees)
  - Keys can be anything hash-able, for instance:
    - a timestamp;
    - a composite of information on a piece of data; or
    - a random number

- **HEAP_READ:**
  - Determine the child node that has the largest key (or smallest for reverse HEAP trees) and promote it to the parent node and pop (read) the original parent node

- **HEAP_WRITE:**
  - Find a leaf node or a not fully populated parent node whose key is greater than (less than for reverse HEAP trees) and push (write) the data at that location.

FIGURE 17.2   Typical Heap Features Relative to Why Usage Causes Nonstable Data-Structure Reads. Binary GDS is a heap. Any verification of GDS needs to understand this and prevent this from becoming a problem.

Although the size of the database does not change because of a heap read, there is no guarantee that the order of the structures in the database will remain the same. This presents a validation problem because reading an entire database of polygons might be necessary to see the one validation token while realizing that the first token in the database during library release probably will not be the first token in any subsequent viewing of the database and does not invalidate the database.

Although there are multiple translation tools that will translate GDS into ASCII format and back again, these are problematic because of the very need for the added steps for translation. Because of the nature of the storage of GDS (note the preceding discussion concerning heaps), the translation between the binary and ASCII format assuredly produces "changed" polygon ordering in the posttranslated database. So how does a validation engineer ensure that the GDS is correct and up to date? Here even the loose validation mentioned in the preceding discussion on compiled liberty is too strict. It is possible to insert text that is readable with a GDS "STRINGS" command into a GDS structure, although it is possible to ensure, because of the size of the database not changing, that new polygons have been added or subtracted. However, it is not possible to ensure that the polygons have not been altered "slightly" (the shifting of an edge, for instance) but insufficiently to change the number of edges of any given polygon. Here, then, is where reliance on the nondevious nature of design engineers must come into play. Any GDS database can be manipulated outside of the library-support environment to such a degree as to make it broken and useless. That manipulation cannot be seen by inspection of the GDS, although it might be seen by means of chip-level verification software that will do design rule checks (DRC) and layout versus schematic checks

(LVS), along with other runs such as layout parasitic extractions (LPE). The validation of this view can only be perfunctory.

- GDS exists.
- It is the same size as that which was released.
- When a STRINGS command is run, the correct text is present.

The real level of validation must then happen at the design-release phase, where DRC comparison to known good PDK-based design rules and LVS comparison to known good library-released carrier-to-interference ratio (CIR) circuit files can occur. Beyond the preceding, the only option available to the design-center manager is periodically to write the GDS out to an ASCII format and do some amount of data sorting and comparing. This is done to see if any polygon has been adjusted. This is doable, admittedly, but it is suboptimal.

17.3   METADATA, DIRECTORY STRUCTURES, AND POINTERS

The term *metadata* has come into the lexicon to mean the token words included in Web-page HTML header sections that help search engines tag those Web pages for reference during customer searches. The more and diverse the collection of applicable token words in the Web-site header, the more likely that matches for various and sundry pertinent user search criteria will occur and the more likely the search engines will direct the user to the Web site.

In terms of library support, metadata have come into use for related although not exactly similar reasons. First, they assist in understanding where certain data are located. Second, they help understand the validity and applicability of those data for the user's needs. Slightly contrary to the location of the metadata in modern Web-site headers, library database metadata are usually in separate files and usually at each level of a directory structure.

Modern library digital design kit (DDK) releases can be exceptionally large data structures covering more than a gigabyte of data, and possibly much more. Many design centers will also require the maintenance of multiple versions of a given library concurrently (because of the need to support or debug previous designs that were built with earlier versions of a library while simultaneously designing new integrated circuits with the latest design-center–accepted versions of libraries). Many design centers do not have enough concurrent and unused disc space available to allow placement of entire library releases without breaking the data structure into smaller chunks. This is true at least until the purchase of additional disc space, which is not always an expense-level item, allows immediate purchase and can require long lead times, whereas the capital expense goes through whatever signoff loops are required. On receipt of a new library release, the design-center manager usually has the local IT personnel load the various directories and files where possible and then include UNIX pointers at appropriate points within the dispersed directory tree to allow what should be seamless navigation through the structure. The bad news is that design-center IT personnel are not necessarily design-center users or library developers. Not all the links and pointers that a typical designer would want are perceived by the IT personnel and immediately properly added. The design-center users can work with the IT people and add what is needed over time. However, this process can take days, is repeatedly error prone, and

TABLE 17.1   An Easily Parseable, Nearly Flat Directory Structure (Top Level). Although the human-readability argument would indicate that flatness of directories is undesirable, modern EDAS tools have no such requirement. Maintaining flat structures, on the other hand, allow for easier directory structure verification and validation, among other testing.

Technology Node (for instance 20 nanometer)/	
	GDS/
	CIR/
	Extracted CIR/
	Physcial LEF/
	LIB/
	DB/
	VERILOG/
	DOC/
	REPORTS/
	Other Uncompiled View (for instance:Celtic)/
	Other Compiled View (for instance; CDB (compiled Celtic))/
	continued ...

can be described at best as ad hoc and haphazard. With slight additions to each directory, and with a little expertise in software development, installation tools can be added to the library-release mechanism that can foreshorten and stabilize this process greatly.

For this stabilization to occur, the first item is to develop a workable basic directory structure for the release of a library. The term is *workable* as opposed to *humanly comfortable. Workable* in this context means "easy to implement, easy to scan and parse, but still human readable." Such a structure is a directory that is as nearly flat as possible. The flatness of it may cause it to be "humanly uncomfortable," it being easier for a human to read and understand—say, five subdirectory titles in a more hierarchical directory structure—and to guess which subdirectory to explore for the desired data file, as opposed to reading and comprehending 20 subdirectory titles. However, each one having only pertinent data files inside, the exact opposite is true for parsing activities. It is easier to k while parsing, for instance, that only uncompiled liberty are in a LIB directory and only compiled liberty are in a DB directory than to guess that perhaps both are in a SYNOPSYS_VIEW directory. As another benefit, it seems more natural to check if a DB directory and a LIB directory have the same number of files, with similar naming as opposed to digging through a directory that has both and determining if all LIB files have matching DB files. With that in mind, one such of a plethora of "nearly flat" directory structure might be of the form shown in Table 17.1.

In the table, the views types are all listed at the same level of the hierarchy. The other uncompiled (ASCII) and compiled (binary) views illustrate how additional specific design-center desired views might be added. In many cases, notice that there are uncompiled and compiled (or at least nonextracted and extracted) versions of views. Walking down through the list, all directly under the technology node, note the following.

- GDS: All polygon-based sources of items that can be placed together on a design should be in the same directory (probably in separate GDS files under the directory). This is to say that if IP from different sources (for instance, internally developed and acquired third party) are desired, then they should be maintained in separate subfiles (which allows for separate revision control as the opportunity arises). There also are instances where a set of technology-node–consistent structures are okay to be put on a device but cannot be matched with devices of the same technology node but slightly different processing (for instance, deep N-well or isolate P-well versus common P-well structures). These should be placed in separate technology node directories.

- CIR: With the exception of glue structures such as gaskets and gapers that do not have extractable structures inside the physical view, a corresponding SPICE netlist for each item appears in the various GDS. If the GDS is in separate subfiles, then it should also be in the CIR, which is also known as *circuit design language* (CDL).
- Extracted CIR: This corresponds to the extracted SPICE netlist that was used in the characterization for each item in the GDS and CIR directories and in the same manner of directory structure as for those two subdirectories with the exception of those gapers and gaskets with no extractable devices within.
- Physical LEF: This corresponds exactly to the GDS, including gapers and gaskets, and in the same directory structure.
- LIB: Uncompiled liberties for timing, power, and noise calculations are found in separate subdirectories for each division as in the GDS mentioned previously.
- DB: Compiled liberties for timing, power, and noise measurement and synthesis are in corresponding separate subdirectories for each division, as in the LIB.
- DOC: This file contains white papers and usage notes for the intellectual property and the technology node.
- REPORTS: This includes validation reports and verification reports on the various views of each of the various pieces of IP in each subdirectory.
- Other compiled view: This includes noncompiled and compiled view subdirectories as the specific design center requires.

Not shown but present in both levels of the preceding directory structure are hidden (for instance with UNIX DOTFILES) metadata files that describe what should be in each subdirectory and the path to those subdirectories. These paths, on release point within the immediate directory but with proper software development, can be adjusted as desired as the various subdirectories are moved to other locations on a disc farm by the IT personnel. Notice that there is not yet a version or date identification nor is there any IP source identification in either level of the directory structure. Having these at a minimum of at least one more level down in the hierarchy is a personal preference because doing so allows the potential to easily mix and match releases of subdirectories (revision 2.1.0 of a Verilog may not need to be updated when revision 2.2.0 of a LEF occurs). With that in mind and with a division of process, voltage, and temperature data, a layering might look as in Table 17.2. In it, a CURRENT pointer has been added to point to the most recent release of a view.

Note that there are separate verification and validation subdirectories and these are under the IP source level. This may be a splitting of minutia, but there is a difference between the two forms of quality data, and this is explored in another section of this book.

In addition, as noted for the first of these tables, hidden metadata files exist at each level of the directory tree.

TABLE 17.2   An Easily Parseable, Nearly Flat Directory Structure (Three Additional Levels). Design-center–required EDA tools can and will cause some changes (in terms of both additions and deletions) to this structure.

Technology Node (for instance 20 nanometer)/			
GDS/			
	IP SOURCE 1/		
		CURRENT» REV 1.0.0/ through REV X.Y.Z/	
	through IP SOURCE N/		
		similar as above	
CIR/			
	similar as above		
		similar as above	
Extracted CIR/			
	similar as above		
		similar as above	
			BESTCASE_PVT/ TYPICALCASE_PVT/ WORSTCASE_PVT/ other PVT as desired by design center
physical LEF/			
	similar as above		
		similar as above	
LIB/			
	similar as above		
		similar as above	
			BESTCASE_PVT/ TYPICALCASE_PVT/ WORSTCASE_PVT/ other PVT as desired by design center
DB/			
	similar as above		
		similar as above	
VERILOG/			
	similar as above		

(Continued)

TABLE 17.2   (Continued)

Technology Node (for instance 20 nanometer)/			
		similar as above	
DOC/			
	similar as above		
		similar as above	
REPORTS			
	similar as above		
		similar as above	
			VERIFICATION/ VALIDATION/
Other Uncompiled View (for instance: Celtic)/			
	similar as above		
		similar as above	
Other Compiled View (for instance; CDB (compiled Celtic))/			
	similar as above		
		similar as above	
continued . . .			

17.4 CONCEPTS FOR FURTHER STUDY

1. Exploration of various EDA open-source Backus-Naur formats (for instance, Verilog and liberty):

   • To greatly increase the user community for their various tools, many EDA companies have released the BNF description for their view-description languages (two such open-source languages are Verilog and liberty). The student or interested engineer might be inclined to investigate the BNF for these two (and other) formats versus how they are actually used within a released stdcell library.

2. Minimum-length cyclic-redundant checksums:

   • Guaranteed minimum length cyclic-redundancy checksum feedback patterns that ensure minimum chances of aliasing are known for all word bit-count widths up through the ranges useful to a stdcell-library support engineer. Many Web sites can produce such a list. The student or interested engineer might find it useful to attempt to create a tool that will use such an LFSR checksum by passing the various ASCII-format stdcell views through it, thus producing a checksum for that particular view. The student or interested engineer could demonstrate how changing a line of ASCII description within such a view (or even reordering some lines within the view) demonstrably breaks the original checksum thus produced.

3. Investigate tagging for view data security purposes with a series of released stdcell-library views:

- The ASCII checksum approach to stdcell-library ASCII-view tamper resistance is not the only such method used to ensure that some stdcell-library views are not adjusted. The student or interested engineer might be inclined to investigate what is done within any vendor-released stdcell library to which the student or engineer has access. If none exists, the student or engineer might be inclined either to use that described within the chapter or to invent an alternative.

4. Investigate metadata and pointers for revision-control purposes with a series of releases stdcell-library views (especially for a stdcell library that has gone through multiple releases):

- The metadata and pointer approach to stdcell-library revision-control and view-consistency assurance is not the only such method used to ensure that some stdcell-library views do not get out of date with reference to other views, stdcell-library release to stdcell-library release. The student or interested engineer might be inclined to investigate what is done within any released stdcell library (by third-party vendor or internally developed) to which the student or engineer has access. If none exists, then the student or engineer might be inclined either to use that described with the chapter or invent an alternative.

CHAPTER 18

RELEASING AND SUPPORTING

18.1 LESSON FROM THE REAL WORLD: THE MANAGER'S PERSPECTIVE AND THE ENGINEER'S PERSPECTIVE

"There is never time to do it right, but there is always time to do it over." This old truism in the integrated circuit design world applies more to actual design of integrated circuits than to stdcell library development. At deeper submicron technology nodes, although it remains possible in terms of both time and resource to always "do it over," the shear cost of even a single mask makes a mistake virtually too expensive to recover from. In the future, do not be amazed that only the highest volume devices will "fix" poor design whereas other "mistakes" are either resolved by a software fix or by removal of the proposed feature for the marketing of the device. The truism for stdcells is more along the lines of "It has to be done right, and it has to be done now."

A couple of years ago, I was working in a digital design center, supporting its foundation intellectual property (IP). That foundation IP, which included external vendor devices of various types, included internally developed stdcells from another central library organization. The stdcell library for one of those technology nodes was optimized for density "at all cost." That library, among other issues, had removed the output buffers from the latch sequential cells (in order to remove as much added area as possible from the resulting placements). Figure 18.1 gives the basic circuit of these unbuffered latches. Unfortunately, removing the output buffer meant that the external to the cell route was now partially responsible for the performance of the latch. The liberty for the latch, however, did not consider this output loading. As a result, the library

---

Engineering the CMOS Library: Enhancing Digital Design Kits for Competitive Silicon,
First Edition. David Doman.
© 2012 John Wiley & Sons, Inc. Published 2012 by John Wiley & Sons, Inc.

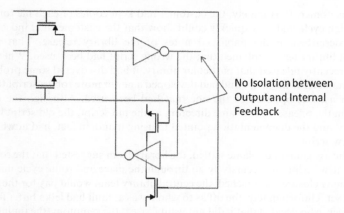

No Isolation between Output and Internal Feedback

FIGURE 18.1 An Unbuffered Latch. The fact that there is no isolation between the output of the latch and the internal latching feedback loop means that any output load will affect the performance of that latching mechanism and render the cell effectively useless for common ASIC place and route usage.

became an expert-use-only library. Haphazard usage of these unbuffered latches could easily cause broken silicon.

Of course, the central library team was aware of this previously described possibility and had developed a methodology to resolve the issue. The team had a script that would read the parasitic load on the output of the latch, calculate how much that load would affect the performance of the latching, and then adjust the input parasitic of the route enough for it to emulate that affect. Realizing how dysfunctional this workaround really was, the team took great pains to communicate the absolute need for the use of this script during any automated place and route (APR) of netlists in this particular technology, which contained such unbuffered latches.

The design center in which I then worked had an external customer who did indeed use that particular library and had synthesized a netlist with several hundred of such unbuffered latches (which amounted to a significant percentage in both instances and in the area of the design center's design). The application engineer in the design center who was involved with the external customer was well aware of the hazard of such cells and had strongly stressed to the customer the absolute need to adjust the parasitic file by means of the script before timing closure could be completed. Duly forewarned, the customer had run the script over repeated APR timing closure attempts; although the process was difficult, the customer had reached timing closure, and the design went to mask and processing. When the design came out, it was dead. This was okay, however. The technology node was such that the entire mask sets were fairly cheap as compared to the price of the design, and the date (of a trade show) when the design had to be ready was months off. All that was needed was some circuit debug in order to find the issue, fix it, and rerelease it to new mask and processing. The engineers in the design center, along with the customer engineers, quickly narrowed the issue down to an unbuffered latch not grabbing the appropriate data at the appropriate time.

The central library team was notified that there was an apparent issue with the unbuffered latch. Its response was to say that we must have not run the adjustment script as we had been instructed and that it was not the team's fault. The cost of the new mask and processing would have to be either attributed to the design center or passed onto the customer. Fortunately, the customer had kept copies of each file for each step of the design cycle and very quickly could show that the customer had indeed used the script as described in the published notes of the library release. That was when the central library team explained that the wrong script had been used. When the team had most recently released that particular library, it had discovered some problems with previous versions of the script and had developed a newer more robust script that should have been used as opposed to the one that had been used. Although the new script had been in the release, in the same directory as the old script, the old script was indeed still present and the documentation pointed to using it and, in fact, had nowhere said to use the new script.

With the assignment of blame settled, the library team suggested that the solution was to use the new script and recursively go through the place and route cycle until the new current timing closure was reached; the central library team would pay for the cost of the new mask set. Unfortunately, the tiff as to whom was at fault had left a bitter feeling with the customer, who decided that it did not want to treat the symptom (the timing around a poorly performing circuit being adjusted) but instead wanted to treat the problem (and get the poorly performing circuit actually redesigned). The design center went back to the central library team and asked how long it would take. It was told that it would take months because this was an old technology node and the central team was off developing deeper technologies. That is when I got involved. The actual layout of the unbuffered latch, even though its stated reason for existing was for density support, was designed with a loose layout of its circuit. I sat down with a library layout engineer, and we determined that an inverter could be added to the output of the cell at the cost of just one grid. We built the new cell on a Friday afternoon. I wrote the adjusted Verilog for the new circuit by adding an inversion to the output (actually removing the intrinsic one that was already there). The parasitic layout parasitic extraction (LPE) was extracted, and over the weekend we characterized the new cell. Monday morning the customer had the new cell in the library offering. It was not happy that the new cell was one grid larger than the previous cell (recall that the customer had used this latch significantly in the design, and the one extra grid per latch caused a significant amount of growth). In addition, the customer did require an entire resynthesis because of the inversion on the output. However, the customer completed that and APR and trimming closure by the end of the day that Monday. It went to mask shop that same week. When the part came out of fabrication, it passed all testing and the customer made its release date. Since that time, that particular customer has used that buffered latch in more than a dozen additional designs, all of which have been first-pass successes.

Meanwhile, on the day that the good silicon came, the central library team announced that it was going to add a buffered latch of its own design to the library that would be out in about 6 more months. Although it demonstrated the lack of customer support that a central organization can exhibit, the design actually caused issues. Specifically, that particular company had a policy of not certifying a design unless all circuits had been released in an "official" library. The "design center special" buffered latch that was by then demonstrably working and would be in several more of the soon to be dozen additional designs was not going to be part of the central library official release. When all was said and done, the design center released its own version of the library as its own "official" release. The central library team never did release its own version of the buffered latch, but it did update the documentation of the library to point to the correct unbuffered latch adjustment script and removed the older version of the script from future central releases.

Releases of internal databases by central development organizations can be clean and exhibit remarkable degrees of synergy. However, supporting such released libraries—especially when they contain poorly developed cells or views or pressure, volume, and temperature (PVT)—can rather quickly and easily develop into painful customer-support situations (for both internal and external customers). Just as modern design flow requires that multiple libraries exist within a design center (see Chapter 14), development and support of the library is required to be much closer to the final user (whether internal design centers or external third-party customers).

One final comment should be made about proper levels of circuit design in relation to design for manufacturing (DFM). Although the extra grid in the buffered latch caused the design to grow, it produced a working piece of silicon as opposed to the smaller but less well-designed unbuffered latch.

18.2 WHEN IS TEST SILICON NEEDED FOR VERIFICATION?

In one of the very first CMOS library-development efforts in which I was involved, I was asked to build a high-powered inverter out of several smaller library-based inverters placed in parallel. Much to my relief, that inverter happened to be used in a test circuit for use in a test chip. When the test device came out of processing, it first appeared to not function correctly. The first thing that I did was go into the lab and place test probes down onto the input and output of the inverter to make sure that it wasn't something that I had done. Although I was quickly relieved to find that my inverter was not an issue, this story does illustrate a viable answer as to when test silicon is required for proper verification (and validation). Test silicon is required whenever the functionality and performance of a circuit is called into question and its verification is desired. What level of circuit personalization of a stdcell or view or PVT or what level of addition of a cell or view or PVT that a test circuit is required depends highly on the technology node at which the modification or addition is being added. The addition of a higher drive strength inverter at a 1.25-micron VHSIC level technology node, especially when it is based on parallel connections of previously supplied inverters and probably does not rise to the need for silicon verification. However, the creation of a combination level shifter and metastable double FLIP-FLOP for a high-speed communication channel, especially if that circuit involves analog structures, might require such a test device.

Table 18.1 provides an incomplete list of the situations (not circuits) that might require a test structure.

Test silicon is probably justified for flight-critical circuitry, for more than minimally complex circuitry, for anything that is not straight digital circuitry, and for strongly performance-critical circuitry. This is especially true if the performance limit is based on "pushed SPICE" (that is, using non–3-sigma SPICE models, possibly even self-modified).

Conversely, for this last category, if the circuit is for a testable path in actual silicon then test structures might not be required, the actual part becoming the "real" test circuit. This is true because any customer part that does not pass the final test, where such performance testing of the timing critical part of the circuit could and should be tested literally, is not shipped to the customer. In effect, the performance issue is actually a yield issue. Badly performing silicon on the actual device forces it to be "broken." Remember the clock-buffer ground-plane illustration in Chapter 14. The "unfixed" ground plane effectively is converted into a performance yield issue as opposed to a technical issue as

TABLE 18.1   Situations That Might and Might Not Require Test Silicon. This list is not meant to be all-encompassing. All such entries can be influenced by the technology node for which the design in question is implemented and by the amount of estimated risk versus historically demonstrated ability of the designer of the circuit in question Your design center's list of structures that might require test silicon will vary from this one, perhaps significantly.

Probable for Test Device
  –MSI or Larger Circuit
  –Custom Physicals at technology nodes below 130 nanometer
  –Anything on Airframe Flight Critical Designs
  –Performance Critical Structures (such as High speed Communication Channels)
  –Analog or Mixed Signal

Less Probable for Test Device
  –SSI Logic (especially above 130 nanometer)
  –Circuits with ONLY Modified Logical or Temporal Views
  –Yield Improved (or Reduced) Modified Physicals
  –Consumer Product Circuitry
  –Derived Physical Structures

long as there is a production test capable of finding the performance limiter. That is what could and should occur in this similar condition.

On the other hand, for simple digital logic for most technology nodes, any structure at the small-scale integration (SSI) level, possibly even some simple structures at the medium-scale integration (MSI) level, especially those derived from similar structures already in the library, building test circuitry is probably not necessary. Similarly, for modified logical or temporal views, assuming that those modifications are viable and faithful renderings of the actual logical or temporal performance of the cells, test silicon will not be required. In addition, any circuit that is only modified to improve yield probably do not require test devices. Even one of these few basic categories can be nullified for technology nodes deep enough to cause lithographic considerations. Anything in the realm of technology requiring that optical proximity correct (OPC) should be at least reviewed by the fabrication house before it is committed to real customer silicon.

Aside from "the customer requirements insist on it," which should trump all other considerations, the decision as to whether test silicon should be produced comes down to a cost–benefit decision. In terms of engineering resources, time, and money to produce the test structure and ensure it lands on a timely test shuttle circuit and then actually test the resulting silicon for proper operation, is the cost offset by the benefit of potentially finding silicon hindrances? One way to analyze the cost aspect of the issue is the same as the need for logical test structures within a given piece of digital logic within a system on a chip (SOC). In the niche business of sales of logical test analysis circuits that ensure observability and reachability for circuits that may be deeply embedded within a modern SOC device, various engineering design automation (EDA) vendors try to sell their product to SOC system-integration design customers with attempts to prove that certain embedded circuits might unknowingly not be functioning correctly. Because these circuits are so embedded inside the SOC in question, without the use of the embedded circuit analysis tools being offered, the circuit will go on with undetected (and undetectable) failures for the life of the design (causing at minimum a potentially significant performance hit). These vendors suggest that the permanent addition of their logic-analysis tools within the SOC, with the area penalty that results, make detection of these faults much more likely. In theory, this is true. In actuality, it can be too intrusively costly to make it a useful addition (for either the EDA-based analysis circuit or the time- and resource-based costing test silicon).

18.3 SENDING THE BABY OUT THE DOOR

18.3.1 Validation Reports

Recall that the definition used in this book for validation is how well the cell, view, PVT, or library (or other SSI structure) meets its stated design goals. Effectively, validation is the justification of the library or other structure. It gives the potential user in the design center, the information that demonstrates whether the library will meet his or her goals for it. This in turn allows the design center to better estimate the feasibility of chip designs done by using that library. With that in mind, there is an obvious need to report on the level of matching of those design goals. This is because the design user community, both internal and external, will be rather interested in how well the released item actually meets the goals given for it.

Part of the documentation that should be released is the result of active validation of the library against specific validation measures. These measures should include such items as the following.

- *Uniform density testing circuits*: These netlists contain large percentages of the entire library release, very strongly preferring all cells in the entire library, in roughly uniform distribution so that the cells can be automatically placed and routed in a given density, thus demonstrating the libraries' place and route capability (including the ability to connect to all pins on the cells).
- *Multiple representative density testing circuits*: This collection of typical netlists, most likely supplied by the various design centers being supported, represents typical usages of the library, including typical distributions of cells in the library, such that these netlists can be automatically placed and routed in "stressed" conditions (with some significant portion of the available router resource on several router layers held back, representing actual chip-level router resource that would be reserved for chip-level integration, thus demonstrating the density capability of the library in typical design-center designs.
- *Multiple representative timing closed high-performance netlist*: This is a collection of netlists again supplied by the various design centers being supported that represents a library's typical performance capability. The closer these timing closed high-performance netlists are to reality, the more likely that they can be directly used—as soft netlists, firm netlists with placement restraints, or even hard blocks—by the design-center user community;
- *Multiple representative large circuits*: Again supplied by the design-center user community, these represent a library's typical power usage and noise nonsusceptibility capability. These probably are not directly useful by the various design centers unless they represent useful hard IP such a microprocessor cores.
- An *all-cells test case*: Together with a series of performance rings of various lengths and various composite cells (such as INVs, NAND2s, NAND3, NOR2, NOR3, NAND2–NOR2 combinations, and NAND3–NOR3 combinations) this is used

with the actual circuit being added to any test vehicle that might occur so that the functionality and performance and power usage of the cells can be manually tested on the test bench by those interested.

- Even *representative collections* of the cells of the release, all pin-accessible in a package-capable test-chip test circuit, are included such that stress lifetime yield studies can occur.

Documentation in the form of validation reports as to the results of all of the preceding is probably not only a good addition to the release but also necessary. The various design centers to which the release is targeted will be extremely interested in this level of information (and library justification).

For all of these structures, the actual results should be compiled in as fine a granularity as the engineering resource and time limits available will allow and released in the documentation section of the full library delivery. In such a manner, the design-center user community can be best enabled to take full advantage of the release and simultaneously make the fewest "please explain" bug requests back to the library-development-and-support staff.

If the three-cycle development flow discussed in a Chapter 17 is used (see Figure 15.1), the third cycle is continuously running these various design centers' supplied netlists in the background. There is no added engineering effort in generating these validation reports other than actually compiling the data from the last round of validation from release.

Another piece of validation documentation that is important to include is the trend analysis between the cells of the release demonstrating how:

- each drive strength has output load regions in which it is most applicable as opposed to the other drive strengths for the given cell family;
- the cell areas and power usages follow consistent trends (for instance, where INV are smaller and consume less power than BUF, NAND functions, and NOR functions and smaller and consume less power than AND functions and OR functions, and similarly for inverting Booleans functions being smaller and consuming less power than noninverting Booleans functions, etc);
- to give individual cell documentation such as density and power and estimated DFM yield increase over some baseline;
- to demonstrate noise susceptibility baselines per cell or cell family; and
- per-cell "quality estimates" match what is in the per cell tagging text strings, which should give a reasonable estimate as to the general quality and previous usage of the cell in earlier designs at the same technology node. These allow design managers to judge the overall quality of a chip during stages of the design cycle by polling the databases in order to see numbers of cells that have been previously successfully used in previous designs versus new renditions (see the chapter on tagging for a more extensive description).

Please note the strong commitment to the active acquiring of test structures from the various customer design centers. This is critical for at least two reasons.

First, the circuits supplied will be closer to the future actual usages of the library by the design centers. These are more likely to show issues early during the validation cycle of the library, before actual release while still in the hands of the library-development staff as opposed to later once the release has occurred and the library is in the hands of the design-center user community. This not only saves time in developing corrected views but also saves face for the library-development staff and reduces the otherwise useful workload of "finding library bugs" for the design-center engineer.

Second, it more closely engages the design center into the development phase of the library, allowing closer ties between what the library-development team can supply and what the design-center user community requires and expects to receive.

18.3.2   Verification Reports

Verification of a library, as stated in the earlier chapter on the subject, is the actual testing of the different library cells and views against each other and against the design-center methodology-required EDA tools. This is done to see if these various views are accurate, mutually supportive, and consistent with the correct version of the correct tool.

Conceptually, the first of the purposes of the verification report section of the documentation for a release is nothing but a collection of reports showing:

- what the resulting reports reveal for each cell and for each view of that particular cell against the EDA tool (or tools) for which it is designed to operate within;
- how the particular views interrelate with each of the other views;
- how the individual cells interrelate;
- which version of each EDA tool the particular view was tested against; and
- any exception or issue that any particular view of any particular cell has with any particular pertinent EDA tool or version

The generation of these reports should not be difficult. Indeed, they should be readily available as part of the cell and view development process. Each view of each cell must go through a prescribed list of verification activities. This list runs from SPICE simulation of the carrier-to-interference ratio, both pre- and postextraction, in order to determine functionality, performance, power consumption, and noise immunity, through design rule check (DRC) and layout versus schematic extractions (and LPEs), and through logic and test view simulations, including chip level (or at a minimum, intercell level placement and routing experiments).

All of these various steps of the testing process generate reports. Once those test reports come up either completely clean or "clean enough" and the cell (or particular view of the cell) is declared to pass that test, then the report can be archived for later compilation into a final library-release verification report document (or series of report documents). If resource allows, then just prior to release it probably makes sense to rerun the entire test on all the views of all the cells, compiling the result. However, this can be resource intensive (both human and hardware) and time consuming if the number of tests, views, or cells is extensive. Therefore, the archived good reports might suffice (assuming sufficient care is given to forcing the rerunning of any particular previously passed test if a fix for a later test causes the previous test to be called into question). Which of these two methods is chosen is really a library-development-team decision, in conjunction with the customer-design-center management approval.

The purpose of these reports is the demonstration of the viability of the various cells of the library, together with the demonstration of the interrelatedness of the various views of the cells with each other and with the pertinent EDA tool that is being used within the design-center methodology to the design-center engineer. In fact, it is saying that the design engineer should not see any issue with any particular cell or view beyond any items that are already listed as exceptions somewhere within the verification documentation. If there are any exceptions, and if those exceptions are limiting in terms of how the cell can be expected to behave (for instance, in terms of performance), then it would be useful to include "issue work-around" methods and ideas for the benefit of the design engineer. As a secondary benefit, it is partial protection of the library-development engineer from too many design-center engineer questions. Such documentation allows the library-development-and-support engineer the luxury of responding with "Read the documentation."

A second goal of verification reporting is to list the characterization PVT corners and LPE and characterization methodology that was employed in the generation of those PVT corners. A listing of available liberty files should go beyond "just the PVT" and include the following:

- the PVT;
- the thresholds and edge rates within each PVT;
- the percentage roll-off for both the setup and hold constraints that was assumed during the SPICE runs;
- the monotonic or nonmonotonic nature of each PVT liberty and whether this is different for the timing or the constraint or the power or the noise arrays in the particular PVT corner;
- the nonnegative value or negative allowed nature of the each PVT liberty;
- any added margin into the various timing or power or noise arrays as agreed to by the design-center management;
- the assumed usage (for instance, setup measurement versus hold measurement versus power measurement versus noise-susceptibility measurement);
- the top-plate versus non–top-plate nature of the LPE extraction that was used for each PVT liberty;
- the assumed minimum, typical, or maximum parasitic backend stack that was used in the LPE for each PVT liberty;
- the extent of resistance capacitance (RC) accuracy for each LPE for each PVT liberty, or even if it was RC versus "lumped C";
- any cells internal to the liberty file that have been declared "Don't use" or "Don't touch," together with an explanation as to why (and in what situations those tokens might be safely removed; and
- anything else that can be questioned on a resulting liberty.

The reason for this secondary goal is to document the correctness of the PVT views of the library. It tells of the extent of the appropriateness of the PVT toward the requirements for them as given by the design-center user community.

All of these pieces of information are or should be both readily produced by the characterization flow and a priori agreed on by the design-center management. Production of this documentation is both of minimal effort (just compiling the extant information) and of high value, proving to the design-center team the previously mentioned correctness of the PVT views.

A third goal of verification documentation is to list possible demonstration circuits, including soft register-transfer language (RTL), firm netlist, or hard blocks, illustrating the use of various new or innovative cells and views.

The reason for this third goal for verification is readily apparent. New cells or extended (or new) views will be called into question by the user community. Indeed, such additions will be typically viewed as "risk items" with no history of resulting good silicon. Demonstration of the usefulness of those cells and views answers such questioning, at least partially, in an effort to demonstrate the value of adding them versus any associated risk.

Unfortunately, the level of effort for the documentation for this third level of verification report is the only real labor-extensive part of the entire effort. It requires design resource. True design resource within a library-development organization is unlikely. Demonstration of the usefulness of new and innovative library additions will require a priori approval by the design-center user community to the extent that it is willing to supply the resources needed to produce the verification.

18.4   MULTIPLE QUALITY LEVELS ON THE SAME DESIGN

Every piece of IP used at any design center in the world has some level of quality attached to it, even if it is not defined in any official or view-detectable manner. Every engineer can tell a quality design when he or she reviews it. That quality can be thought of as representing the degree to which that IP fulfills the goals for its design. At the cell or library level, the quality may include the density and performance (and power consumption and noise immunity) of the cell. It may include some way of determining its likelihood of surviving usage over the proposed lifetime of the part in which it is used (because of electromigration protection). It may include some way of determining its ability to maintain its performance capability over that lifetime of the part (because of its ability to withstand hot carrier injection HCI effects) or its ability to yield across hundreds or thousands of instantiations on parts that are produced in the hundreds of thousands or millions. At the MSI or LSI IP level, it may have to deal with its proven conformity to some standard or another. It may include information on the circuit's demonstrable ability to interact with other vendors' designs of similar functionality during "plug-fest" (events where various vendors of IP, usually based in communication channels, get together to demonstrate their IPs are usable in systems and networks that are comprised of collections of various vendors' functional versions of the particular IP). In all instances, the quality is a measure of how likely (or unlikely) it is that the IP in question, if used in a design, will cause some future issue with that design. Defining a metric for that quality level and somehow attaching that metric in the tagging of the various views (as described in the tagging section of this book) and updating that metric tag for every release is a valuable addition to any library and should be included in the release flow.

As mentioned in Chapter 14, in the modern SOC design world it is likely that a SOC design will need to contain cells (or entire libraries) that are of divergent levels of quality (in whatever manner it is defined and measured). These various cells will not only have the same function but also possibly even the same lineage (one cell or library having been developed from an older cell or library). In that chapter, the reason given was that the inclusion in the SOC of previously developed blocks of potentially hard (or firm) IP that contain previous versions of the internal library releases or refer to externally developed and released libraries. These "other" libraries, either the earlier released and less quality-improved internal library or the externally acquired and quality uncontrollable external library, will have varying degrees of quality from the blocks of the SOC that will use the more up-to-date versions of the library (again, either internally developed and controlled or externally acquired). The same can be said for the other various blocks of IP on a

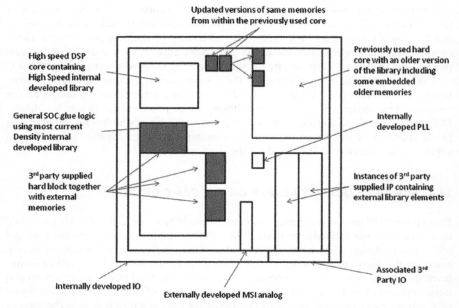

Updated versions of same memories
from within the previously used core

High speed DSP
core containing
High Speed internal
developed library

Previously used hard
core with an older version
of the library including
some embedded
older memories

General SOC glue logic
using most current
Density internal
developed library

Internally
developed PLL

3rd party supplied
hard block together
with external
memories

Instances of 3rd party
supplied IP containing
external library elements

Internally developed IO

Externally developed MSI analog

Associated 3rd
Party IO

FIGURE 18.2 A Typical Modern SOC Requiring Various Levels of Quality IP. The modern SOC can contain IP from multiple internally developed and externally acquired sources. The ability to allow some metric to judge the quality level of each such piece of IP is of utmost importance to the management of risk of such a device. Development of this metric, the active updating of same, and its inclusion in the library-support-team–released views of the various pieces is a valuable *internally generated* addition to the views of the IP and its regular polling by the design-center customers a valuable risk-mitigation method during the design of the modern SOC.

modern SOC, from memory arrays to IOs to SSI- or MSI-level analog. Figure 18.2 gives a representative illustration of such a modern SOC. The various blocks represented in the illustration are indicative of the types of IP that could easily be found on the modern SOC. They include multiple versions of the library (because of older versions instantiated in hard-block IP). They include multiple versions of memories (again, because of older versions embedded within hard-block IP). They include third-party libraries, IOs, memories, and analog sections. They even include several versions of the most current internally developed libraries (some sections using high-density versions of the library versus other sections using high-performance or low-power versions of the library).

How does a library-support team enable such multiple divergent levels of IP quality? The definition of the metric itself has to be something that is defined in consultation with the design-center user community. However, the preceding discussion gives a decent list of possible items that should be considered, at the least for a starting point.

- some actual measure of the density and performance (and power consumption and noise immunity) of the cell (and the library);
- some way of determining its electromigration protection;
- some way of determining its ability to withstand HCI effects;
- some way of determining its ability to yield;

- some way of measuring its demonstrated acceptable usage in earlier designs;
- some manner of measuring its proven conformity to some standard or another; and
- others as the design-center user community deems fit for addition.

The metric also needs to be flexible enough to allow updates of the metric in the same manner that the IP on the design is updated. If the only revision of IP within a design is at the block and library level, then the metric is not as difficult to maintain and only requires modification when major IP releases occur. However, if individual cells within a design can be modified, than the metric has to be modifiable at this level. If various views of cells (or PVT of library) are modifiable, then that is the required level of active quality-level measurement and tagging.

In addition, it is important to enable the design-center user community actually to use this added information within a SOC development (otherwise, the effort at developing and maintaining such quality tags is wasted). Some scripting of polling software tools that are capable of automated reading of the tags (in both the ASCII and binary views of the various pieces of IP) and compiling the results into human-readable updates might have to be developed or bought. The usage of such scripts will need to be encouraged. It is not just sufficient to grab the quality tags of the various IP "during the final design release to tapeout" phase of the project. Any quality issue that is realized at that point is either going to cause unrealized delay in order to allow it to be fixed or going to remain in the design, thus limiting that design in some manner or another. Instead, it is important to develop a regular quality-review schedule within the SOC development cycle. It should occur when the design's quality tags are determined, thereby allowing any discovered quality issues to be resolved as early as possible and as often as necessary during the design-center development process.

Proper development of this quality-tagging metric, in combination with the usage of the automated polling schedule, can allow it to morph into an entire design-center tagging metric. As an example, it can be used to tag the RTL developed by the design-center engineers so that those RTL blocks can be placed on some IP shelf within the design center (or across the company) so that the IP developed within the design center or company can more easily be judged valid for reuse.

18.5 SUPPORTING "BUG FIXES"

In library IP design, mistakes happen. Usually, these mistakes occur in one of two manners: random or endemic. *Random mistakes* are easy to understand. Specifically, the mistake is the problem. Fix the mistake and the problem is resolved. In *endemic mistakes*, the mistake is a symptom. Fix the mistake and you have treated the symptom but possibly have not solved the problem. For endemic issues, the mistake can keep recurring. Unfortunately, during both the library-development and library-support phases, recognition of reoccurring issues might not occur easily enough. As a result, such reoccurring mistakes tend to get out to the end customer repeatedly (either internally in the design center or externally at a third-party design house). When this happens, the library-development-and-support team can very quickly become known as a poor-quality organization. This is why tracking issues (or "bugs") during both development and support is both a benefit to the customer base and a benefit to the library IP organization. Regular and systematic review of incoming bugs, tracking of extant bugs as they get resolved, and, most important, postmortem dissection as to the underlying cause of the bug on closer examination (including directed analysis toward finding similar or related previously reported and possibly closed, bugs) are important. Such activity can uncover such endemic issues before they besmirch the fine name of the library organization.

However, engineers, just like anybody else, do not want to attend more meetings, especially those that are not directly related to their design and development activities, and a regular meeting to review the current cumulative issue list will not be viewed any differently. Indeed, because the issues in such a cumulative list will be concerning problems uncovered in their own previous work output, they will easily take attending such a meeting as a painful disincentive toward continuing to work in the library organization. A regular (weekly or otherwise) review is not going to be received well. Alternate means of accomplishing the goal of exhaustive tracking and review will need to be established. That is where the bug-reporting mechanism comes in.

A proper bug-reporting mechanism will force continuous cataloging and categorizing of issues into bins so that any recurrent endemic concerns become apparent. A bug- or issue-reporting tool (for instance, on a company's intranet) should be more than just a mechanism to record the name and phone number of the requestor combined with the issue at hand. It must have a certain amount of expert-system–level questioning such that the user of the tool is led to a particular bin, depending on how appropriately that bin might "fit" the question. The more refined (and probing) the questioning, the more likely the mechanism will easily uncover endemic issues. Certainly, the questions do need to cover the name and contact information of the requester, as well as the technology node and an attempt to get as close to the specific piece of IP to which the requester is referring. In addition, a brief description of the issue or question, plus a separate and more detailed peeling of the concern (including any attachments of pertinent files at the user's behest), should be part of the reporting system. In addition, however, it should attempt to determine the area of concern (lack of compilability in a tool, lack of meeting a perceived goal, lack of a view to represent some feature accurately or correctly, and so on). If the user is requesting an update (for instance, to support a new version of a design-center approved tool), then the bug-reporting mechanism should be able to determine the basis of the need for this update. As mentioned, the more extensive the capability of the reporting mechanism, the more likely that valuable information on the quality of the IP as perceived by the user community can be gleaned. This can then be fed back into the library-development-and-support organization in order to attempt to fix the underlying endemic "problems" while the individual members of the organization resolve the "symptoms." One caveat: the level of questioning can become too aggressive. If the users of the reporting system start to complain about the need to answer so many questions, and especially if they stop using the system, then the *perceived value* of the bug-reporting system is lost to the user community, as is the *actual value* to the IP development-and-support team.

Having said that, a second leading cause of bugs in a view of some such piece of IP can be viewed as a pure communication problem such as the following issues.

- The purpose or goal of the view, cell, library, PVT, or other piece of IP was not originally adequately communicated by the customer to the IP development organization.
- That purpose or goal was not adequately communicated to the responsible design or layout engineer.

- The correct time and resource requirement was not communicated back to the IP development management in a timely manner to allow for proper resourcing.
- The method of verification and validation of the IP was not communicated to the verification and validation staff.
- The resulting verification and validation reported issues were not adequately communicated back to the development engineer for resolution.
- The proper release of the clean views of the clean IP was not communicated to the end customer.
- The end customer did not communicate changes to the original specified goals.
- The end customer did not use the IP in the manner in which it was designed even though it was designed to the correct (original or otherwise) specified goals.

Although it is impossible to prevent fully such miscommunication (or complete lack of communication) a priori, the bug-reporting system can also be used to alleviate such issues in both a priori and a posteriori fashion. As mentioned at the beginning of this section, humans make mistakes. Anybody can miss a comment, either oral or written, be they the engineer responsible for the generation of the IP, the engineer requesting it, or the manager attempting to adequately resource its development. At a minimum, all of these should be pushed toward resolution by the bug-reporting system. Although the system cannot *make* one understand the list of information being communicated, it should be able to prompt the owner of that particular stage of the communication to stress particulars, thereby making the receiver of the communication more likely to *perceive* that information.

As an example, when a design-center user inputs a request for a piece of IP to be developed, the tool could ask that the requirements be cut and pasted into a field (or even a reference to a standard), or it could specifically and repeatedly ask "give me another specific requirement" until the list of requirements is complete. The second of these two options is the better. A detailed checklist can be developed that can be used to prompt the developer of the IP along the lines of "Have you verified this requirement against you IP yet—and, if so, how?" Such questioning can be viewed as another form of the "issue questioning" mentioned earlier in this section. Although the goal there was proper categorization of the issue, and the goal here is preparation for "checklist prompting," the mechanism useful for both goals is remarkably similar. This level of automated prompting also can be useful at the resourcing level. An assigned development engineer should be telling the tool how much time and resourcing will be required to complete the task. The tool then can prompt the manager about support and can prompt the development engineer as time goes by as to how the effort is or is not meeting schedule. Finally, such an extensive bug or issue-reporting mechanism can be reporting to the original requester on the development as well. In that manner, before-the-fact communication is institutionalized, thus reducing the likelihood of miscommunication, and after-the-fact communication is forced, thereby increasing the likelihood of early detection and easier resolution of problems that might arise from any earlier misconstruance.

All of these features add value to the bug-reporting tool and should be part of the system. Several good EDA productivity tools that certainly contain the hooks needed to allow such internal company-based issue-reporting systems exist on the market.

Identifying the *actual* mistake, as opposed to the *perceived* mistake, is important. In the opening section of this chapter, in the story of the unbuffered latch, that particular central library organization immediately but incorrectly perceived the problem. The customer, or the design-center supporting application engineer, had "messed up" by not using the standard parasitic format (SPF) modification script correctly or possibly not at all. Added time and effort had to be wastefully brought into the situation in order to demonstrate to the library team that neither of these was the case. The central library organization, again perceiving incorrectly, then eventually came around and admitted that the release of the incorrect script was the real problem. Wastefully added time and effort had to be brought to the situation, this time by management, to demonstrate that the script was just a patch and the underlying problem was the actual cell. As mentioned, that particular central library organization has still not fully and correctly identified, let alone resolved, the actual issue. It is an often overused tautology, but the following old adage remains a legitimate and valuable way of viewing customer-based bug reports:

- Rule number 1: The customer is always correct.
- Rule number 2: When the customer is incorrect, refer to rule number 1.

As with all truisms, following this one 100% of the time is improper. Although the customer (that is, the design-center engineer) knows the market into which the final design will go better than the library developer, the library developer knows the capabilities of the technology (and possibly the EDA tools) better than the design-center user. Sometimes it is important to merge the two views into a compromise result as opposed to continuously "letting the customer have his way." A word of caution, however. When such is the case, the library developer had better come to the table with legitimate compromise solutions. Just saying "You can't get that" will be viewed by the design center as obstructionism.

18.6 CONCEPTS FOR FURTHER STUDY

1. Unbuffered outputs of transmission gates in stdcells that go directly to those outputs:
   - There are instances where the use of pure transmission-gate output circuits in a design is the correct approach. Most of the time, such instances involve custom or semicustom techniques. A potentially viable APR usage might be for development of a given (and well characterization capable) clock-tree multiplex structure that is easily isolated from the remainder of the APR. The typical stdcell library usually makes stdcells available to the user, typically in a separate "Don't touch, don't use" library and directory structure. The student or interested engineer might find the development of some "naked transmission-gate output" stdcells of interest. Care should be given in developing a set of restrictions as to where and how such stdcells can be used without allowing significant risk for dead silicon as a result. Checker scripts will need to be developed that would check for proper usages and would be automatically run in such a manner as to not allow devices to reach tapeout with having gone through these checker scripts successfully.
2. Unbuffered outputs of stdcells with internal feedback from those outputs:
   - There are instances where the use of stdcells with unbuffered outputs that feedback into the stdcell circuit is viable. These usually occur in custom and semicustom techniques. The almost never occur in APR techniques. The student

or interested engineer might be inclined to demonstrate the risk of such circuits by building SPICE netlists that demonstrate how some loading can force any such circuit, even with large amounts of setup and hold requirements built in, to fail. The student or interested engineer might develop a script to determine, for designs that use stdcell libraries in which such structures have been delivered by a third-party vendor, when they are used within a design-center or customer design. This could be accomplished so that the design center or customer can be educated in redeveloping those sections of its netlist so that those cells are not used.

3. Investigation of a stdcell library's validation documents:

   * The student or interested engineer should develop a habit of *always* intensively reading a stdcell library's released *validation document* set. Analysis of a stdcell libraries release, in an attempt fully to understand it, should be the result.

4. Investigation of a stdcell library's verification documents:

   * The student or interested engineer should develop a habit of *always* of intensively reading a stdcell library's released *verification document* set. Analysis of a stdcell libraries release, in an attempt fully to understand it, should be the result.

5. Proposal for support of multiple stdcell library releases with multiple quality levels on a single design:

   * As mentioned in the chapter, eventually the library-support engineer will have to support an SOC design that has multiple releases of any given library and has multiple vendor IP elements. The student or interested engineer should develop a methodology of library element name-space modification and IP tagging and directory structure definition that allows for such.

6. Investigation into a vendor's bug-reporting system:

   * Being a library-development-and-support engineer means that sooner or later one will have to use some issue (or bug) reporting system, either a third party's to which one enters issues or one's own to which others enter issues that the library-development-and-support engineer must resolve. The student or interested engineer who has access to such a bug-reporting system should explore its intricacies. The student or interested engineer who does not have such access should develop such a tool (or assist in its specification) for use by the library-development-and-support organization.

CHAPTER 19

OTHER TOPICS

19.1 LESSON FROM THE REAL WORLD: THE MANAGER'S PERSPECTIVE AND THE ENGINEER'S PERSPECTIVE

In Chapter 14, Gertrude Stein's quotation "A rose is a rose is a rose is a rose" was slightly parodied as "A NAND2 is a NAND2 is a NAND2." There it was used in negation, claiming that this is not necessarily true. I should like to point out that sometimes "A library is a library is a library is a library." Several years ago, I was working in the library-development-and-support organization for a large semiconductor company where I produced what still remains the densest [and because of intellectual property (IP) sharing through various industry consortiums] the most widely used 90-nanometer libraries in the industry. Within that company, my development team was known for producing dense libraries. Located in the same building was the company's high-performance microprocessor group. All was well between the two organizations because there was so little overlap. My team and I designed bulk libraries for other ASCI-oriented divisions within the company.

Meanwhile, the microprocessor organization designed silicon-on-insulator (SOI) processors. We designed dense libraries, doing such obvious low-performance techniques as allowing active and polysilicon "snake routes" for those other ASIC-oriented divisions within the company. The microprocessor organization was interested in performance libraries that did not have such parasitics causing and performance robbing layout techniques (see Figure 19.1). We designed stdcells without well and substrate ties because the design rule check (DRC) rules on distance to such ties could allow them to be placed at far lower frequencies than adding them within each cell would produce, thereby allowing

Engineering the CMOS Library: Enhancing Digital Design Kits for Competitive Silicon,
First Edition. David Doman.
© 2012 John Wiley & Sons, Inc. Published 2012 by John Wiley & Sons, Inc.

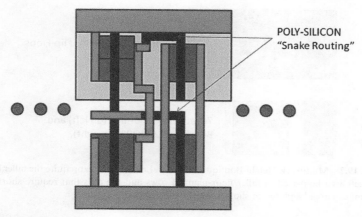

POLY-SILICON
"Snake Routing"

FIGURE 19.1   A Snake Route. Such routing can help density but cause increased parasitic performance-robbing capacitance.

density increases for those other ASIC-oriented divisions within the company. The microprocessor organization needed SOI technology that, somewhat serendipitously, do not have biased substrates and do not require well and substrate ties. When that realization came to the microprocessor organization, it suddenly realized that it could use my bulk 90-nanometer-dense library, polygon for polygon, for the dense sections of its 90-nanometer SOI designs. The company did not need to develop its own versions of a dense library. Its library-development organization could concentrate on high performance (i.e., GHz-range performance). Meanwhile, even if it could *only* get a couple hundred MHz performance with an SOI version of my "bulk" dense 90-nanometer library (which the company started to refer to as a *low-performance* library), then it could significantly reduce the size of many of the blocks used in much of its then-upcoming microprocessor designs. The company took my library as it was, extracted the parasitic netlist by its approved methodology, ran the characterization against its SOI SPICE models, and used the results in benchmark place and routes of several designs, hoping to ensure "a couple hundred megahertz" operation. The company concluded that my so-called low-performance library could actually run those benchmark designs at clock frequencies up to 600 MHz. The company immediately stopped calling my library *low performance* and started referring to it as *middle performance, high density*. What is more, electromigration analysis said that most of the cells could operate in this range without modification. That revelation did cause some consternation with the bulk designers in the ASIC divisions, because they started to wonder if the library could get smaller if they removed the obvious extra widths in the various interconnect within the cells that enable such high-performance electromigration (EM) resilience. However, that consternation vanished when it was reviewed that the majority of the cells used minimum dimension internal interconnections.

To maintain complete consistency between the bulk designs and SOI designs that used my dense library, the few remaining cells that did not meet EM limits were duplicated and renamed within the library, with the duplications having the required EM changes. In that manner, as long as a bulk design did not mind using the EM-resilient versions of the few cells, then the resulting block could be placed on an IP shelf for use in future IP in either bulk or SOI, subject only to any performance issues in SOI that did not show in

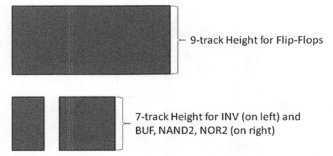

9-track Height for Flip-Flops

7-track Height for INV (on left) and
BUF, NAND2, NOR2 (on right)

FIGURE 19.2 Multitrack-Height Boutique Libraries. Let the cells that require the taller height for efficient cell area be placed in taller floor-planned rows and the cells that require shorter height for efficient cell area be placed in shorter floor-planned rows.

bulk. Likewise, any SOI block completed in the so-called middle-performance, high-density library could be placed on the IP shelf for use in future designs subject to only any performance restrictions in bulk that did not show in SOI.

With the preceding successful entry into the SOI market, the microprocessor team started coming to me about refinements to the support model. My bulk 90-nanometer library was a 9-track library. The real reason for this was that the efficient internal routing of the FLIP-FLOPs required that much height. However, many of the simpler functions (INV, BUF, NAND, and NOR), which were high-use cells in SOI (as opposed to the more complex Boolean functions that were additional high-use cells in bulk), had loose layouts (see the section on cell architecture and footprints for the reasoning on this). The compromise that I developed for pure SOI blocks was multitrack-height libraries. The concept was as shown in Figure 19.2). The FLIP-FLOPs were designed as 9-track cells, whereas the INVs, BUFs, NANDs, and NORs were designed as 7-track cells (and even smaller track height versions were proposed).

The resulting library was then used with the following methodology:

- Synthesize a design using the library (synthesis could be in regular Synopsys DC_shell or in Synopsys "area-aware" PC_shell).
- Determine the total horizontal length of all nine- and seven-track cells.
- Increase these totals by some use factor consistent with typical place and route utilizations.
- Create a floor plan with the proportional number of rows, "combined cell length" percentage-wise, interspersed between each other, that allows such "utilized space" (see Figure 19.3).
- Rerun area-based synthesis to ensure easier performance-based place and route and timing closer.
- Finally, automatically place and route, the combined design with the nine-track cells landing only in the appropriate nine-track rows and the seven-track cells landing only in the appropriate seven-track rows.

This method has a certain amount of placement inefficiency. Synopsys PC_shell or any given synthesis engine assumes that it can place a cell near where its exact optimal location is determined to be, and it chooses the drive strength of the cell with this assumption.

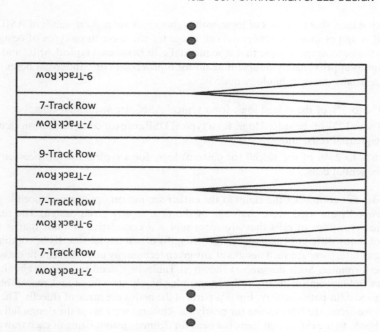

FIGURE 19.3 Place Rows for a Multitrack-Height Library-Based Design. The proportions of each type of row are determined from the synthesis (this example assumes two-thirds seven-track versus one-third nine-track cells in the netlist). After a round of refinement in PC_shell, the resulting synthesized multitrack-height netlist is allowed to place, route, and timing closed.

However, when the cell is actually placed using whatever placement tool the design center currently uses, the cell will be placed in the closest available location in a legal placement row for that particular cell. In designs with abnormally large or small numbers of nine-track cells compared to seven-track cells (actually, the ratio of their combined lengths is abnormally large or abnormally small), the ratio of nine-track rows to seven-track rows will be dramatically lopsided. For high-performance usages, this usually happens not to be the case. For high-density applications, it unfortunately occurs more often.

What this story shows is that support for such items as high performance (or low power) with a library that is not originally geared toward that goal can be viewed as creative extension of the original goals of the library as opposed to hard distinctive realignment of resources. Much of this chapter is about such creative extension.

19.2 SUPPORTING HIGH-SPEED DESIGN

High speed is a relative term. Which frequency constitutes high speed is not just a function of the technology node at which the demarcation is made but also depends on the logic staging style in which the register-transfer language (RTL) netlist is written. High-performance cores will require more sequential memory stages, thereby breaking the cones of logic into more divisions and allowing faster clock rates than digital signal processing (DSP) cores, and those DSP cores will likewise have more such sequential memory stages, shorter cones of logic, and higher clock rates than standard ASIC logic. A good rough estimate of sequential cell usage for the these three types of design is to double the percentage of logic that is sequential cells between a typical ASIC and a DSP core and to approximately double it again for a processor core. Indeed, a decent rough estimate at any given technology node is:

- 4% to 5% of the stdcell logic for a typical ASIC are sequential cells,
- 8% to 12% of the stdcell logic for a typical DSP core (or communication channel) is sequential cells, and
- 16% to 25% of the stdcell (or custom) logic for a typical microprocessor core is sequential cells.

This does not contradict the claim in the earlier section on optimizing stdcell logic that very few cells are used more than 1% of the time in any given design. The reason is that the percentage of cells that are sequential is a cumulative amount that is usually made of many different types and drive strengths of sequential stdcells in combination.

At any rate, there are such designs at any given technology node that can be classified as high performance. Such designs, as shown in Table 19.1, tend to contain circuitry that includes full-custom (or nearly full-custom) plus highly structured and repeatable semi-customized data-path circuitry. Finally, parts of the netlist are made of stdcells. The first of these three areas, the full-custom (or nearly full-custom) sections of the design fall outside of this book, to the extent that there is a need for defined custom sizing of each transistor in defined custom and creative functions within the netlist. Although it is true that many of the techniques described here can be applied to such custom design, it would be incorrect to assume that these techniques would be adequate or sufficient for such an effort. But the second and third types of circuitry that occur within a high-speed design, those of repeated data-path style designs and full stdcell design, do fall within the realm of this book.

To handle the simplest of the three types first—design by means of the normally fully synthesizable and automatically placeable and routable stdcells—although this *can* still be the case, there are high-speed design cases in which netlist must be custom built rather than synthesized. There are cases such that a netlist has to be hand-placed and

TABLE 19.1 Types and Characteristics of High-Speed Circuitry. Many of the techniques described in this book can be applied to each of these three categories, especially the second and third.

High-Speed Circuit Type	Characteristic	Clock Rate
Custom, semicustom, nearly custom	Creative definition of functions Custom drive strength staging Netlist hierarchy at the transistor level	Highest in design
Structured data path	Unique semicustom design per bit Repeated across multiple bits Netlist hierarchy at the bit level	Can range from highest in design to lowest in design
True stdcell	Usually designed at RTL level and synthesized Random Netlist hierarchy at the stdcell and block level	Lowest in design

hand-routed (as opposed to automatically placed and routed), and there are cases where a netlist has to be *both* custom built and hand-placed and hand-routed. In general, however, such hand manipulation of the typical stdcell library is completely supportable by the normal views in a stdcell release; after all, the lineage of most of these views have come down from the 1980s when such design practice was common. There is, of course, the need to ensure that the various polygons in the actual cells meet EM limits for the higher operating frequencies, but this is just a stricter version of an already-defined requirement. The one caveat might be what has become known as "pretty pictures."

Back in the 1980s, hand generation of schematics extended down to the early generation stdcell libraries. These hand-drawn schematics tended to have common and easily readable formats. Specifically, inputs came in from the left of the schematic, and outputs tended to be on the right. The direction of data tended to flow left to right. When a schematic of a NAND gate was drawn, it usually had a characteristic style that was easily recognizable as a NAND gate. Specifically, there were horizontally placed P-channel transistors near the top of the schematic sheet and vertically placed N-channel transistors along the right side of the schematic sheet. Similarly, each of the other functions tended to be drawn in rather structured and easily recognizable templates. With the advent and increased use of synthesis engines in the last 20 years, many current RTL-aware design engineers have downplayed the need to continue such generation. Many are much happier with a netlist representation of the transistors in a stdcell as opposed to a schematic representation. As a result, most modern library-released schematics tend to be unreadable collections of randomly placed P-channel and N-channel transistors, whose signals are connected "by name" as opposed to by drawn wire on the schematic. This is not the desired case as expressed by the typical high-speed design engineer. Such an engineer still tends to be interested in being able to see how a particular stdcell might be constructed out of the individual transistors. An additional task for the library-development-and-support organization is required to support a high-speed design organization, that of producing and maintaining a complete set of clean and *readable* schematics.

Finally, in the intermediate case between these two extremes (that of structured data paths), the components may be thought of as being members of a small boutique library, with an otherwise awkward stdcell architecture and footprint. The typical data-path element footprint will tend to be designed to allow easier stacking of cells representing the functions in the data path (usually, although not necessarily) sequentially down the data path in one direction and the stacking of like functions for adjacent bits across the data path in the orthogonal direction. The footprint tends to be almost like a bit cell in nature, allowing the arraying of the individual cells into a grid, although with the bit cell replaced by a one-bit logical function. Indeed, in many instances, automated data-path stacking is accomplished through *data-path compilers* that function not too differently from those memory compilers described in Chapter 4. The data-path footprint tends to be designed for data to "flow" through and over (and sometimes around) the individual cell. This is accomplished with the following methodology:

- Data inputs to the function tend to have routing access from one side of the cell, and data outputs tend to have routing access from the exact opposite side of the cell, allowing easier data flow through the stacking of such structures.

- Occasional additional open routing channels within the cell in the same direction as the data-input and data-output routing access accommodate easier data flow through the same stacking.

- Control inputs with routing access in the orthogonal direction are sometimes directly over the cell and sometimes adjacent to it.

These data-path footprints tend to be longer in the direction of data flow (taller and narrower for vertical data paths, wider and squatter for horizontal data paths) such that adjacent stacking of bits of the data path can be most compressed. Many data-path footprints are designed to connect signals by abutment, eliminating any variance in timing because of the randomness of automatic place and route (APR). This, of course, means that these types of cells tend to be physically designed to the specific instance of the data path in question, in almost a custom fashion. That boutique data-path library will consist of just a few stdcells; most, if not all, have but one drive strength (Figure 19.4).

The caveats for such a boutique data-path library are twofold:

1. It must meet all EM requirements placed on it by the higher operating frequencies expected to be seen, as is the case for the high-performance stdcell library previously mentioned earlier.
2. It must be dense "at all cost." Any extra space, in the case shown in Figure 19.4), in the horizontal direction, within a cell will be multiplied across the entire final circuit. If it occurs within each bit, it will cost wasted area around any denser data-path element elsewhere in the stack, and any extra space in the vertical direction (again, in the case shown in Figure 19.4) will cause unnecessary additional delay in the performance of the overall circuit.

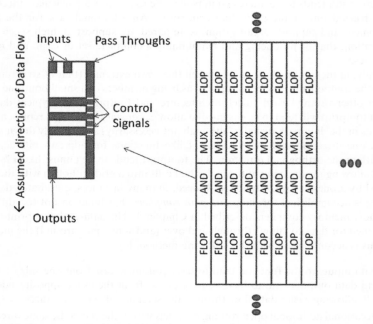

FIGURE 19.4 A Typical Data-Path Footprint and Representative Stacking. Such a set of custom cells can be treated as a special purpose library.

19.3 SUPPORTING LOW-POWER DESIGN

Support for low-power consumption designs with a stdcell library has been, so far in this book, treated in a piecemeal manner and across several sections. However, just as in dense libraries and high-performance libraries, there are several key characteristics of low-power libraries that should be highlighted in a consistent and contiguous fashion.

As in the high-performance case, *low power* should be thought of as a relative term. The amount of static leakage power and total dynamic power change sometimes drastically over technology nodes. So also the extent of effort brought to modify these two components is design-center dependent.

Dynamic current, although increasing in total across designs as technology nodes shrink, is constant per transistor. The reason for this is that the load to which a transistor must switch shrinks with the square of the scale factor, whereas the transistor width shrinks with the scale factor. At every deeper step in the technology-node progression, the transistor has to drive less proportional load from one rail to the other on any given switching (which may or may not be after every clock event). The typical growth in dynamic power defined in Table 19.2 results from the deeper the technology node, the more transistors per device, and the faster frequency of the device. The numbers of transistors per device and the frequency of the clock per device are consistent with those given in Tables 6.1 and 6.2.

Static leakage power *per transistor*, for many reasons—including shorter transistor channel lengths, thinner gate oxides, subthreshold currents, and several other components, none of which vary with frequency of the signal controlling the gate of the transistor—is growing exponentially as technology nodes shrink. Table 19.3 gives a scaled representation

TABLE 19.2 Dynamic Power Trends Over Multiple Technology Nodes. Although dynamic power per transistor at a given frequency tends to be nearly constant, the total dynamic power of devices continues to increase.

Technology Node (Nanometer)	Representative Total Chip Power in W (All ASIC Stdcell Designs)
250	5
180	18
130	50
90	90
65	160

Representative Total Chip Power in W
(all ASIC stdcell designs)

TABLE 19.3 **Static Power Trends per Transistor. The exponential increase for "static leakage power" consumed by a transistor is exacerbated by the increase in the number of transistors per design.**

Technology Node (Nanometer)	Percentage Increase in Dynamic Power Over 250 Nanometers
250	1
180	3
130	10
90	50
65	150
45	500

Percentage increase in dynamic power over 250 nanometer

of the amount of such increase. The vertical axis of the chart is percentage increase over 250-nanometer technology.

Starting at technology nodes around 90 nanometers, there is a dramatic increase in the need for some kinds of power management on designs. Libraries are integral in the various kinds of power management that becomes more and more required at such technology nodes. The actual extent of power management that a given technology node requires, however, should be defined by the design center that the library organization supports. The various types addressed elsewhere through the book in approximate increasing order of difficulty are as follow.

- Higher drive strengths: This is somewhat counterintuitive. Higher drive strengths mean wider transistors per function. Although wider transistors definitely mean increased static power consumption, they also mean harder switching of nodes between rails. This in turn decreases the dynamic power consumption because inputs to the various gates remain within the switching threshold ranges of those gates for less time per each clock cycle of the succeeding stage P-channel, N-channel stack, reducing the time that both transistors are active, which reduces the short-circuit current (also known as *crowbar current*) through the stack. Library support for this is relatively simple: build higher drive strength cells. Of course, this power-saving technique is of limited use as steps through the technology nodes progress.

- High-Vt design: The higher the threshold voltage becomes, the lower the leakage and dynamic current. Unfortunately, it is also true that the higher the threshold voltage becomes, the slower the resulting performance of the stdcell function. Library support for this is relatively simple. If the fabrication house can support it, then it comes down to just recharacterization using a different SPICE model. The lower performance, however, makes this a method of limited usage.

- Multi-Vt design: As previously noted, high-Vt design can reduce both major power-consuming components. But if the fabrication process can allow it, using this technique on some of the design and allowing lower Vt fabrication on other parts of the design to allow those other parts to not be hindered in performance allows for low power consumption and adequate performance. Library support for this depends on the extent of the desired intermixing of the various Vt options. If the various Vt cells are segregated into regions, then it becomes just a slight increase in complexity in support from that given previously. Specifically, the characterization effort remains, and a pair of recognition layers, one for the high–Vt-region P channels and one for the high–Vt-region N channels is required to be added to a copy of the library. These recognition layers can then be used during design fracture to build the proper processing plates for the alternate Vt implants. Depending on the fracture flow, it may even be possible to use just one recognition layer in combination with methodology that realizes the difference between the P-channel and N-channel implants. If, however, the various multi-Vt cells are intermixed, or if the cells themselves have different transistors of different Vt, then the effort is more complex. Layout of the circuitry within the cell must ensure that there is proper space between the transistors of divergent Vt implants to allow producible processing plates. The most notorious possible problem is that illustrated in Figure 2.8, the "four corners" lithographic issue. Multi-Vt design techniques remain a useful addition to any set of lower-power-design methodology.

- Voltage–frequency trade-offs: Power is proportional to the square of the voltage. By lowering the voltage on the rail, it is possible to reduce significantly the power consumed in a design. The unfortunate side effect, however, is that lower voltage on a circuit also tends to reduce the performance of that circuit. This remains a possible tool within the power-reduction toolkit. Library support for this technique starts to cause a rapid increase in the number of required PVT characterization points as each different rail voltage will need to be checked for how it affects the overall performance of those blocks being so scaled. This may not seem an issue at first, just the addition of a few more PVT corners, but as the desired voltage corners increase, it can become untenable. I spent a computer year on a large processor array producing the total number of PVT corners that a single design center required.

- Well isolation: This can be viewed as a variation of the preceding voltage frequency trade-off. If a well can be isolated, then it can be turned off (that is, the circuitry contained on it can be powered down) by removing power to it during those times it is not required to operate the device. Library support for this includes not only the characterization effort previously mentioned but also the need for logic isolation circuitry that prevents "switched off" signals from propagating noise-induced random logic levels into those areas of the design that remain in operation. Other logic-isolation devices are required to prevent validly driven signals from reaching the switched-off region of the design and partially repowering the isolated and switched-off well through the connected transistors. This requires circuit design (and layout) of these isolation functions as well as logical view design in order to allow proper simulation of the surrounding circuitry during such powered-down events. Some level of potential isolation endcap cells may be required to be developed. If required, these endcaps must be such as to be APR on the ends of the various isolated wells.

- Well pumping: This can be viewed as a half step between the two immediately preceding techniques but a more complicated version. Here cell power consumption (and, unfortunately, cell performance) can be reduced as the isolated well is biased away from the common "well and rail are equal" voltage of normal CMOS. Library support for such an effort is the cumulative effort of the immediately preceding two techniques but with the added need for development of analog well pumps, usually capable of being automatically placed within the various cells on the pumped well.

- State-retention sequential cells: This process includes developing sequential cell functions that retain knowledge of the state they are in when much of the function and all of the surrounding circuitry is shut down. They can be used then to retain the state of a block of logic while that block is in a reduced power mode and can thus allow for quick restarts of the logic when the power is restored. Library support for such is the cumulative "all of the preceding" list and the addition of such state-retention sequential cell circuits. As mentioned in the stdcell section, various companies have patented different design techniques and circuits for various versions of these types of cells. If your company wishes to develop alternative implementations, then proper patent searches should be made to ensure that your version is not similar to those patented elsewhere.

19.4   SUPPORTING THIRD-PARTY LIBRARIES

Lucky is the library-development-and-support organization that actually gets the opportunity to develop the internal and complete stdcell library, memory compiler, IO library, and analog IP at a given technology node. Depending on the technology node being used by the design-center user community, much of the actual effort by the library-support team can be one of supporting external libraries as opposed to developing internal libraries. Still, this level of effort does not mean that it becomes one of shuffling bug reports and information requests from the design center to the third-party library vendor and releasing updates to the design center. The reason for this is that the typical external library vendor, be it stand-alone or integrally connected with a given fabrication house, will do any amount of modification and view development and characterization releasing that a customer desires. It just does it for a price—which is often exorbitant. In addition, because the library vendor's financial position is on the line (it not wanting to absorb the cost of dead silicon from a customer), it usually adds guard-banding margin around the views and characterizations that it produces for the price that it asks. Although there is certainly a usually nontrivial amount of shuffling of bug reports and information results from the design center to the third-party vendor as well as a similar amount of update release effort from the third-party vendor to the design center, the library-support staff usually gets the chance to develop special views of special cells. The views of these cells are consistent with special characterizations that represent the third-party library for the design-center user of that IP. Much of that level of activity is chronicled throughout other sections of this book.

In addition to the extra design activity, there exists a level of activity that needs to be addressed when dealing with third-party IP vendors.

- First, rare is the design center that accepts the directory structure in which each of its individual third-party vendors supplies data. This is not necessarily because the design center believes that its internal directory structure is somehow better than that of any individual supplier's directory structure but because of the design center having many such IP suppliers who have divergent concepts on how to present data to the user. The effort required of the library-support organization is of two types. The first type is actually to adjust the directory as delivered by the vendor into one that is acceptable by the design center. Depending on how open the internal design center is to the external IP vendor, the library-support organization should be cautious to do this adjustment with the full knowledge of the IP provider. The reason for this is that the IP provider usually sends an application engineer to the design center at some point to address one or another issue or in other ways assist the design-center user base. If that application engineer finds the data structure not as expected, there can result a level of "justification" activity by both the vendor's engineer and the design-center engineer, usually in front of one's own upper management. That upper management may wish to understand its level of financial exposure for changing the vendor's IP (even though the IP, in such a case, remains intact and just the directory structure is adjusted). This can cause significant risk mitigation rework within the design center. The second camp is to point to the various files in the extant third-party IP vendor's directory structure from the locals for them within a more design-center–familiar directory structure. Here is one place where the metadata tagging of directories, as described in Chapter 17, can certainly help. Note that in either the case of the directory structure adjustment or of metadata tagging, pointers should be automated as much as possible. Assurance that directories are moved properly or pointed at properly should not rely on human error-prone hand manipulation. That having been said, however, the result of the automated manipulation must be hand checked and reviewed. It is rare enough for a vendor to release two sequential releases consistently in which at least some files are not moved correctly.

- Second, as an addition to the discussion on tagging, the design engineer is responsible for tracking the internal measure of the quality of the third-party IP. This is done so that once it is known to be good ("known good" in the methodology of the design center), then the tag of the particular piece of IP (or the particular view of that particular piece) should be updated. Once that IP is known to have produced functioning silicon within the design-center driving design, then its tag should be updated. When that IP is deprecated within the design center, its tag should be updated. Although it would be nice to have this tagging actually placed within the view, there is the same chance of having to justify modification of the vendor's IP, thereby potentially removing its fiduciary obligation as previously described. The tagging mechanism and the reading and reviewing of such tags must allow for the possibility of requiring parallel files with the tagging present external to the IP.

- Third, any bug reports registered within the internal issue-reporting system or any information request for the IP needs to be in some manner, preferably automatically, mirrored in the third-party issue-reporting system. Likewise, any resulting feedback or other deign modification activity by the third-party vendor should be registered, again preferably in an automated manner, back into the internal issue-reporting and issue-resolution system. Communication is why it is important for the automated mirroring of bugs and questions from the internal system to the third-party vendor's system. As an illustration, consider the old party game of telephone. At the end of a line, one participant whispers something to his or her neighbor, and then each participant shares the information with his or her neighbor down the line until the last person repeats out loud what he or she has heard. Usually, much to the amusement of the sequence of individuals, the information is significantly different from what the first person in the sequence originally whispered. This game illustrates the hazard of the library-support engineer who sits between the questioning end user and the perplexed external vendor. Such communication chains can all too easily result in several rounds of information being changed between all the parties involved before adequate resolution is found. However, in addition, the reason for the automated mirroring from the third-party vendor system back to the internal system is the propensity of vendors to solve symptoms as opposed to problems. If a vendor "fixes" a piece of IP that is used in multiple places throughout a design center to the pleasure of a particular customer design engineer but that "fix" breaks the design for a second customer, possibly in a second design center, then that vendor has broken the IP. That vendor has just managed to split the original IP into two independent pieces of IP, each requiring its own support. If those two divergent pieces of IP are integral in the two design-center projects that the one original piece used to support, then the IP pieces can easily justify the need to double the amount of resource that they can charge to the design center, thereby increasing their cash flow at the expense of the customer. A library-support organization should be in a position to prevent such eventualities.

- Unless the design-center methodology is one of being purely within the tool suite of one of the "big three" engineering design automation (EDA) houses, internal view or cell or PVT development will be needed. So-called Frankenstein flows usually break some view or another that one of the EDA vendors assumes to be one way but a competing vendor assumes should be another. That internal view or cell or PVT development is generally but not necessarily specifically described within this book. The real keys to such development are the verification and validation of the view against the tool for which it is built against the other views of the third-party IP for which it is descriptive, as well as against the defined goals of the piece of IP. That verification and validation is also described in Chapter 18.

There is a complication in supporting external third-party IP vendors. Rare is the environment in which just one third-party IP vendor exists for a given design, let alone an entire design center or an entire company. This is just as rare as the environment in which an internal library-development-and-support team gets to develop all of the IP used within the design center or company. Not only will most design environments require input from multiple external vendors, but also there probably will be a need to merge internally developed IP. In such instances, it certainly isn't the responsibility of any of the individual vendors to ensure that the various pieces of IP integrate nicely together, and it certainly is not in the knowledge of the design-center user engineer to ensure such either. The responsibility of the library-design-and-support team extends to here. If the thresholds of one device do not match the assumed thresholds of another to which it is to connect and the design-center methodology assumes that they must, then the library team needs to ensure that one or the other (or both) is adjusted appropriately. If one piece of IP contains a structure labeled "NAND2" that is divergent from one with the same label in another piece of IP, then it is the responsibility of the library-design-and-support organization to resolve the issue before it affects (i.e., delays) a critical project or, worse, goes undetected and causes dead silicon. If a particular view of a particular cell or subcell in a vendor's IP is insufficient for the design-center methodology, then it is the responsibility of the library-design-and-support organization to ensure proper resolution, either internally or externally.

19.5 SUPPORTING BLACK BOX THIRD-PARTY IP (INTELLECTUAL PROPERTY) DESIGN

The previous section discussed the support of third-party IP, but the IP being described tended to be small- or medium-scale integration (SSI and MSI). This particular section takes that discussion a step further. As opposed to discussing SSI- and MSI-level IP—which is what most stdcell library, IO library, basic analog, and memory compilers could be considered—there are some basic extensions. Such extensions include the need for some level of system-level validation and verification against initial (and tracked) IP requirements that need to be addressed when referring to designs of a slightly larger level of complexity. This section addresses the need to support blocks of IP at the large-scale integration (LSI) level in general and third-party–generated LSI-level IP in particular.

Over the last couple of decades, the Federal Aviation Administration (FAA) has more strictly interpreted what it means to build passenger-capable airplanes that are allowed to fly over the country. Most of the rest of the world has followed suit, usually taking FAA requirements word for word as requirements for airplanes flying over their respective countries. Currently, to comply with FAA strictures, all airframe manufacturers must now comply with *requirement tracing*. In addition, many other manufacturers of human safety-critical designs—for instance, automobile manufacturers and heart-pump manufacturers—are moving toward the same type of compliance stricture. What does requirements tracing imply? Specifically, it is the strict tracing of (1) what pieces of the design are in place to fulfill what specific specification requirement and (2) what part of the verification and validation test bench exists to test for that specification. In the FAA's strictures, nothing is allowed in a design except that which is present specifically to fulfill a requirement. A specified requirement is both sufficient and necessary for a component to be present, and that specific component is both sufficient and necessary for a test-bench component to be present. In practice, what this means is that every requirement is the specification must have a direct mapping (although not necessarily a one-for-one mapping) with a piece of the design and every piece of the design must have a direct mapping (again, not necessarily one-for-one) with a test-bench vector. There must be an auditable trace capability between requirement, component, and verification functions. This is true for the hardware that makes up and otherwise manipulates the various flight controls of an airplane, the brake components of an automobile, the human body-monitoring sensors of a health device, and the electronics that control those hardwares and report on their operational statuses. Those design centers that have designs that are going into airframes, automobiles, health devices, or many other human-critical designs must be able to conform to these traceability requirements.

As just mentioned, most stdcells, IOs, and basic SSI-level analogs (phase-locked loops, power-on resets, and oscillators) can be described as individual pieces of hard IP, whereas memory compilers might be thought of as a special cross between soft, firm, and hard IP (as mentioned in Chapter 4). LSI-level IP, whether externally generated by a third party or internally developed and used off the IP shelf, tends to fall into one of the three following categories. How that LSI-level IP maps into the three categories affects how that IP accomplishes its requirement tracking.

First, it is important to delineate the difference between three types of IP.

1. Soft IP tends to be synthesizable RTL netlist that has no direct mapping to individual pieces of polygons that will eventually represent the function described in that RTL on silicon. For LSI-level IP, this is the simplest of the three categories. In many senses, this is the easiest of the three categories to handle. Because the netlist is written as RTL code, which is much more human readable than a large sequence of seemingly random stdcells that constitute the netlist in the next two categories, it can be easily understood. More so, it probably has some level of embedded vendor comments. There should be a means of requirement tracking from the initial published specification of what the LSI as viewed as a component for a design is suppose to fulfill in that design to what the third-party IP vendor has actually designed and produced in RTL to the sections of the RTL code that implement the feature. A well-written, commented, and requirement-tracked test bench should also be present. The IP vendor may even be able to deliver a mapping file that indicates which requirement is mapped to which line (or lines) of code and mapped to which test-bench vector (or vectors). Admittedly, this level of documented requirement tracking has been unusual in the past, but it is becoming increasingly common over time because all airframe manufacturers and most other human-safety–critical manufacturers are starting to require this absolute tracking of specification to netlist to test bench. The various IP vendors see a differentiator for their products in the marketplace. If the third-party IP vendor does not supply such requirement-tracking scripts, several EDA tools can be used within the library and IP support organization to facilitate the internal development of this capability. This may mean that special tags must be added to the specification, the RTL code, and the test bench, but this is little penalty in terms of adding to a schedule if such tracking is required later.

2. Firm IP tends to have device-specific, postsynthesized gate-level netlist that directly maps to the individual pieces of polygons that will be eventually patterned on silicon to represent the function described. But firm IP has no final descriptive relationship as to how those various patterns relate to each other outside of the stdcell or how they connect via routing with each other. For many reasons, this type of delivery is a little more problematical to the library-support organization. Although such a delivery eliminates the need to do the actual synthesis, it does require that a specific stdcell library be employed (at least for that block of IP). Depending on the size of the netlist, especially if delivered without hierarchy, there may be serious difficulty in accomplishing any meaningful level of rigorous requirement tracing. If the netlist is hierarchical and the various subblocks in that hierarchy have schematic views, then there is a chance that a specific subblock, in its entirety or piecemeal, can be mapped to original specification requirements, especially if there are printed comments on the schematic. There is a chance that the test bench will remain partitioned enough to allow some level of requirement tracing. If either of these is not the case, then this type of delivery is completely useless for any future requirement tracing. If such design exists in the design-center

IP shelf, whether third-party acquired or internally generated, it should be considered antiquated and be replaced before further use allows it to limit the market of future design-center designs.

3. Hard IP tends to be made of device-specific postsynthesized gate-level netlist that also has the placement and routing of those patterns hard coded. This type of third-party IP delivery, although it remains highly appealing to the support organization with the lack of any need of constraint files (it is delivered timing closed) and with the presupposed level of test suites, it also is nothing but an even more highly structured for a the previous firm IP delivery. It suffers from the same requirement traceability limitations. Depending on the perceived value of the hard IP by the vendor of that hard IP, there may be a perceptual problem with it delivering significant amounts of detail concerning the inner workings of that design. There is a perplexing trend of such hard IP vendors even to deliver the layout versus schematic verification netlist in some level of encryption. The obvious reason is that the third-party vendor wishes to limit exposure of its perceived valuable IP to any potential competition, but such encryption immediately forces it out of the human-critical–design market because such encryption effectively prohibits requirement tracing.

In addition to requirement tracing, a second feature of LSI-level IP support is the required level of system IP validation and verification. Although most to all IP vendors are conscientious engineers and providers of functional IP, a foolish design-center organization would accept a LSI-level block at face value without some level of verification and validation. All IP needs to interface with that produced within the design center. The design of that interface may be either in the internal design center or in the third-party IP vendor's facility, but it too needs a level of verification. These are the realms of system-level IP verification, and it is a requirement even for IP not destined for human-critical projects.

So what constitutes system-level IP verification? Figure 19.5 gives a general flow diagram of a typical system IP verification effort. The steps constitute the same level of validation through which a design center would put a piece of internally developed IP.

- The RTL is put through each of the EDA tools in the methodology flow against the stdcell library to wring out any compiling issues.
- Once the netlist compiles in each EDA tool in the flow, it is tested against any supplied test suite.
- It is then (usually concurrently) tested against the specification for feature coverage (and for completion of test suite coverage).
- It is validated as to whether it is synthesizable.
- The synthesized netlist is validated against fault-coverage requirements.
- Finally, test place and route and timing closure is attempted, followed by any substitution-required formal verification.
- If any step fails, the IP is usually returned to the vendor for some level of remediation. If it passes all steps, it is accepted within the technology IP shelf.

This system IP activity is accomplished inside the design center for internally generated IP (even by the individual designer responsible for the initial effort). This activity

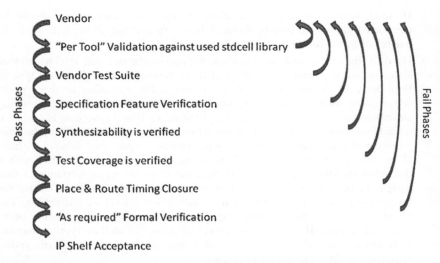

Pass Phases

- Vendor
- "Per Tool" Validation against used stdcell library
- Vendor Test Suite
- Specification Feature Verification
- Synthesizability is verified
- Test Coverage is verified
- Place & Route Timing Closure
- "As required" Formal Verification
- IP Shelf Acceptance

Fail Phases

FIGURE 19.5   A High-Level Representation of a System IP Flow. Each step mirrors, for external IP, what internal design-center–developed IP goals through.

may also occur for third-party IP at the using design center. However, because the third-party LSI-level IP block is, in fact, a piece of IP just like anything else delivered by the library-development-and-support team, it is usually assumed that the responsibility for this system-level IP-verification effort stay within the library-and-support organization.

19.6   SUPPORTING MULTIPLE LIBRARY DESIGN

Figure 18.2 shows a representative SOC development. Various components on that device use various pieces of IP sourced from many places. Although this is an illustration, it certainly represents the valid modern SOC. There will be many designs that contain source libraries and source IP with multiple ancestries. There will be many designs that contain source libraries and source IP with the same lineage but with divergent versioning (and divergent density, power consumption, performance, yield improvement, or noise-immunity features, some of which are critical to the IP blocks in which they are embedded).

When this occurs, there will be a need to allow for changes to the various labels of the various cells in the various IP to be changed. As implied in Chapters 14 and 15 on design for manufacturing (DFM) and quality, a NAND2 in an internally released library version (e.g., A.B.C) might have different physical views than another NAND2 in an internally released library version (e.g., X.Y.Z), which will have different physical views than an externally acquired library version (e.g., L.M.N). An internally released NAND2 version A.B.C will have different quality levels than those in internal release (e.g., X.Y.Z). They all will have different liberty files, logic files, and test files.

1. In some manner, in the release and support mechanism there has to be the ability to both rename the cells of the various libraries "on the fly" such that name clashing

becomes impossible and to keep the names similar enough that name recognition remains tantamount. Automatically renaming all of the views within a particular NAND2 cell to ZIPPY2 might be able to resolve the first issue but completely obfuscates the second. It is common to name the cells by some combination of source, revision, and function. A NAND2 in vendor JKL release version A.B.C might become JKL_A_B_C_NAND2. The caveat here is that many older-generation versions of various EDA tools limit the absolute size of cell names.

2. The tracking and tagging of cells' quality levels within the various views of the cells needs to be such that these can be automatically identified and polled. The reasons for this are twofold. First, externally acquired library will not have pre-instantiated tags. They will need to be generated. Because the IP vendor may have issues with actual modification of the various views of the various cells in its library, the tags might need to be in alternate added files within the vendor's delivery directory. Second, both internally generated and externally acquired libraries will have to have such tags uniquely updatable in a probably random manner (because not all views of all cells will be updated at any given time).

3. The various cells various views may not contain merge-capable features. Some libraries might have liberty characterizations, for instance, with different thresholds and edge rates than other libraries. Some libraries might have different delivered logic views (Verilog versus VHDL). Some libraries might have different placement restrictions—for instance, for noise-immunity reasons—that force them to be phys-ically guard banded from each other on silicon. Some libraries might have DRC restrictions. Some libraries may even have divergent graphical display standard layer lists.

4. The directory structure will need to accommodate having multiple different librar-ies, each with multiple different "current" revisions or multiple different views of the multiple different cells in the multiple different PVT characterization corners present.

5. The bug-reporting mechanism must be able to handle the division of issues among the various library sources and revisions. This is especially cumbersome for those times when multiple revisions of the same library are required on the same design (for instance, when a previously closed piece of now hard IP is included on a new design). Several previously closed issues may have to be reopened, causing new revision branches off otherwise antiquated libraries. In such instances, the eventual merging of the branches back into a single trunk is notoriously difficult. It may mean that this eventuality is impossible in many of those instances.

All of these issues must be handled within the release and support mechanisms of the library-support function.

19.7 CONCEPTS FOR FURTHER STUDY

1. Multitrack resource calculation tool:
   - The chapter described a possible multiple-height library together with a methodology for its usage in a real design. This methodology required knowledge of the relative amounts of the stdcells of the various heights that are synthesized. The student or interested engineer who has access to RTL might

be interested in constructing a fake library exchange format (LEF) and fake liberty file of a fake stdcell library. Some of the stdcells in this fake library are of one type and height, and some of the stdcells are of another type and height and use this fake LEF and liberty in order to do synthesis. Before APR, it is important to build an APR area calculator tool that reads the resulting synthesized netlist and calculates how much area is required for each type of row height. This tool can then proportionately allocate rows in an APR region and actually interdigitally place those rows. Finally, the APR tool can attempt to place the fake LEF represented stdcells in the netlist.

2. Data-path–structured stdcell layout:

   • The student or interested engineer who has a desire to investigate data-path stdcell layout and has access to a netlist that has such a requirement can identify the pertinent cells that would be required by the data path and attempt to build GDS layouts so that they can best fit together.

3. Data-path stdcell APR tool:

   • The student or interested engineer who has accomplished the layout of the required data-path stdcells in the preceding concept can develop an automated data-path compiler placement tool that will, by means of abutment, correctly place the developed data-path stdcells appropriately for the netlist.

4. State-retention sequential cell design:

   • The student or interested engineer who has a desire to support low-power design might be interested in first doing a literature search on the various patented state-retention sequential cells that exist today and then attempt to find unique alternate circuit designs for them.

5. Black-box IP support survey:

   • The third-party IP vendor that has a desire to protect its IP has collectively developed a plethora of means to accomplish this task. The student or interested engineer who has an interest in the field should contact multiple such IP vendors and develop a working knowledge of these various means and methods.

6. System IP development and verification methodology:

   • The chapter described a system IP verification and validation methodology. The student or interested engineer could be inclined either to expand on this methodology or to develop a working flow of his or her own.

7. Multiple library design-support methodology revisited:

   • In the previous chapter, one concept for further study included the idea of developing a methodology to allow for multiple vendor libraries on the same design. The student or interested engineer who has access to two or more compatible stdcell libraries of the same technology node should use the methodology and naming conventions developed for that effort in the actual develop of a multilibrary design.

CHAPTER 20

COMMUNICATIONS

20.1 MANAGER'S PERSPECTIVE

This book has discussed various means and techniques that are useful in intelligently
reducing unneeded margin (or in intelligently adding user-defined margin) in the various
views of the elements of a design. However, little has been said about what, in the eyes of
the management of the library-support organization and the customer design center,
constitutes enough margin reduction or addition. Such a determination is a decision that
should not be made at the level of either the design-center engineer or the library-support
engineer. It really should be made at the management level, which is financially
empowered to accept the risk of such margin reduction. Such a change in a view,
the addition of a view, or the choice of a view should never be done without explicit
management approval, even when the need for it is obvious to the library-support
engineer or the design-center engineer. As implied, all such change or addition involves a
change in the risk level of the resulting view. This means that information on those risks
and the reasons for the need for a view change will need to be communicated up to the
level of the organization where that particular financial accountability lies. This section
deals with that communication.

What type of information needs to be communicated in order to justify a change of a
view? At any level of management, it will be the automatic reaction for that particular
manager to minimize risk. It is the common first reaction to reject an engineering request
to change a library vendor-supplied view, no matter whether that vendor is internal or
external. The use of a vendor-supplied view, no matter how insufficient (or how incorrect)

Engineering the CMOS Library: Enhancing Digital Design Kits for Competitive Silicon,
First Edition. David Doman.
© 2012 John Wiley & Sons, Inc. Published 2012 by John Wiley & Sons, Inc.

that view may be, is the low-risk decision. If a design fails because of an issue with some vendor-supplied view, then the manager is "not at fault." He or she was just following what the vendor provided. However, if a design fails because of an issue that can be traced back to some change to a view made by some engineer within the organization, done with the manager's approval but not with the vendor's approval, then the financial responsibility for that decision will be judged by the upper management of the company as being squarely the manager's. Such a risk can destroy (or at minimum derail) a manager's career. It is the extreme rarity when a manager is rebuked for not taking a risk. As mentioned, the automatic reaction to any request by an engineer to change a vendor-supplied view is likely to be met with little support. Even when the need for a change in a vendor's view is obvious or has been successfully communicated to an appropriate level of the organization, the second-tier knee-jerk reaction is to accept the vendor view "as is." This is because the view "as is" can be justified as having been placed into service by the vendor and potentially used by many other of the vendor's customers on other designs. It can be the result of a request sent to the vendor asking for it to change its view (thus, offsetting the risk for such a change from the manager back to the vendor). Therefore, unless there is significant reason for a manager to trust an internal library-support engineer's intent, judgment, and effort, then that decision to allow an internal change is probably unlikely. This is the case no matter if the relationship between the engineer and the management level that is empowered to make the decision is direct or multitiered.

However, to increase the chances of getting approval for a margin-reducing, risk-increasing change, the following pieces of information probably should be made available to the responsible management tier:

- What is the actual underlying issue that requires the change? This is straightforward enough. There is an identified problem with a view of a supplied library element. Describe what that identified problem actually is.
- What would the change actually be? (It is probably better to use before-change and after-change examples in the question.) This too is a straightforward set of data. Report what the view currently looks like versus what the view would look like once the change has been approved. It makes sense to show before-and-after examples of how the view will affect synthesis or place and route or timing closure or test insertion or whatever else is affected in the final design in which the view change will be used.
- Why is it beneficial to the organization to make the change and what would be the resulting competitive advantage to the organization? Here the information is one of marketing. If a change to a view is to be beneficial, then it should be easy to indicate how it would be so. What does it give to the organization that a competitor either does not have (giving an advantage to the organization) or does have (closing a competitive disadvantage).
- Why is it better to make the change than not make the change? Here the idea of cost versus benefit probably first arises. Specifically, what is the cost of making the change versus what is the missed opportunity cost of not making the change? However, this may or may not include actual monetary calculations. This really is more along the lines of "why" the change should be made.
- Why is it better to do so in house as opposed to going to the vendor (or, what did the vendor do when previously approached on the issue and why)? Again, this is a fairly easy and straightforward level of communication. It amounts to "When we approached the vendor on this issue, it...."

- What does the cost of making the change entail? Here, finally, any estimated actual monetary cost calculations concerning the determination of the data that is to be changed or added (or the cost of the view that is to be added) should be identified. Here, for instance, is where the amount of the time and hardware resource for a recharacterization should be made.
- What would the verification process entail, including its cost? Just as previously noted, the cost should be calculated for the verification of the data in time and hardware resource requirements for the change or addition. In addition, yes, verification does cost.
- What are the legal risks to the organization making the change (in acceptance of both design risk and possible legal risk from the vendor)? As mentioned in the book's introduction, vendors may not fully appreciate the need or desire of a customer to enhance their various library-offering views. As a result, there may not just be technical risk but also a fiduciary risk in making a margin enhancement change to a vendor's particular view. There may also be a legal risk. Some views that leave no telltale evidence on the resulting device are inherently immune to this level of risk. But the views that actually change the physical dimensions or locations of one or more polygons not only can cause physical technical risk but also can be identified by the vendor "on silicon" to such an extent that they could suggest that such an effort abrogates any legal responsibility to the organization to make such a physical change actually work. This could be as simple as adding another power rail over the top of an existing layout supplied by the vendor (possibly adding capacitive coupling to other elements of the vendor-supplied circuit).
- What would the financial benefits of making the change mean? This is another marketing issue. All of the potential costs (technical, legal, and monetary, and well as time and resource) have been identified. What is the potential benefit of the change?

The following do not bear directly on the need for proper communication of potential margin-reducing view adjustments, but they do encompass the effort.

- Regular weekly status meetings and status reports should be held. Any manager who is not in a minimum of weekly (or, even better, daily) communication with an engineer is not fulfilling the roll of manager. This is not to say that the manager should minutely direct the effort of the engineer, but he or she should be able to completely understand and generally describe the activity of each engineer in the organization.
- Regular monthly status reports should be presented by the lead engineer of each project on which an organization is working. Open presentation by the engineering staff to upper management, together with fostering similar knowledge among the rest of the engineering staff and fostering discussion of the efforts with the internal customer design-center engineers and management is the goal of any such effort.

Finally, as an example of how to formalize this level of regular monthly information flow, I have copied a form that I have used in multiple previous positions (all with slight modifications per the desire of management at those organizations at which I have worked) into Appendix IV.

20.2 CUSTOMER'S PERSPECTIVE

Along with the need for proper communication of margin-reduction view changes to management, it is also highly important to make sure that:

- the design-center customer understands the increased risk of such reduced-margin view, and
- the end-user customer understands that such views may be used in the design and development of the device.

For the intermediate design-center organization, there is a need to understand the extra risk that is to be assumed in building a device with reduced-margin views as well as the benefit of doing so. The end user of the resulting design must understand those parts of the design that may be marginally less robust because of the design-center engineers using some of the reduced-margin views. For instance, adding fewer rail connections to the power supply may cause slightly more ground bounce and power droop, which in turn may cause slightly slower that otherwise expected performance or possibly slightly higher bit-error rates within a memory.

In the first case, much of the same amount and types of communication that is required by management to justify the choice of building such reduced-margin views is also required by the design center in order to justify the use of this reduced-margin view set. The design-center engineering staff and management will need to address the following.

- The need for the change and exactly what the change actually entails should be elaborated.
- As in the previous section on management communication, it is probably beneficial whenever possible to present actual before and after views so that the design center can better visualize what the actual change entails.
- In addition, it might be beneficial to demonstrate how the change actually can improve the performance, power, noise immunity, design for manufacturing (DFM), or whatever else of the design the reduced-margin view change is meant to accomplish.
- The expense of the change (to the organization as a whole and top the design center specifically) should be explained in terms of lost opportunity cost from the time spent and resource expended by the library-development organization in developing, testing, and supporting the changed views. Here the expense to the organization probably does not directly mean money. It is more important to make sure that the design center understands that human engineering and hardware and software and time (and expense) will be expended to build, test, and support what becomes a custom view that is internal to the organization. Custom views may be the correct answer by allowing for internally developed core competence and marketing competitive edge, but they are not free and the design center must understand this.
- The risk of using the change must be clear. This includes the increased possibility of producing dead silicon and the possibility of not being able to pass the expense of that dead silicon along to the fabrication house, even when it is provable that the issue is because of a library view that was not changed.

- All reduced margin of any amount, even that accomplished in nonbroadside manners as described within this book, entail increasing the chances that the reduced margin will cause a design that is otherwise barely within the required specifications to fail.
- The marketing risk of not using the reduced-margin views should be specified. In particular, the end-user customer might be tempted to move to the competitor that is willing to take a smaller margin for a design that it produces without the margin reduction possible.
- This is the basic premise of this book. If your design center starts with the same initial library offering from the same manufacturing house as the next design center down the block and you go after the same customer proposal, then there are no magic bullets hidden from the competition, and the winning bid goes to whichever company is willing to take the smaller margin. If you happen to have engineers capable of using the vendor-offered library better and producing a slightly better (tighter, faster, whatever) design, then eventually the competing design center "down the street" will get some of its customers to change companies or get better engineers of its own. There are noncompetitive advantages in not reducing the margin within the vendor-delivered library views. On the other hand, if your organization shaves some of that margin from the various views of the library vendor, then tangible competitive advantages exist that can be protected as the engineering staff is enticed away to the competition. This tangible competitive margin can provide for production of smaller, faster, cheaper, less-power-hungry, more-noise-immune, and better-yielding products, allowing your organization to undercut the competition.

In the second case, the end-user customer must be made aware of the time risk on requiring a second or even a third cycle in order to get a design properly tuned with the new views. As mentioned earlier in this section, using reduced-margin views can increase the likelihood of a first-pass failure. As a result, the end user must be made aware that the part, although it may be better in the end (smaller, faster, cheaper, or whatever), might take a longer time between development and market introduction.

Knowledge of the risk and reward for such development should be broached with the potential customer from very early on in the sales cycle. This level of communication needs to be in an extended and continuous format as in the following:

- Sales and marketing must be made aware of risks and rewards. The attendance of both departments at the monthly project reviews mentioned in the previous section is important.
- Engineering must understand what the sales and marketing organizations are implying about the abilities of the engineering organization. In that manner, not only do the potential risks and benefits from using the reduced-margin views become known to the customer, but also any misinterpretation of those risks and benefits that might develop can be realigned as quickly as possible.
- Once a customer is engaged and from the earliest weeks of the design development cycle, regular conference calls to discuss weekly progress and issues, plus regular face-to-face status meetings between the customer, the design center, and the library-development-and-support staff must be maintained. In that manner, any potential issue can be more easily and earlier seen, allowing for a more expedited resolution.

20.3 VENDOR'S PERSPECTIVE

Two categories of vendors count when considering this perspective of the communication issue:

1. the vendor of the initial intellectual property (IP) with views that are being modified in house, and
2. the vendors of the various engineering design automation (EDA) tools with which the modified views of the IP will have to interface.

For the first group, depending on the relationship that the vendor has with the internal design organization, this communication could be open and free flowing or it could as easily be minimal and terse communication.

- If the first of these types of communication is the case, then the reason for the internal cell-development organization doing the actual work of building reduced-margin views of the IP is one of two possibilities. It could be that the IP vendor either does not have the tools and means or does not have the knowledge required to accomplish the actual build. Conversely, it could be that the internal cell-development organization wants to maintain the knowledge of the reduced-view development inside the organization. This could be the case when an organization wishes to build such knowledge within as opposed to possibly or probably losing it when the relationship with the outside IP vendor ends. It could be the case that the internal cell-development organization is protecting what it considers to be a competitive advantage and desires to minimize exposure of that competitive advantage outside the organization. This level of open discussion on the features and restrictions of the vendor's IP that the vendor can supply is ideal. Fostering such open communication, if possible, is a value-adding function of the management of the internal cell-development organization and should be prioritized. There should be frequent conversations, possibly even on-site visits to the vendor's offices. The goal is to acquire as much knowledge of the IP as possible. If the communication is open but protects the development organization's competitive advantage, then information back to the vendor can be just as free and open.
- If the relation with the IO vendor is the second possibility, then the reason for the internal cell-development organization doing the actual work of building the reduced-margin views of the original IP could be that the IP vendor is not willing to risk the chance of the IP failing. As a result, the IP vendor is much less likely to have open discussions on the areas where the design is "overconstrained" and the margin in the view is reduced. This means that much of the knowledge on the actual internal function and performance of the IP will have to be gleaned from internal analysis. Here, although the information flow will be minimal and terse, it remains a good idea to foster as much accumulation of information as possible. What will probably be required is to have some amount of formal design review of the delivered IP.

The second groups of vendors who require open communication channels in both directions are the various EDA vendors whose tools must contend with the reduced-margin views. This communication is developed over time. On the good side, the EDA vendors are highly incentivized to develop such a open relationship (at the least for larger companies) because doing so could mean more return sales of licenses for their EDA products. As a result, the likelihood of such open communication is significant. This is important because the EDA vendor though its assigned application engineer (AE), whether full-time or part-time, is more likely to understand the intricacies of its tools and specific information and information formatting that is best for allowing optimal interface with and use of that particular set of EDA tools.

- It then behooves the cell-development organization to foster a regular even daily on-site interfacing with the AE. This will allow the best means of acquiring the knowledge needed for the proper presentation of the reduced-margin data within the tool. It will also foster knowledge within the organization that is useful in future cell and view development of the vendor's tools and how best to present reduced-margin data to the tool.

The use of informal and formal training that can be supplied by the EDA vendor should be of significant importance to the cell-development organization. This is for tools used for the development of views and for tools used by the design-center engineer in conjunction with the views delivered by the cell developer. There should be a regular requirement that the individual cell-design engineering staff attend some amount of vendor-supplied training at least yearly if not more often.

20.4 ENGINEER'S PERSPECTIVE

Communication with the engineers on the topic of margin-reduction techniques is usually simple and straightforward. Engineers in general are interested in following whatever directives they are given as long as those directives make sense. A manager's role here is to define the "what and how and why" of the margin reduction, give the benefits (and possibly the risk) to the design-center project of such margin-reduction techniques, and then get out of the way of the professional engineering staff as it accomplishes the building of the views. Getting out of the way means facilitating the effort as well as not interfering with it. If a recharacterization effort can be speeded up with the addition of more hardware computing resources, then the manager has the duty to find all of the requested hardware resources (or as much as can be brought within reason).

Going the other way on this communication channel, the engineer has the duty to:

- keep the first-line manager aware of the progress of the view-generation and view-verification process;
- keep the first-line manager aware of any found issues in the mechanics of generating the appropriate reduced-margin views;
- keep the first-line manager informed, in as timely a manner as possible, of any mistakes made along the way so that they can be remedied appropriately as early on as possible;
- let the first-line manager become aware of potential better methods and techniques of generating and verifying the reduced-margin views;
- let the first-line manager become aware of reduced-margin limits that may be unintentionally crossed if a reduced margin is taken too far;
- let the first-line manager become aware of additional margin that can be effectively recovered by other view generation;

- allow the first-line manager to redirect unused hardware and software resources toward other efforts;
- inform the first-line manager about when the engineer can take on additional tasks as currently assigned tasks are completed; and
- keep the first-line manager aware of the resource needs already mentioned.

It makes sense to have such communication on a regular basis (daily informal and weekly formal staff meetings, for example). As with all manners of communication, the engineer and the first-line manager should keep the communication level, be it formalized or not, as efficient as possible. Nobody enjoys attending meetings.

20.5 CONCEPTS FOR FURTHER STUDY

1. Communication survey:
   - The student or interested engineer should be encouraged to discuss "information flow and progress and status reports" for various purposes and with as many sources as can be found.

20.6 CONCLUSIONS

This is a dangerous book. It has shown the reader where to look in order to find the places in vendor-delivered IP where extra margin is hidden. It has suggested ways to determine how much extra margin is present. It has suggested means on how to efficiently and intelligently rewrite and re-create the views for the various pieces of IP in such a way as to reduce that margin so that the reader's organization can maintain a competitive edge over the design center down the street or around the world. This is true even if that competitor is using the same IP delivered by the same third-party vendor in order to design the same type of product with similar EDA tools in order to compete for the same customers in the same marketplace.

You have seen illustrations for the creative addition of a large amount of additional Boolean stdcells that are typically not included within a third-party vendors' library offering, together with the knowledge needed to actually construct and add them to the vendor's library.

You have been given insight into the same means for adding special IOs, whether the same footprint or a different footprint, to a vendor's general-purpose IO offering.

Descriptions on various memory compilers and small-scale-integration–level analog have been given so that these can be used, be they internally developed or acquired from alternate third-party IP vendors, in conjunction with the stdcell and IO vendors' libraries.

Alternate physical architectures have been described that allow the reader insight into how alternate internally developed libraries can be used in conjunction with vendor third-party IP in various regions of a design.

The limits of vendor supplied logical views have been described, and the reason for rewriting them when they have proven themselves inadequate have been given.

The means of ripping apart a SPICE model has been given, together with ideas on how to determine when such models, again supplied by third-party vendors, can be "tweaked" to produce increased accuracy and performance.

The techniques required for parasitic extraction and for timing, power, and noise recharacterization has been extensively shown, enabling the reader to confidently reextract and recharacterize third-party vendor IP internally in order to save the cost (which can be very large) and time of having the vendor do the recharacterization.

The test views required to evaluate the logical models and assist the development of the chip-level test vectors have been explored.

Beyond this, discussions on consistency between views and the verification and validation of the same have been given, thus facilitating the creation of accurate and non-contradictory views of the various pieces of IP.

An extended discussion on various aspects of DFM has been given such that the reader is more aware as to when and how DFM improvement to a library or other piece of IP is and is not advantageous.

Discussions on how to best interface the resulting internally re-created digital design kit with the intact third-party manufacturing house's physical design kit were given.

A completely workable means for proper IP tagging and revisioning, followed with proper releasing, supporting, and (inevitable) bug fixing and expanding was provided.

You were then shown a series of discussions in support of various special-purpose libraries and how these become extensions of the aforementioned topics.

Finally, how best to generate and maintain proper but open discussion of the usage of the techniques illustrated in this book has included.

If you are happy to compete against those other design houses in what would otherwise be a commodity market on just price alone, then the reading of this book has been a waste of time (and money) for you. If, on the other hand, you have developed the idea that perhaps there is extra margin that a vendor has taken to better ensure that its IP is "clean" but whose action in doing so now prohibits your ability to compete on performance, power, signal integrity, yield, test cost, or package cost, and that you might be able to retool the vendor-supplied views in such a way as take back some of that vendor-absconded margin, and as a result you might reap the reward of a smaller, faster, and cheaper design that competes as opposed to competing purely on reduced profit margin, then this book is a success!

I am reminded of a staff meeting I had to attend several years back. A team of engineers of which I was a member was called into the meeting. We were developing a next-generation high-end microprocessor. The development of the microprocessor was not going according to the schedule that the company's upper management had set for it. The various reasons for the delays are not important to this illustration, but they were many in number and of varying severity. This is the usual case for any design project. There are always previously undiscovered issues that make themselves known and must be resolved before a project can proceed. If the average users of integrated-circuit (IC) design (whether that IC goes into cell phones, laptops, televisions, cars, trains or planes, or any of a plethora of other devices) understood the design process of the typical integrated circuit, then they would be appalled. At any rate, the company happened to be contracting with a construction firm to construct another building adjacent to the one in which the microprocessor design was being done. Senior managers for the electronics company had the insight to give an incentive to the construct firm for on-time delivery of the completed new building. Unfortunately for the construction firm, the weather had been unseasonably wet for several months and the schedule for construction of the new building had started to suffer. To facilitate the construction and to recover lost time from the continuing wet weather, the construct firm brought in a paving subcontractor that literally laid a layer of asphalt down on top of the cleared ground. The building started to be built on top of the asphalt as opposed to having to wait for the otherwise muddy ground to dry out. At any rate, the microprocessor team management saw an opportunity to teach the engineering staff a lesson. They brought the entire staff into a meeting. The vice president of the division got up and told the story of the construction firm to everybody. He then informed us that he felt that a microprocessor could not be much more complex that an entire building with all of its electrical, plumbing, and HVAC requirements, especially because that building was for an electrical assembly plant that would require even more intricate handling of such. He mentioned that the construction firm had found a means of resolving the hindrances to its schedule, and he felt that the engineering staff should take the lesson offered and creatively find ways to recover the original schedule for the development of the microprocessor design. One engineer, not me, then raised his hand and asked if one of the light switches in one of the offices in the new building happened to be placed two inches out of alignment, would the construction firm have to throw the new building out and start over?

The reader can use the information in this book in a manner that allows for the building of smaller, faster, and less power-hungry design, but the reader can use the information in this book a little too aggressively and a little too uncontrollably and, as a result, create, as I said in the introduction, expensive sand.

APPENDIX I

MINIMUM LIBRARY SYNTHESIS VERSUS FULL-LIBRARY SYNTHESIS OF A FOUR-BIT FLASH ADDER

To show the value of a richer stdcell library offering, I have attached these two netlists plus the equations from the original RTL. The following are two synthesized versions of a four-bit full adder. The first was synthesized using a minimum library: one FLIP-FLOP, one NAND2, one NOR2, and one INV. The second was synthesized using a fuller and richer library that is representative of those in the stdcell chapter of this book, complete with multiple drive strengths. It would be more beneficial to the reader to attempt to write up a register-transfer language (RTL) and synthesize it as opposed to copying these resulting netlists.

A general review will highlight that the richer library results are dramatically smaller and contain far fewer inversions (can run much faster). In addition, the reduction in the number of cells indicates the reduction in power consumption as well. The richer the stdcell library offering, the better the smaller, faster and less power hungry will be the designs synthesized with said library.

Note that the original RTL from which these came was rather stylized in that it defined the P and G signals and forced a true *flash-add* design. In that way, no full-adder or half-adder blocks were used. As noted in the stdcell chapter of the book, full-adder and half-adder blocks can be "abused" by the algorithms in current synthesis tools. Without the forcing of a true flash-add design, most synthesis tools will attempt to use full-adder and half-adder blocks in a more serial fashion that, although slightly smaller in general, is usually much slower.

Engineering the CMOS Library: Enhancing Digital Design Kits for Competitive Silicon, First Edition. David Doman.

The equations from the original RTL are as follow:

$$PI = AI + BI, \text{for } I = \{0, 1, 2, 3\}$$
$$GI = AIBI, \text{for } I = \{0, 1, 2, 3\}$$
$$C0 = CIN$$
$$C1 = G0 + P0CIN$$
$$C2 = G1 + P1G0 + P1P0CIN$$
$$C3 = G2 + P2G1 + P2P1G0 + P2P1P0CIN$$
$$COUT = G3 + P3G2 + P3P2G1 + P3P2P1G0 + P3P2P1P0CIN$$
$$SUMI = AI \text{ EXOR } BI \text{ EXOR } CI, \text{for } I = \{0, 1, 2, 3\}$$

The minimal library synthesis is as follows:

MODULE FOUR_BIT_ADD SUMQ0, SUMQ1, SUMQ2, SUMQ3, COUTQ, AD0, AD1, AD2, AD3, BD0, BD1, BD2, BD3, CIND, CLK;
DFFP U1 (.Q(AQ0), .D(AD0), .CK(CLK));
DFFP U2 (.Q(AQ1), .D(AD1), .CK(CLK));
DFFP U3 (.Q(AQ2), .D(AD2), .CK(CLK));
DFFP U4 (.Q(AQ3), .D(AD3), .CK(CLK));
DFFP U5 (.Q(BQ0), .D(BD0), .CK(CLK));
DFFP U6 (.Q(BQ1), .D(BD1), .CK(CLK));
DFFP U7 (.Q(BQ2), .D(BD2), .CK(CLK));
DFFP U8 (.Q(BQ3), .D(BD3), .CK(CLK));
DFFP U9 (.Q(CINQ), .D(CIND), .CK(CLK));
DFFP U10 (.Q(SUMQ0), .D(SUMD0), .CK(CLK));
DFFP U11 (.Q(SUMQ1), .D(SUMD1), .CK(CLK));
DFFP U12 (.Q(SUMQ2), .D(SUMD2), .CK(CLK));
DFFP U13 (.Q(SUMQ3), .D(SUMD3), .CK(CLK));
DFFP U14 (.Q(COUTQ), .D(COUTD), .CK(CLK));
NOR2 U15 (.X(NP1), .A(AQ0), .B(BQ0));
INV1 U16 (.X(P0), .A(NP1));
NOR2 U17 (.X(NP2), .A(AQ1), .B(BQ1));
INV1 U18 (.X(P1), .A(NP2));
NOR2 U19 (.X(NP3), .A(AQ2), .B(BQ2));
INV1 U20 (.X(P2), .A(NP3));
NOR2 U21 (.X(NP4), .A(AQ3), .B(BQ3));
INV1 U22 (.X(P3), .A(NP4));
NAND2 U23 (.X(NG1), .A(AQ0), .B(BQ0));
INV U24 (.X(G0), .A(NG1));
NAND2 U25 (.X(NG2), .A(AQ1), .B(BQ1));
INV U26 (.X(G1), .A(NG2));
NAND2 U27 (.X(NG3), .A(AQ2), .B(BQ2));
INV U28 (.X(G2), .A(NG3));
NAND2 U29 (.X(NG4), .A(AQ2), .B(BQ2));

INV U30 (.X(G3), .A(NG4));
NOR2 U31 (.X(N0S1), .A(BQ0), .B(C0));
NAND2 U32 (.X(N0S2), .A(AQ0), .B(N0S1));
NOR2 U33 (.X(N0S3), .A(AQ0), .B(C0));
NAND2 U34 (.X(N0S4), .A(BQ0), .B(N0S3));
NOR2 U35 (.X(N0S5), .A(AQ0), .B(BQ0));
NAND2 U36 (.X(N0S6), .A(C0), .B(N0S5));
NAND2 U37 (.X(N0S7), .A(AQ0), .B(BQ0));
INV U38 (.X(N0S8), .A(N0S7));
NAND2 U39 (.X(N0S9), .A(C0), .B(N0S8));
NAND2 U40 (.X(N0S10), .A(N0S2), .B(N0S4));
NAND2 U41 (.X(N0S11), .A(N0S6), .B(N0S9));
NOR2 U42 (.X(N0S12), .A(N0S10), .B(N0S11));
INV U43 (.X(SUMD0), .A(N0S12));
NOR2 U44 (.X(N1S1), .A(BQ1), .B(C1));
NAND2 U45 (.X(N1S2), .A(AQ1), .B(N1S1));
NOR2 U46 (.X(N1S3), .A(AQ1), .B(C1));
NAND2 U47 (.X(N1S4), .A(BQ1), .B(N1S3));
NOR2 U48 (.X(N1S5), .A(AQ1), .B(BQ1));
NAND2 U49 (.X(N1S6), .A(C1), .B(N1S5));
NAND2 U50 (.X(N1S7), .A(AQ1), .B(BQ1));
INV U51 (.X(N1S8), .A(N1S7));
NAND2 U52 (.X(N1S9), .A(C1), .B(N1S8));
NAND2 U53 (.X(N1S10), .A(N1S2), .B(N1S4));
NAND2 U54 (.X(N1S11), .A(N1S6), .B(N1S9));
NOR2 U55 (.X(N1S12), .A(N1S10), .B(N1S11));
INV U56 (.X(SUMD1), .A(N1S12));
NOR2 U57 (.X(N2S1), .A(BQ2), .B(C2));
NAND2 U58 (.X(N2S2), .A(AQ2), .B(N2S1));
NOR2 U59 (.X(N2S3), .A(AQ2), .B(C2));
NAND2 U60 (.X(N2S4), .A(BQ2), .B(N2S3));
NOR2 U61 (.X(N2S5), .A(AQ2), .B(BQ2));
NAND2 U62 (.X(N2S6), .A(C2), .B(N2S5));
NAND2 U63 (.X(N2S7), .A(AQ2), .B(BQ2));
INV U64 (.X(N2S8), .A(N2S7));
NAND2 U65 (.X(N2S9), .A(C2), .B(N2S8));
NAND2 U66 (.X(N2S10), .A(N2S2), .B(N2S4));
NAND2 U67 (.X(N2S11), .A(N2S6), .B(N2S9));
NOR2 U68 (.X(N2S12), .A(N2S10), .B(N2S11));
INV U69 (.X(SUMD2), .A(N2S12));
NOR2 U70 (.X(N3S1), .A(BQ3), .B(C3));

NAND2 U71 (.X(N3S2), .A(AQ3), .B(N3S1));
NOR2 U72 (.X(N3S3), .A(AQ3), .B(C3));
NAND2 U73 (.X(N3S4), .A(BQ3), .B(N3S3));
NOR2 U74 (.X(N3S5), .A(AQ3), .B(BQ3));
NAND2 U75 (.X(N3S6), .A(C3), .B(N3S5));
NAND2 U76 (.X(N3S7), .A(AQ3), .B(BQ3));
INV U77 (.X(N3S8), .A(N3S7));
NAND2 U78 (.X(N3S9), .A(C3), .B(N3S8));
NAND2 U79 (.X(N3S10), .A(N3S2), .B(N3S4));
NAND2 U80 (.X(N3S11), .A(N3S6), .B(N3S9));
NOR2 U81 (.X(N3S12), .A(N3S10), .B(N3S11));
INV U82 (.X(SUMD3), .A(N3S12));
INV U83 (.X(NCARRY1), .A(CINQ));
INV U84 (.X(C0), .A(NCARRY1));
INV U85 (.X(NCARRY2), .A(G0));
NAND2 U86 (.X(NCARRY3), .A(P0), .B(CINQ));
NAND2 U87 (.X(C1), .A(NCARRY2), .b(NCARRY3));
INV U88 (.X(NCARRY4), .A(G1));
NAND2 U89 (.X(NCARRY5), .A(P1), .B(G0));
NAND2 U90 (.X(NCARRY6), .A(P0), .B(CINQ));
INV U91 (.X(NCARRY7), .A(NCARRY6));
NAND2 U92 (.X(NCARRY8), .A(NCARRY4), .B(NCARRY5));
NAND2 U93 (.X(NCARRY9), .A(P1), .B(NCARRY7));
INV U94 (.X(NCARRY10), .A(NCARRY8));
NAND2 U95 (.X(C2), .A(NCARRY10), .B(NCARRY9));
INV U96 (.X(NCARRY11), .A(G2));
NAND2 U97 (.X(NCARRY12), .A(P2), .B(G1));
NAND2 U98 (.X(NCARRY13), .A(P1), .B(G0));
INV U99 (.X(NCARRY14), .A(NCARRY13));
NAND2 U100 (.X(NCARRY15), .A(P2), .B(P1));
NAND2 U101 (.X(NCARRY16), .A(P0), .B(CINQ));
NAND2 U102 (.X(NCARRY18), .A(P2), .B(NCARRY14));
NOR2 U103 (.X(NCARRY17), .A(NCARRY15), .B(NCARRY16));
INV U104 (.X(NCARRY19), .A(NCARRY17));
NAND2 U105 (.X(NCARRY20), .A(NCARRY11), .B(NCARRY12));
NAND2 U106 (.X(NCARRY21), .A(NCARRY18), .B(NCARRY19));
NOR2 U107 (.X(NCARRY22), .A(NCARRY20), .B(NCARRY21));
INV U108 (.X(C3), .A(NCARRY22));
INV U109 (.X(NCARRY31), .A(G3));
NAND2 U110 (.X(NCARRY32), .A(P3), .B(G2));
NAND2 U110 (.X(NCARRY23), .A(P2), .B(G1));

INV U111 (.X(NCARRY24), .A(NCARRY23));
NAND2 U112 (.X(NCARRY25), .A(P3), .B(P2));
NAND2 U113 (.X(NCARRY26), .A(P1), .B(G0));
NAND2 U114 (.X(NCARRY28), .A(P2), .B(P1));
NAND2 U115 (.X(NCARRY29), .A(P0), .B(CINQ));
NOR2 U116 (.X(NCARRY27), .A(NCARRY25), .B(NCARRY26));
NOR2 U117 (.X(NCARRY30), .A(NCARRY28), .B(NCARRY(29));
NAND2 U118 (.X(NCARRY33), .A(P3), .B(NCARRY24));
INV U119 (.X(NCARRY34), .A(NCARRY27));
NAND2 U120 (.X(NCARRY35), .A(NCARRY31), .B(NCARRY32));
NAND2 U121 (.X(NCARRY36), .A(NCARRY33), .B(NCARRY34));
NAND2 U122 (.X(NCARRY37), .A(P3), .B(NCARRY30));
NOR2 U123 (.X(NCARRY38), .A(NCARRY35), .B(NCARRY36));
INV U124 (.X(NCARRY39), .A(NCARRY38));
NAND2 U125 (.X(COUTD), .A(NCARRY39), .B(NCARRY37));
ENDMODULE

The expanded library synthesis is as follows:

MODULE FOUR_BIT_ADD SUMQ0, SUMQ1, SUMQ2, SUMQ3, COUTQ,
    AD0, AD1, AD2, AD3, BD0, BD1, BD2, BD3, CIND, CLK;
DFFPX1 U1 (.Q(AQ0), .D(AD0), .CK(CLK));
DFFPX1 U2 (.Q(AQ1), .D(AD1), .CK(CLK));
DFFPX1 U3 (.Q(AQ2), .D(AD2), .CK(CLK));
DFFPX1 U4 (.Q(AQ3), .D(AD3), .CK(CLK));
DFFPX1 U5 (.Q(BQ0), .D(BD0), .CK(CLK));
DFFPX1 U6 (.Q(BQ1), .D(BD1), .CK(CLK));
DFFPX1 U7 (.Q(BQ2), .D(BD2), .CK(CLK));
DFFPX1 U8 (.Q(BQ3), .D(BD3), .CK(CLK));
DFFPX1 U9 (.Q(CINQ), .D(CIND), .CK(CLK));
DFFPX1 U10 (.Q(SUMQ0), .D(SUMD0), .CK(CLK));
DFFPX1 U11 (.Q(SUMQ1), .D(SUMD1), .CK(CLK));
DFFPX1 U12 (.Q(SUMQ2), .D(SUMD2), .CK(CLK));
DFFPX1 U13 (.Q(SUMQ3), .D(SUMD3), .CK(CLK));
DFFPX1 U14 (.Q(COUTQ), .D(COUTD), .CK(CLK));
NOR2X1 U15 (.X(NP1), .A(AQ0), .B(BQ0));
INV1X2 U16 (.X(P0), .A(NP1));
NOR2X1 U17 (.X(NP2), .A(AQ1), .B(BQ1));
INV1X2 U18 (.X(P1), .A(NP2));
NOR2X1 U19 (.X(NP3), .A(AQ2), .B(BQ2));
INV1X2 U20 (.X(P2), .A(NP3));
NOR2X1 U21 (.X(NP4), .A(AQ3), .B(BQ3));

INV1X2 U22 (.X(P3), .A(NP4));
NAND2X1 U23 (.X(NG1), .A(AQ0), .B(BQ0));
INVX2 U24 (.X(G0), .A(NG1));
NAND2X1 U25 (.X(NG2), .A(AQ1), .B(BQ1));
INVX2 U26 (.X(G1), .A(NG2));
NAND2X1 U27 (.X(NG3), .A(AQ2), .B(BQ2));
INVX2 U28 (.X(G2), .A(NG3));
NAND2X1 U29 (.X(NG4), .A(AQ2), .B(BQ2));
INVX2 U30 (.X(G3), .A(NG4));
EXNORX1 U31 (.X(NS1), .A(AQ0), .B(BQ0));
EXNORX1 U32 (.X(SUMD0), .A(C0), .B(NS1));
EXNORX1 U33 (.X(NS2), .A(AQ1), .B(BQ1));
EXNORX1 U34 (.X(SUMD1), .A(C1), .B(NS2));
EXNORX1 U35 (.X(NS3), .A(AQ2), .B(BQ2));
EXNORX1 U36 (.X(SUMD2), .A(C2), .B(NS3));
EXNORX1 U37 (.X(NS4), .A(AQ3), .B(BQ3));
EXNORX1 U38 (.X(SUMD3), .A(C3), .B(NS4));
INVX1 U39 (.X(NCARRY1), .A(CINQ));
INVX1 U40 (.X(C0), .A(NCARRY1));
AOI21X1 U41 (.X(NCARRY2), .A1(P0), .A2(CINQ), .B(G0));
INVX1 U42 (.X(C1), .A(NCARRY2));
AOI321X1 U43 (.X(NCARRY3), .A1(P1), .A2(P0), .A3(CINQ), .B1(P1), .B2(G0), .
    C(G1));
INVX1 U44 (.X(C2), .A(NCARRY3));
NAND2X1 U45 (.X(NCARRY4), .A(P2), .B(P1));
NAND2X1 U46 (.X(NCARRY5), .A(P0), .B(CINQ));
AOI321X1 U47 (.X(NCARRY7), .A1(P2), .A2(P1), .A3(G0), .B1(P2), .B2(G1), .C(G2));
NOR2X1 U48 (.X(NCARRY6), .A(NCARRY4), .B(NCARRY5));
NAND2X1 U49 (.X(C3), .A(NCARRY6), .B(NCARRY7));
AOI321 U50 (.X(NCARRY8), .A1(P3), .A2(P2), .A3(G1), .B1(P3), .B2(G2), .C(G3));
NAND2X1 U51 (.X(NCARRY9), .A(P3), .B(P2));
NAND2X1 U52 (.X(NCARRY10), .A(P1), .B(G0));
NAND3X1 U53 (.X(NCARRY11), .A(P3), .B(P2), .C(P1));
NAND2X1 U54 (.X(NCARRY12), .A(P1), .B(CINQ));
INVX1 U55 (.X(NCARRY13), .A(NCARRY8));
NOR2X1 U56 (.X(NCARRY14), .A(NCARRY9), .B(NCARRY10));
NOR2X1 U57 (.X(NCARRY15), .A(NCARRY11), .B(NCARRY12));
NOR3X1 U58 (.X(NARRY16), .A(NCARRY13), .B(NCARRY14), .C(NCARRY15));
INVX1 U59 (.X(COUTD), .A(NCARRY16));
ENDMODULE

APPENDIX II

PERTINENT CMOS BSIM SPICE PARAMETERS WITH UNITS AND DEFAULT LEVELS

Parameter	Units	Default
AF		1
CBD	Farads	0
CBS	Farads	0
CGDO	Farads/meters	0
CGSO	Farads/meters	0
DELTA		0
ETA		0
FC		0.5
KAPPA		0.2
KF		0
L	meters	DEFL
MJ		0.5
MJSW		0.33
NEFF		1
RB	Ohms	0
RD	Ohms	0

(Continued)

Engineering the CMOS Library: Enhancing Digital Design Kits for Competitive Silicon,
First Edition. David Doman.
© 2012 John Wiley & Sons, Inc. Published 2012 by John Wiley & Sons, Inc.

Parameter	Units	Default
RG	Ohms	0
RS	Ohms	0
THETA	1/Volts	
TOX	meters	[infinity]
TPG		1
UCRIT	Volts/centimeter	1.00E+04
UEXP		0
VMAX	meters/seconds	0
W	meters	DEFW
WD	meters	0
XJ	meters	0
XQC		1

APPENDIX III

DEFINITION OF TERMS

APR	Automated place and route.
ASCII	A human-readable way of representing information. Most digital design kit (DDK) views are ASCII based.
ASIC	Application Specific Integrated Circuit; the majority of modern integrated circuits, other than custom-designed devices such as microprocessors.
ATPG	Automatic test pattern generation.
BER	Bit error rate.
BIST	Built-in self-test.
BNF	Backus-Naur format. Formal definition of a computer language.
BSIM3/BSIM4	Berkley short-channel IGFET model versions 3 and 4
CDM	Charged device model of ESD testing
CMOS	Complimentary metal-oxide semiconductor.
CRC	Cyclic-redundancy check. An unsecure error-detection coding scheme often used in communication channels that can be used for checksum operations within library file deliveries.
DDK	Digital design kit. Usually a collection of stdcells, IOs, possibly some SSI-level analog functions (such as PLLs) and some memory compilers, together with a set of separate views compatible with engineering design automation (EDA) tools for use in logical, physical, temporal and test development of integrated circuits.

Engineering the CMOS Library: Enhancing Digital Design Kits for Competitive Silicon, First Edition. David Doman.
© 2012 John Wiley & Sons, Inc. Published 2012 by John Wiley & Sons, Inc.

DDR	Double data rate.
DRC	Design rule check (physical verification deck).
Ebers-Moll	Early bipolar SPICE model
EDA	Engineering design automation. An industry geared to the creation of design tools for use in the integrated-circuit design environment.
ESD	Electrostatic discharge.
FDDI	Fiber-distributed data interface.
GDDR	Graphics double data rate.
GDS	Graphic display system. A physical representation of a structure; binary in format.
GPIO	General-purpose IO
Gummel-Poon	Second-generation bipolar SPICE model.
Hamiltonian circuit (or Hamiltonian path)	Minimal path that crosses every arc in a network once and only once and returns to the initial position.
HBM	Human body model. Model of ESD testing.
HCI	Hot carrier injection. One of several processes that can cause transistors to decrease in performance over time.
Heap	Tree-based data structure.
IDDQ	Quiescent supply current.
IEEE	Institute of Electrical and Electronics Engineers.
IO	Input–output.
JTAG	Joint Test Action Group.
LEF	Library exchange format. An open-source format (originally Cadence) for the description of APR information for a stdcell, IO, or other SSI block, together with information on the router's technically viable choices.
LFSR	Linear feedback shift register.
Liberty file	Also known as LIB. An open-source (originally Synopsys) timing, power, and noise view.
LPE	Layout parasitic extraction. For use extraction parasitic resistors and capacitors of physical designs at the cell or chip level.
LSI	Large-scale integration.
LVS	Layout versus schematic. Schematic verification deck.
MM	Machine model. Model of ESD testing.
Mocron	1/25th of an inch, almost a micron. Unit used by a company that was once afraid of the metric system.
Moore's law	The ability to place transistors on a piece of silicon doubles every 18 months.
MSI	Medium-scale integration.
MUX	Multiplex.
NRX	Nonreturn-to-zero coding.
NVM	Nonvolatile memory.
Nyquist rate	Required frequency of sampling of a signal of a given maximum frequency, which is twice that maximum frequency.
OTP	One-Time programmable memory.

PDK	Physical design kit. Technology files and verification decks for a given fabrication house's process. Can include SPICE decks, DRC, LVS, LPE decks, as well as other files for use in various EDA tools.
PLL	Phase-locked loop (circuit).
POR	Power on reset (circuit).
PROM	Programmable read-only memory.
PVT	Process, voltage, and temperature. The three axes that most liberty files are defined against (best-case, typical-case, or worst-case corners on each axis).
Raw gate density	Density of a cell or a library based on 100% use.
RTL	Register transfer language. A logical or equation-based description of a function that can be translated by EDA tools into a gate-level netlist for that function.
SKILL	Cadence database manipulation coding language.
SNR	Signal-to-noise ratio.
SPICE	Simulation program with integrated circuit emphasis. A circuit-simulation tool.
SSI	Small-scale integration.
Stdcells	Usually SSI- or MSI-level logic functions that are capable of being connected together, usually by EDA tools, to accomplish some larger logic function. The collection of views for each function is usually referred to as a *library*.
Unate	Description of the input-to-output relationship of an input–output pair in a logic function. If positive unite, the output can rise if the input rises and fall if the input falls, but cannot do opposite (AND cells and OR cells are positive unite). If negative unite, the output can fall if the input rises and rise if the input falls, but cannot do opposite (NAND cells and NOR cells are negative unite). If non-unate, the output can switch in either direction as an input switch (EXOR cells and MUX cells are non-unate).
Uniform distribution weighting	Method for determining the raw gate density of an entire library based on a uniform distribution of the cells of the library.
UNIX	Computer operating system that ran many to most of workstations used in the integrated-circuit design world for much of the last 40 years.
Verilog	A Cadence-based logical simulator
VHDL	VHSIC hardware descriptive language. An IEEE-based logical simulator. Originally descended from Texas Instruments HDL.
VHSIC	Very high speed integrated circuit.
Vt	Threshold voltage of a transistor. Determined by the dopant levels added during processing. High-Vt transistors are slower but consume less power; low-Vt transistors are the opposite. Most processes allow no more than two to three different Vt across all transistors on a chip; some processes allow none.
Weighted distribution	Method for determining the raw (and sometimes actual) density of a potential design that would use a given library.

APPENDIX IV

ONE POSSIBLE MEANS OF FORMALIZED MONTHLY REPORTING

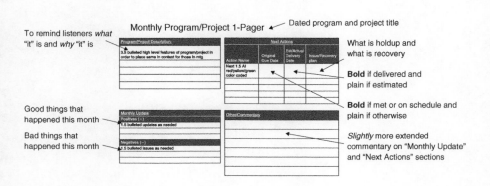

Engineering the CMOS Library: Enhancing Digital Design Kits for Competitive Silicon,
First Edition. David Doman.
© 2012 John Wiley & Sons, Inc. Published 2012 by John Wiley & Sons, Inc.

INDEX

Engineering the CMOS Library: Enhancing Digital Design Kits for Competitive Silicon,
First Edition. David Doman.
© 2012 John Wiley & Sons, Inc. Published 2012 by John Wiley & Sons, Inc.

Printed and bound by CPI Group (UK) Ltd, Croydon, CR0 4YY

16/04/2025

14658417-0004